STUDY GUIDE
ORGANIC AND
BIOLOGICAL CHEMISTRY

JOHN R. HOLUM

John Wiley & Sons, Inc.

New York • Chichester • Brisbane • Toronto • Singapore

ISBN 0-471-13756-1

Printed in the United States of America

10 9 8 7 6 5 4 3 2 1

Contents

1   Organic Chemistry. Saturated Hydrocarbons 1

2   Unsaturated Hydrocarbons 23

3   Alcohols, Phenols, Ethers, and Thioalcohols 41

4   Aldehydes and Ketones 55

5   Carboxylic Acids and Esters 64

6   Amines and Amides 81

7   Optical Isomerism 95

8   Carbohydrates 100

9   Lipids 108

10 Proteins 114

11 Enzymes, Hormones, and Neurotransmitters 125

12 Extracellular Fluids of the Body 136

13 Nucleic Acids 149

14 Biochemical Energetics 157

15 Metabolism of Carbohydrates 164

16 Metabolism of Lipids 170

17 Metabolism of Nitrogen Compounds 178

18 Nutrition 184

Answer Book 190

1

ORGANIC CHEMISTRY. SATURATED HYDROCARBONS

The most important *concepts* in this chapter are functional groups, isomerism, the idea of a structural "map sign," and the basic principles of naming organic compounds.

The *skills* that must be mastered involve the writing and interpreting of structural formulas, naming alkanes and alkyl groups, and using the two following structural "map signs:"

1. All substances whose molecules are wholly or mostly hydrocarbons are insoluble in water, and
2. Alkanes or alkane-like regions of molecules of other compounds do not react with strong alkalis, strong acids, water, reducing agents, or most oxidizing agents.

OBJECTIVES

After you have studied this chapter and worked the Practice Exercises and Review Exercises in it, you should be able to do the following. Objectives 6 through 9 are particularly important because they concern the symbols that we'll use in later work with organic compounds.

1. State what the vital force theory was and how it affected research in its day.
2. Describe Wöhler's experiment and explain how it helped to overthrow the vital force theory.
3. List the main differences between organic and inorganic compounds.
4. Give the ways in which carbon is a unique element.
5. State how a molecular formula and a structural formula are alike and how they are different.
6. Explain why each possible conformation of a carbon chain does not represent a different compound.
7. Give an example (both names and structures) of two compounds that are related as isomers.
8. Write condensed structures from full structures (both open-chain and ring compounds).

9. Examine a pair of structures and tell if they are identical, are related as isomers, or are different.
10. Recognize if a substance is a saturated hydrocarbon from its structure.
11. Write the IUPAC names of all of the straight-chain alkanes (through decane) and give their carbon content (the number of carbon atoms per molecule).
12. Write the name and structure of the simple cycloalkanes through rings of six carbons.
13. Write the names and structures of all alkyl groups through the C_4 set.
14. Write the common names of all alkanes through the C_4 set.
15. Write the IUPAC names of alkanes and cycloalkanes from their structures.
16. Write the structures of alkanes and cycloalkanes from their IUPAC names.
17. Recognize the primary, secondary, and tertiary carbons in a structure.
18. Write an equation for the complete combustion of any hydrocarbon.
19. Write an equation for the chlorination of any given alkane showing the structures of all the possible isomeric monochloro products.
20. Define and, where applicable, give an example of each of the terms in the Glossary.

GLOSSARY

Aliphatic Compound. An organic compound without benzene rings.

Alkane. Any saturated hydrocarbon, one that has only single bonds. A *normal* alkane is any whose molecules have straight chains.

Alkenes. Hydrocarbons with carbon-carbon double bonds.

Alkyl Group. A substituent group that is an alkane minus one H atom.

Alkynes. Hydrocarbons with carbon-carbon triple bonds.

Aromatic Compound. An organic compound with a benzene ring.

Branched Chain. A sequence of carbons atoms to which additional carbon atoms are attached at points other than the ends.

Condensed Structure. (See *Structure.*)

Conformation. One of the infinite number of contortions of a molecule that are permitted by free rotations around single bonds.

Constitutional Isomers. Compounds with identical molecular formulas but different atom-to-atom sequences; structural isomers.

Free Rotation. The absence of a barrier to the rotation of two groups with respect to each other when they are joined by a single, covalent bond.

Functional Group. An atom or a group of atoms in a molecule that is responsible for the particular set of reactions that all compounds with this group have.

Heterocyclic Compound. An organic compound with a ring in which at least one ring atom is not carbon.

Hydrocarbon. Any organic compound that consists entirely of carbon and hydrogen.

Inorganic Compound. Any compound that is not an organic compound.

International Union of Pure and Applied Chemistry System (IUPAC System). A set of systematic rules for naming compounds and designed to give each compound one unique name and for which only one structure can be drawn; the Geneva system of nomenclature.

Isomerism. The phenomenon of the existence of two or more compounds with identical molecular formulas but different structures.

Isomers. Compounds with identical molecular formulas but different structures.

IUPAC System. (See *International Union of Pure and Applied Chemistry System.*)

Like-Dissolves-Like Rule. Polar solvents dissolve polar or ionic solutes and nonpolar solvents dissolve nonpolar or weakly polar solutes.

Nomenclature. The system of names and the rules for devising such names, given structures, or for writing structures, given names.

Nonfunctional Group. A section of an organic molecule that remains unchanged during a chemical reaction at a functional group.

Organic Chemistry. The study of the structures, properties, and the syntheses of organic compounds.

Organic Compounds. Compounds of carbon with other nonmetal elements.

Primary Carbon. In a molecule, a carbon atom that is joined directly to just one other carbon, such as the end carbons in $CH_3CH_2CH_3$.

Ring Compound. A compound whose molecules contain three or more atoms joined in a ring.

Saturated Compound. A compound whose molecules have only single bonds.

Secondary Carbon. Any carbon atom in an organic molecule that has two and only two bonds to other carbon atoms, such as the middle carbon atom in $CH_3CH_2CH_3$.

Straight Chain. A continuous, open sequence of covalently bound carbon atoms from which no additional carbon atoms are attached at interior locations of the sequence.

Structural Isomer. (See *Constitutional Isomer.*)

Structure. A structural formula; an array of atomic symbols and lines (for covalent bonds) that shows how the parts of a molecule are organized.

Substitution Reaction. A reaction which one atom or group in a molecule is replaced by another atom or group.

Tertiary Carbon. Any carbon in an organic molecule that has three and only three bonds to adjacent carbon atoms.

Unsaturated Compound. Any compound whose molecules have a double or a triple bond.

Vital Force Theory. A discarded theory that organic compounds could be made in the laboratory only if the chemicals possessed a vital force contributed by some living thing.

DRILL EXERCISES

I. EXERCISES ON CONDENSED STRUCTURES

The rules for converting from a full structure to a condensed structure are:

1. $H-\overset{\displaystyle H}{\underset{\displaystyle H}{C}}-$ becomes CH_3 (sometimes H_3C)

2. $H-\overset{\displaystyle |}{\underset{\displaystyle |}{C}}-H$ or $-\overset{\displaystyle H}{\underset{\displaystyle H}{\overset{|}{\underset{|}{C}}}}-$ becomes CH_2

3. $H-\overset{\displaystyle |}{\underset{\displaystyle |}{C}}-$ or $-\overset{\displaystyle H}{\overset{|}{\underset{|}{C}}}-$ becomes CH

4. All bonds of hydrogen to carbon, oxygen, nitrogen, or sulfur may be "understood."
5. Any single bond between heavy atoms (C—C, C—O, C—N, C—S, for example) may be shown or it may be understood, provided that it would normally be drawn horizontally and, therefore, is part of the main chain.
6. Any single bond between heavy atoms that hold substituents onto the main chain must be shown. (In other words, we always show a substituent group attached to a main chain by writing the group either above the chain or below it, and by connecting the group to the main chain by a line representing the bond.)
7. All double and triple bonds must be shown. (A modification of this rule will appear in a later chapter when we study carbonyl compounds.)

Practice converting between full and condensed structures by doing these additional exercises.

<u>SET A.</u> Condense each of the following structures.

1. $H-\overset{\displaystyle H}{\underset{\displaystyle H}{\overset{|}{\underset{|}{C}}}}-\overset{\displaystyle H}{\underset{\displaystyle H}{\overset{|}{\underset{|}{C}}}}-\overset{\displaystyle H}{\underset{\displaystyle H}{\overset{|}{\underset{|}{C}}}}-H$ _____

2. $H-\overset{\displaystyle H}{\underset{\displaystyle H}{\overset{|}{\underset{|}{C}}}}-\overset{\displaystyle H}{\underset{\displaystyle H}{\overset{|}{\underset{|}{C}}}}-\overset{\displaystyle H}{\underset{\displaystyle H}{\overset{|}{\underset{|}{C}}}}-\overset{\displaystyle H}{\underset{\displaystyle H}{\overset{|}{\underset{|}{C}}}}-H$ _____

3. $H-\overset{\displaystyle H}{\underset{\displaystyle H}{\overset{|}{\underset{|}{C}}}}-\overset{\displaystyle H}{\underset{\displaystyle H}{\overset{|}{\underset{|}{C}}}}-\overset{\displaystyle H}{\underset{\displaystyle H}{\overset{|}{\underset{|}{C}}}}-\overset{\displaystyle H}{\underset{\displaystyle H}{\overset{|}{\underset{|}{C}}}}-\overset{\displaystyle H}{\underset{\displaystyle H}{\overset{|}{\underset{|}{C}}}}-H$ _____

4. $H-\overset{\displaystyle H}{\underset{\displaystyle H}{\overset{|}{\underset{|}{C}}}}-\overset{\displaystyle H}{\underset{\displaystyle H}{\overset{|}{\underset{|}{C}}}}-\overset{\displaystyle H}{\underset{\displaystyle H}{\overset{|}{\underset{|}{C}}}}-\overset{\displaystyle H}{\underset{\displaystyle H}{\overset{|}{\underset{|}{C}}}}-\overset{\displaystyle H}{\underset{\displaystyle H}{\overset{|}{\underset{|}{C}}}}-\overset{\displaystyle H}{\underset{\displaystyle H}{\overset{|}{\underset{|}{C}}}}-H$ _____

with $H-\overset{\displaystyle |}{\underset{\displaystyle |}{C}}-H$ groups as substituents on the chain

```
             H
             |
         H—C—H
             |
         H   |   H   H   H
         |   |   |   |   |
5.   H—C—C—O—C—C—C—H   _____
         |   |   |   |   |
         H   H   H   |   H
                     |
                 H—C—H
                     |
                     H
```

```
             H
             |
         H—C—H
             |                   H
         H   |   H   H   H   O   H
         |   |   |   |   |   |   |
6.   H—C—C—C—C—C—C—C—H
         |   |   |   |   |   |   |
         H   |   H   H   H   H   H      _____
             |
         H—C—H
             |
             H
```

```
         H   H   H
         |   |   |
7.   H—C—C—C—O—H
         |   |   |        _____
         H   H   H
```

```
         H   H   H   H   H
         |   |   |   |   |
8.   H—O—C—C—C=C—C—H
         |   |       |
         H   H       H        _____
```

```
         H       H   H   H   H   O
         |       |   |   |   |   ‖
9.   H—C—O—C=C—C—C—C—O—H
         |       |       |   |
         H                   |        _____
                         H—C—H
                             |
                             H
```

```
         H   H   H   H   H   H
         |   |   |   |   |   |
10.  H—C—C—C—C—C—N—H
         |   |   |   |   |
         H   O   H   H   H          _____
             |
         H—C—H
             |
             H
```

SET B. Write the full structures that correspond to these condensed structures. Show all single bonds.

1. CH_3CH_3

2.
$$\begin{array}{c} CH_3 \\ | \\ CH_3CCH_2CH_2CH_3 \\ | \\ CH_3 \end{array}$$

3.
$$\begin{array}{c} CH_3 \qquad CH_2CH_3 \\ | \qquad\qquad | \\ CH_3CH_2CH_2CH_2CCH_2CH_2CHCH_2CH_2CH_2CH_3 \\ | \\ CH_3CCH_2CH_3 \\ | \\ CH_3 \end{array}$$

4. $CH_3CH{=}CH{-}CH_3$

5. $HOCH_2CH_3$

6.
$$\begin{array}{c} O \\ \| \\ CH_3OCCH_2CH_2NHCH_2CH_3 \end{array}$$

7.
$$\begin{array}{c} O \\ \| \\ HC{\equiv}C{-}CH_2CH \end{array}$$

8. $NH_2CH_2CH_2NH_2$

II.   EXERCISES ON DEVISING STRUCTURES FROM MOLECULAR FORMULAS

Review Exercise 1.12 in the text asks you to devise a structural formula from each molecular formula given. An answer cannot be correct unless each and every atom in a molecule has exactly the correct number of bonds going from it—four from carbon, three from nitrogen, two from oxygen (or sulfur), one from any halogen, and one from hydrogen. If the question in the text gives you trouble, see how the following semi-systematic approach works. The number of rules may seem a bit awesome, but you will quickly advance beyond needing them.

Rule 1.   Identify those atoms in the given molecular formula that must have two, three, or four bonds—all of the atoms other than a halogen or hydrogen. (We'll work with these to establish a "skeleton" for the structure upon which the hydrogen and halogen atoms will be hung. Think of each allowed bond of every atom (4 for carbon, etc.) as an unused arm. A correct structure will be one in which all the arms are in some way linked. (There can be "no unjoined arms" could be another way of stating it.)

Rule 2.   The number of unused bonds ("free arms") left over on the skeleton—**all being single bonds**—must equal the number of hydrogens or halogens to be appended.

Rule 3.   Don't put a halogen on anything but carbon—not on oxygen, not on sulfur, and not on nitrogen. (Although such possibilities do exist, we'll never encounter them.)

Rule 4.   In writing a structure, assemble the carbon atoms along a horizontal line as much as possible and pin the substituents to the resulting chain of carbon atoms. (This rule has nothing to do with rules about numbers of bonds, only with looks.)

When atoms have more than one bond, options exist as to how they may occur in structures. All of the bonds may be single, or some may be single and some double or even triple. Here are all the possible options for carbon, nitrogen, oxygen, the halogens, and hydrogen.

For carbon   —C—   ＼C═   —C≡
                  ｜   ／

For nitrogen   —N—   —N═   N≡
                  ｜

For oxygen   —O—   O═

For halogen   —X   (X = F, Cl, Br, or I)
For hydrogen   —H

Now let's work a few sample exercises.

Sample Exercise 1: Write a structural formula for CH_4O.

Step 1.   The multivalent atoms are C and O. (Rule 1)
Step 2.   Assemble the options:

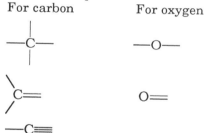

For carbon   For oxygen

—C—   —O—
  ｜

＼C═   O═
／

—C≡

Step 3.   Try the possible combinations of those in the first column with those in the second.

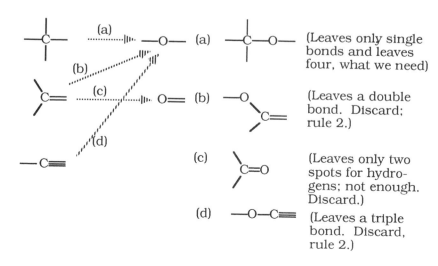

(a) —C—O—   (Leaves only single bonds and leaves four, what we need)
      ｜

(b) —O＼C═   (Leaves a double bond. Discard; rule 2.)
         ／

(c) ＼C═O   (Leaves only two spots for hydrogens; not enough. Discard.)
      ／

(d) —O—C≡   (Leaves a triple bond. Discard, rule 2.)

Step 4.   Identify the skeleton, or skeletons, (rules 1 and 2) and append the hydrogens. In this sample exercise only one skeleton meets the rules, the one generated by combination (a)

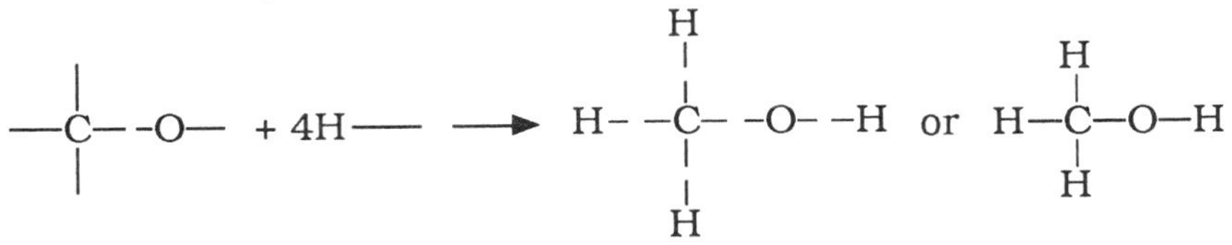

$$\overset{|}{\underset{|}{-\text{C}}}\text{--O--} + 4\text{H} \longrightarrow \text{H--}\overset{\overset{\text{H}}{|}}{\underset{\underset{\text{H}}{|}}{\text{C}}}\text{--O--H}\ \text{ or }\ \text{H--}\overset{\overset{\text{H}}{|}}{\underset{\underset{\text{H}}{|}}{\text{C}}}\text{--O--H}$$

(the answer)

We may condense the answer to CH_3OH.

Step 5.   Run a mental check; does the carbon have 4 bonds; the oxygen, 2; and the hydrogens 1 each? They do; therefore, the structure obeys the rules of covalence.

Sample Exercise 2: Convert C_2H_4 to a structural formula.

Step 1.   The multivalent atoms are just the two Cs.

Step 2.   Assemble the options:

First Carbon   Second Carbon

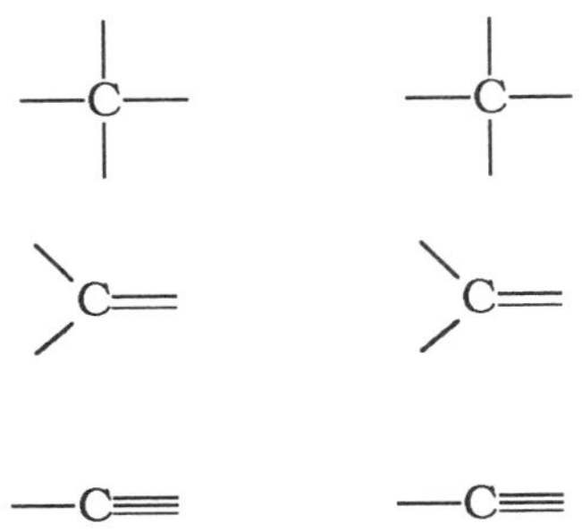

Step 3.   Try combinations from each column. Remember that the remaining unused bonds must equal four in number, for the four hydrogens, and that these remaining bonds must all be single bonds. Mentally discarding all combinations that leave double or triple bonds "open," we have these possibilities:

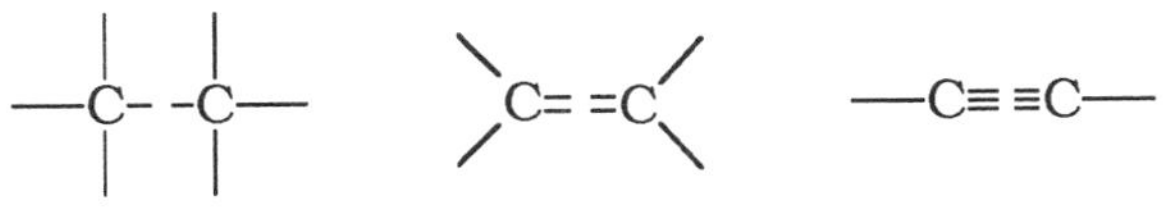

The first leaves room for six hydrogens, too many; the last for two hydrogens, too few. The middle will work; room is left for four hydrogens.

Step 4.   Put the pieces together:

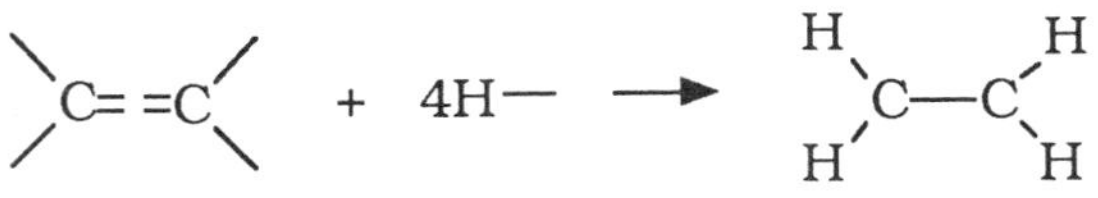

The condensed structure is: $CH_2=CH_2$.

Step 5. Check to see that each atom has the right number of bonds; that each carbon has four bonds, and each hydrogen has one.

Sample Exercise 3: Convert C_2H_4O into a proper structural formula.
Step 1. The multivalent atoms are two carbons and one oxygen.
Step 2. Assemble the options:

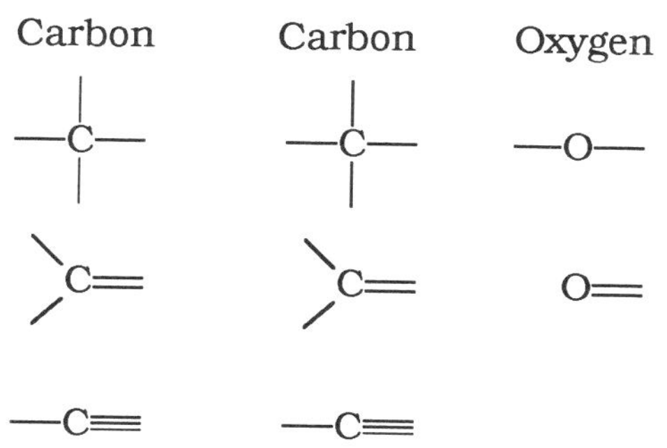

Step 3. Try the possible combinations, remembering that the skeleton must have no double or triple bond left "hanging" and must have, via single bonds only, room for four hydrogens. Here are the possibilities that have only single bonds remaining. (Beneath each is a number equaling the spaces for hydrogens.)

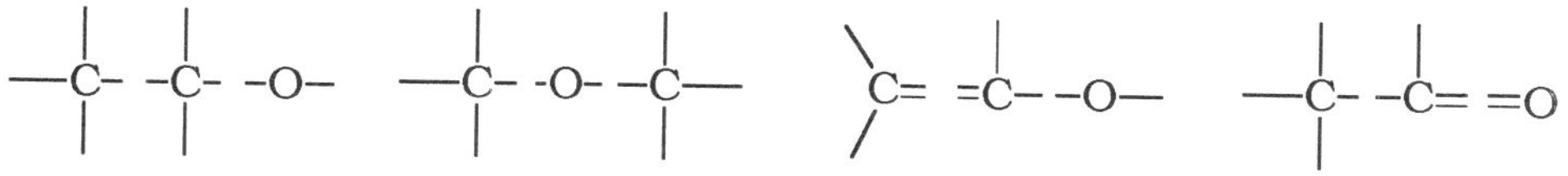

(6)   (6)   (4)   (4)

Any others you may have drawn, that avoid leaving open double or triple bonds, will be duplicates of one or more of these. (For instance,

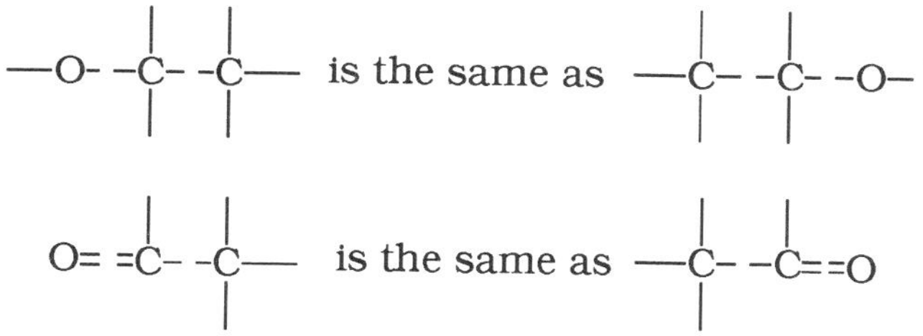

is the same as

is the same as

There are two combinations with four spaces for hydrogens. We must consider them both.
Step 4. Put the pieces together:

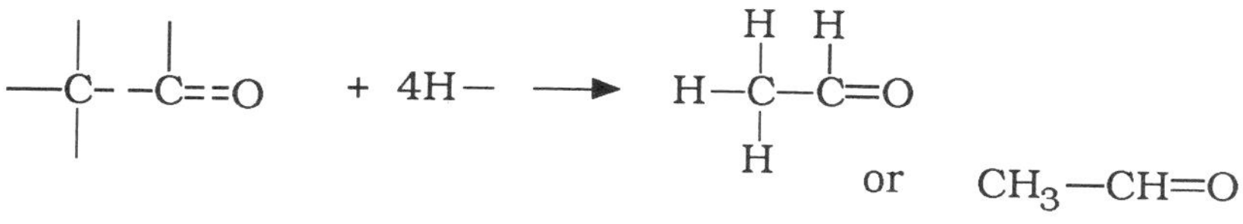

or $CH_3-CH=O$

Usually, you will see carbon-oxygen double bonds aligned vertically, and the last structure will usually be seen as

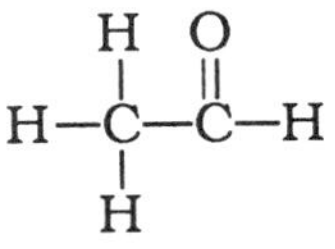

$$\begin{array}{ccc} H & O \\ | & \| \\ H-C-C-H \\ | \\ H \end{array}$$

The second combination gives CH_2=CHOH.
(Remember that the single bonds in the main chain can be "understood.")
Step 5.   Check the answers: four bonds from the carbons; two from the oxygens. Either structure is correct; the two are isomers.

Write structural formulas for the following molecular formulas.

1. H_2O_2 _____

2. N_2H_4 _____

3. CH_4S _____

4. C_2H_5Br _____

5. C_2H_3Cl _____

III. EXERCISES ON STRUCTURAL FEATURES OF HYDROCARBONS

The following set of structures illustrates several features discussed in the text. Supply the information requested by writing the identifying number(s) of the structure(s).

A. $CH_3CH_2CH_2CH_2CH_2CH_3$

B. 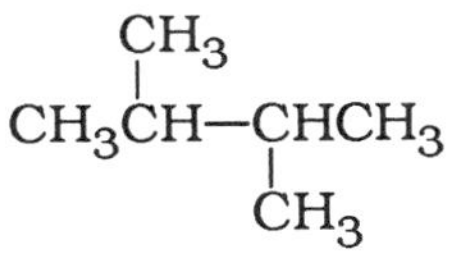 $CH_3CH-CHCH_3$ with CH_3 above first CH and CH_3 below second CH

C. CH_3CH=$CHCH_2CH_2CH_3$

D. $CH_3CH_2CH_2OH$

E. 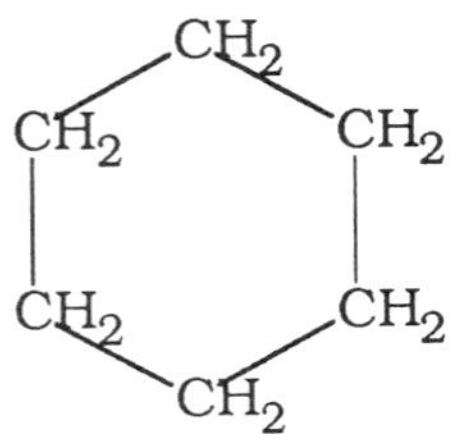 (cyclohexane ring of CH_2 groups)

F. 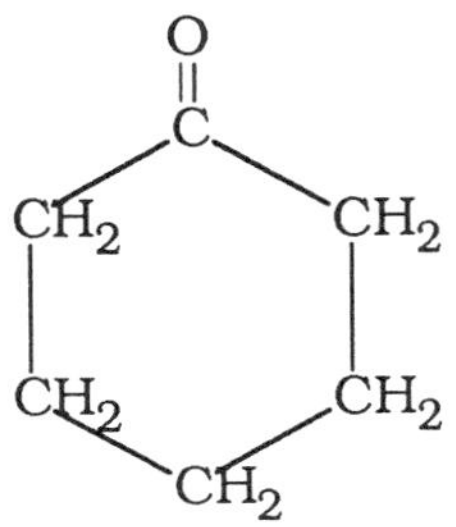 (cyclohexanone-type ring with C=O and CH_2 groups)

SET A.

1. Give the letters of
    (a) the saturated compounds   _____
    (b) the straight-chain compounds   _____
    (c) the branch-chain compounds   _____
    (d) any pairs of compounds related as isomers _____
    (e) any compound with a 3 ° carbon   _____

2. Write the IUPAC name   of A   _____

                               of B   _____

                               of E   _____

3. Write the structures of E and F using the geometric figure method illustrated in the discussion of "Ring Compounds" in Section 1.1 of the text.

   _____   _____
                  E   F

4. Write the structure and give the IUPAC name for the straight chain alkane with one less carbon than A.

   _____

5. Write the full structure of CH_3

   _____

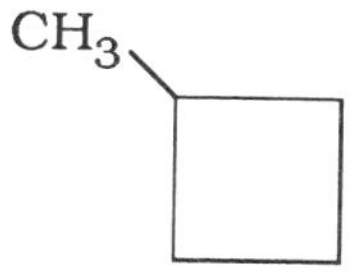

6. Write the full structure of   Cl

   _____

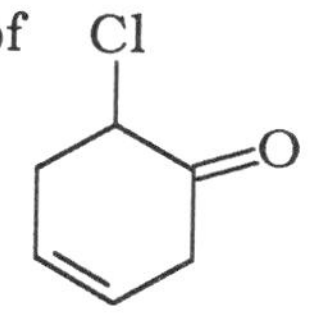

SET B

Examine the following compounds and supply the information requested.

A.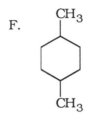

B.

C.

D.

E.

F.

7. Group in sets the letters of structures representing identical compounds.

_____

8. Write IUPAC names   for A _____
                        for E _____

9. To develop skill in the rapid recognition of alkyl groups regardless of how they are oriented in space, place the name of each group on the line beneath its structure.

(a)
$-CH_2$ CH_3

_____

(b)
$-CH_2-CH-CH_3$ CH_3

_____

(c)
CH_3 $CH-$ CH_3

_____

(d) $-CH_2$ CH_3
     CH_2-CH_2

_____

(e)
$CH_3-CH-CH_2-$ CH_3

_____

(f)
CH_3-C-CH_3 CH_3 CH_3

_____

(g)
CH_3-C- CH_3 CH_3

_____

(h)
$-CH-CH_3$ CH_3

_____

(i)
$CH_3-CH-CH_3$ CH_2-

_____

(j)
$CH_3-CH-CH_3$

_____

(k)
$CH_3-CH_2-CH-CH_3$

_____

(l)
$CH_3-CH-CH_2-CH_3$

_____

(m)

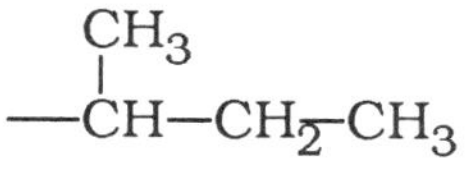

$$CH_3$$
$$-CH-CH_2-CH_3$$

(n)

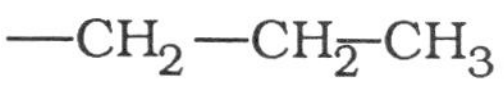

$$-CH_2-CH_2-CH_3$$

(o)

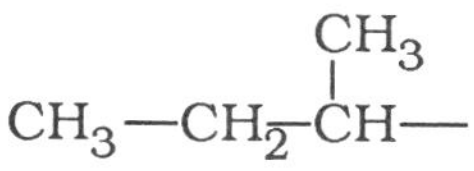

$$CH_3$$
$$CH_3-CH_2-CH-$$

———————

———————

———————

(p)

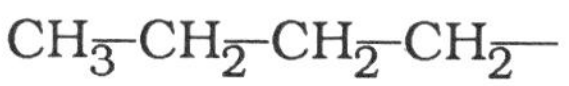

$$CH_3-CH_2-CH_2-CH_2-$$

(q)

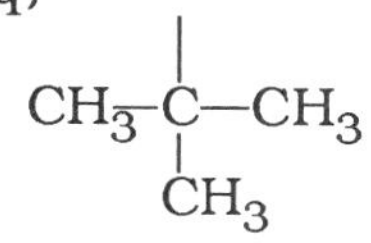

$$CH_3-C-CH_3$$
$$CH_3$$

(r)

$$CH_3-CH-CH_2-CH_3$$

———————

———————

———————

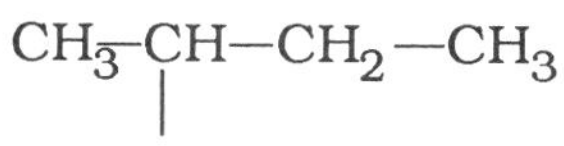

10. To develop the skill of looking at a structure and visualizing it with its longest continuous chain on one horizontal line, write the IUPAC names of each of the following.

(a)

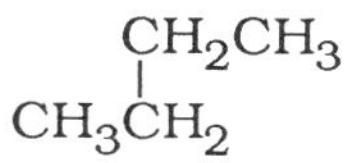

$$CH_2CH_3$$
$$CH_3CH_2$$

(b)

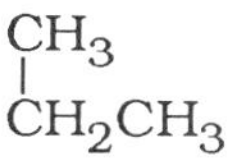

$$CH_3$$
$$CH_2CH_3$$

(c)

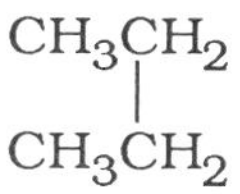

$$CH_3CH_2$$
$$CH_3CH_2$$

———————

———————

———————

(d)

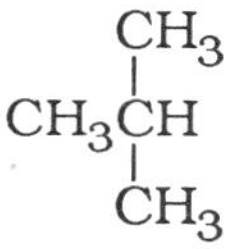

$$CH_3$$
$$CH_3CH$$
$$CH_3$$

(e)

$$CH_3$$
$$CH_2$$
$$CH_3$$

———————

———————

———————

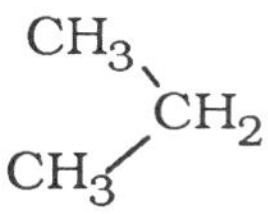

Before going on to the Self-Testing Questions, be sure to work all of the Practice Exercises and Review Exercises in the text first. As always, use the Self-Testing Questions as a "final exam" for the chapter.

SELF–TESTING QUESTIONS

COMPLETION

1. If compound X boils at 176 °C, it is almost certainly in the family of _____ (ionic or molecular) compounds. We would assign it to the organic class if, when analyzed, it was found to contain _____.

2. The structure

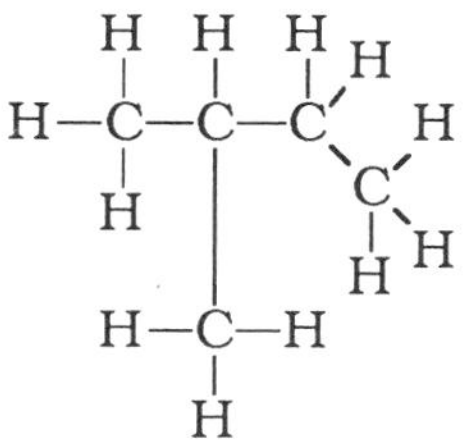

could be most neatly condensed to _____

3. The structure

$$H-\overset{\overset{\displaystyle H}{|}}{\underset{\underset{\displaystyle H}{|}}{C}}-\overset{\overset{\displaystyle H}{|}}{\underset{\underset{\displaystyle H}{|}}{C}}-\overset{\overset{\displaystyle H}{\overset{|}{\overset{\displaystyle H}{\underset{\displaystyle H}{\overset{|}{C}}}}-H}}{\underset{\underset{\displaystyle H-C-H}{|}}{C}}-H$$

could be most neatly condensed to _____

4. Are the compounds whose formulas are given in questions 2 and 3 isomers or are they identical?

_____

5. Write full structural formulas for each of the following:

_____   _____   _____   _____

(a) CH_5N (b) CH_2Br_2 (c) C_3H_8 (d) C_3H_4

6. Write the condensed structure of at least one isomer of $CH_3CH_2CH_2OH$ (C_3H_8O).

_____

7. The IUPAC name of $CH_3CH_2CH_2CH_3$ is _____.

8. The structure of 1-chloro-2,3-dimethylbutane is _____

9. The structure of isobutane is _____

10. Write the structure of each compound on the line above the name.

(a) _____   (b) _____   (c) _____

    isopropyl chloride   *t*-butyl bromide   isobutyl chloride

(d) _____   (e) _____   (f) _____

    sec-butyl bromide   butyl iodide   propyl chloride

11. Give the common names of the following compounds
    (a) the isomer of butane with a 3 ° carbon _____
    (b) the isomer of butane with two 2 ° carbons _____
    (c) the isomer of butane with three 1 ° carbons _____

12. When hexane burns completely in air the balanced equation is _____

13. The IUPAC names and structures of all the monobromo derivatives of pentane are:

_____   _____   _____

14. Pentane has an isomer that can form only one monochloro derivative. Write the structure of this isomer of pentane. _____

15. The structure of 1,3-dimethylcyclohexane is _____

MULTIPLE-CHOICE

1.   The compounds in which of the following pairs are related as isomers?

    (a) CH_2=CH–CH_2–OH and $\overset{\overset{\displaystyle HO}{|}}{CH_2}$–CH–$CH_2$

    (b) H_2 and D_2

    (c) CH_3–CH_2–NH–CH_3 and CH_3–CH_2–CH_2–NH_2

    (d) CH_3–C≡C–CH_2NH_2 and CH_3–CH_2–CH_2–C≡N

2.   According to all the rules about allowed numbers of bonds, which compound(s) should not be possible?
    (a)   AlPO4

    (b) CH_3CH=C=CH_2 (c) $H–O–\overset{\overset{\displaystyle O}{||}}{C}–O–H$ (d) CH=NH

3.   A structural formula for C_5H_{12} would be

    (a) $CH_3(CH_2)_3CH_3$ (b) $CH_3CH_2CH_2CH_2CH_3$

    (c) $CH_3\underset{\underset{\displaystyle CH_3}{|}}{C}HCH_2CH_3$ (d) $CH_3\overset{\overset{\displaystyle CH_3}{|}}{\underset{\underset{\displaystyle CH_3}{|}}{C}}CH_3$

4.   Which (if any) of the choices in question 3 are identical compounds?
5.   Which (if any) of the choices in question 3 are isomers?

6.   A family of organic compounds containing only carbon and hydrogen and having only single bonds are the
    (a) alkenes   (c) alkynes
    (b) alkanes   (d) cycloalkanes

7.   The combustion of butane produces
    (a) butylene and water   (c) an alcohol
    (b) CO_2 and H_2O (d) cyclohexane

8.   The compound CH_3–⬡ is

    (a) methylcyclopentane   (c) cycloheptane
    (b) a saturated compound   (d) cyclohexane

9. What is the name of the following compound?

$$CH_3CH_2CH_2\underset{\underset{\underset{CH_3-CH-CH_3}{|}}{\overset{\overset{CH_2CH_2CH_3}{|}}{CH}}}{C}HCH_2CH_2CH_2CH_3$$

(a) 4-isobutyl-5-propylnonane
(b) 2-methyl-3,4-dipropyloctane
(c) 5-propyl-6-isopropylnonane
(d) 4-isopropyl-5-propylnonane

10. The chlorination of 2-methylpentane would produce how many isomeric monochloro compounds?

(a) 6   (b) 5   (c) 4   (d) 3

11. Because carbon and hydrogen have very similar electronegativities, hydrocarbons are generally

(a) electropositive   (c) polar
(b) nonpolar   (d) hydrogen-bonded

12. The compound shown has how many hydrogens per molecule?

(a) 8   (b) 7   (c) 12   (d) 10

13. The common name of $CH_3\underset{\underset{CH_3}{|}}{-CH}-CH_2-Cl$ is

(a) butyl chloride   (c) *t*-butyl chloride
(b) isobutyl chloride   (d) *sec*-butyl chloride

14. The common name of $\underset{\underset{CH_3 \quad CH_3}{}}{CH_2}$ is

(a) cyclopropane   (c) propane
(b) isopropane   (d) dimethylmethane

15. The compound $CH_3-C\equiv C-H$ is

(a) soluble in water   (c) soluble in gasoline
(b) a hydrocarbon   (d) unsaturated

ANSWERS

ANSWERS TO DRILL EXERCISES

I. Exercises on Condensed Structures

Set A

1. $CH_3CH_2CH_3$ or CH_3—CH_2—CH_3 (Single bonds along the main chain may be shown or they may be understood. You will see both practices often.)

2. $CH_3CH_2CH_2CH_3$ or CH_3—CH_2—CH_2—CH_3

3. $CH_3CH_2CH_2CH_2CH_3$ or CH_3—CH_2—CH_2—CH_2—CH_3

4.
$$CH_3CHCH_2CH_2\underset{\underset{\displaystyle CH_3}{|}}{\overset{\overset{\displaystyle CH_3}{|}}{C}}CH_3$$
$$\underset{\displaystyle CH_3}{|}$$

5.
$$CH_3\overset{\overset{\displaystyle CH_3}{|}}{C}H—O—CH_2\underset{\underset{\displaystyle CH_3}{|}}{C}HCH_3$$

6.
$$CH_3\overset{\overset{\displaystyle CH_3}{|}}{\underset{\underset{\displaystyle CH_3}{|}}{C}}CH_2CH_2CH_2\overset{\overset{\displaystyle OH}{|}}{C}HCH_3$$

7. $CH_3CH_2CH_2OH$ (often seen as $CH_3CH_2CH_2$—O—H)

8. $HOCH_2CH_2CH{=}CHCH_3$ (Do not write this as $OHCH_2CH_2CH{=}CHCH_3$. It implies O—H—CH_2— etc., which is wrong.)

9.
$$CH_3—O—CH{=}CHCH_2\underset{\underset{\displaystyle CH_3}{|}}{C}H\overset{\overset{\displaystyle O}{\|}}{C}OH$$

10. 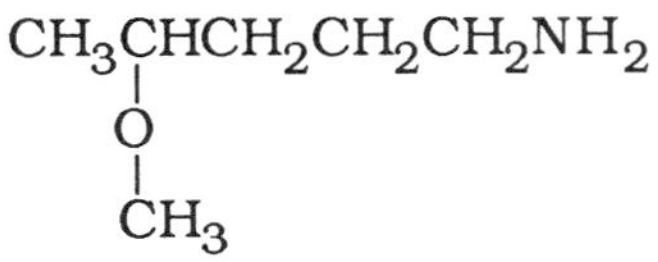 $CH_3CHCH_2CH_2CH_2NH_2$
      |
      O
      |
      CH_3

Set B

1. 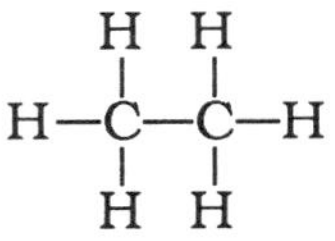
```
    H   H
    |   |
H—C—C—H
    |   |
    H   H
```

2. 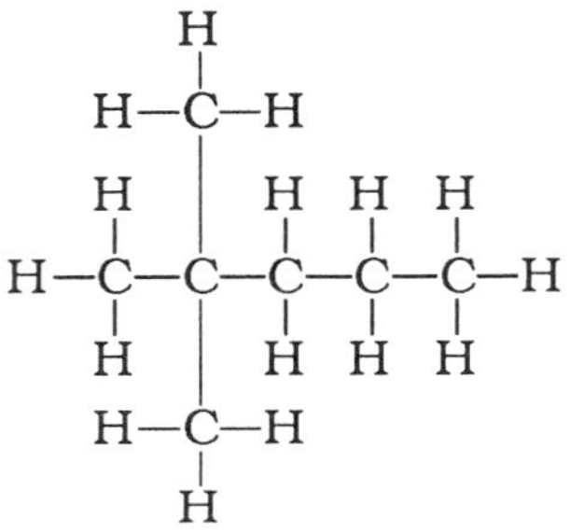
```
        H
        |
      H—C—H
  H   |   H  H  H
  |   |   |  |  |
H—C—C—C—C—C—H
  |   |   |  |  |
  H   |   H  H  H
      H—C—H
        |
        H
```

3. 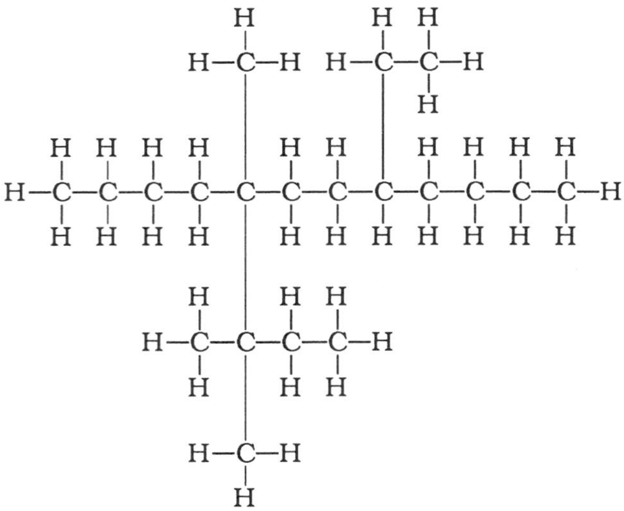
```
              H       H  H
              |       |  |
            H—C—H   H—C—C—H
                          |
                          H
  H  H  H  H     H  H     H  H  H  H
  |  |  |  |     |  |     |  |  |  |
H—C—C—C—C—C—C—C—C—C—C—C—H
  |  |  |  |     |  |     |  |  |  |
  H  H  H  H     H  H     H  H  H  H

              H     H  H
              |     |  |
            H—C—C—C—C—H
              |     |  |
              H     H  H

                H—C—H
                  |
                  H
```

4. 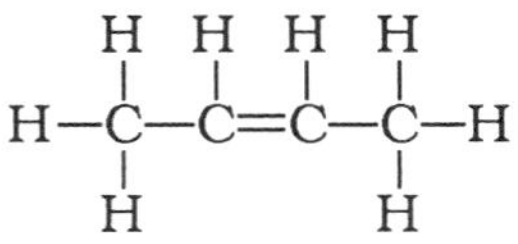
```
  H  H  H  H              H        H
  |  |  |  |              |        |
H—C—C=C—C—H      H—C—C=C—C—H        (The two are equivalent.)
  |     |    or    |  |  |  |
  H     H          H  H  H  H
```

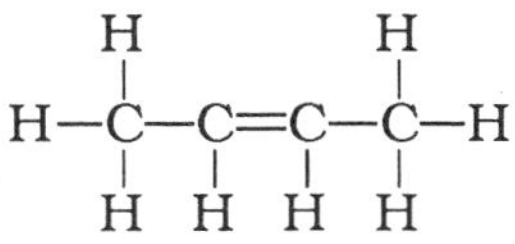

5.
$$H-O-\overset{\displaystyle H}{\underset{\displaystyle H}{C}}-\overset{\displaystyle H}{\underset{\displaystyle H}{C}}-H$$

6.
$$H-\overset{\displaystyle H}{\underset{\displaystyle H}{C}}-O-\overset{\displaystyle O}{C}-\overset{\displaystyle H}{\underset{\displaystyle H}{C}}-\overset{\displaystyle H}{\underset{\displaystyle H}{C}}-\overset{\displaystyle H}{N}-\overset{\displaystyle H}{\underset{\displaystyle H}{C}}-\overset{\displaystyle H}{\underset{\displaystyle H}{C}}-H$$

7.
$$H-C\equiv C-\overset{\displaystyle H}{\underset{\displaystyle H}{C}}-\overset{\displaystyle O}{C}-H$$

8.
$$H-\overset{\displaystyle H}{N}-\overset{\displaystyle H}{\underset{\displaystyle H}{C}}-\overset{\displaystyle H}{\underset{\displaystyle H}{C}}-\overset{\displaystyle H}{N}-H$$

II. Exercises on Devising Structures from Molecular Formulas

1.   H—O—O—H

2. $H-\overset{\displaystyle }{\underset{\displaystyle H}{N}}-\overset{\displaystyle }{\underset{\displaystyle H}{N}}-H$

3. $H-\overset{\displaystyle H}{\underset{\displaystyle H}{C}}-S-H$

4. $H-\overset{\displaystyle H}{\underset{\displaystyle H}{C}}-\overset{\displaystyle H}{\underset{\displaystyle H}{C}}-Br$ (Satisfy yourself that the following,

$H-\overset{\displaystyle H}{\underset{\displaystyle H}{C}}-\overset{\displaystyle Br}{\underset{\displaystyle H}{C}}-H$ and $H-\overset{\displaystyle H}{\underset{\displaystyle H}{C}}-\overset{\displaystyle H}{\underset{\displaystyle Br}{C}}-H$ are identical.)

5. $H-\overset{\displaystyle H}{C}=\overset{\displaystyle H}{C}-Cl$

III. Exercises on Structural Features of Hydrocarbons

Set A

1.   (a)   A, B, D, and E

   (b)   A, C, and D   (The term "straight-chain" cannot apply to rings. Among open-chain compounds, as long as the atoms C, O, N, or S are all in a continuous sequence with or without double or triple bonds, the chain is "straight." If the atoms O, N, or S are appended as part of substituents, the chain is still straight if all the carbons occur in a continuous sequence.)

   (c)   B (Ring compounds are not classified as branched, either.)

(d)   A and B; C and E

(e)   B (Only saturated carbons are designated as 1°, 2°, or 3°.)

2.   A: hexane

B: 2,3-dimethylbutane

E: cyclohexane

3.   E    F O

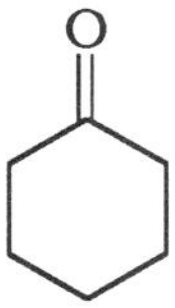

4. $CH_3CH_2CH_2CH_2CH_3$ pentane

5. 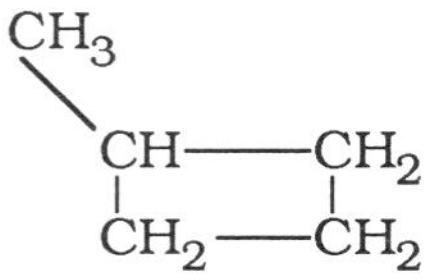
$$CH_3$$
$$CH-CH_2$$
$$CH_2-CH_2$$

6.

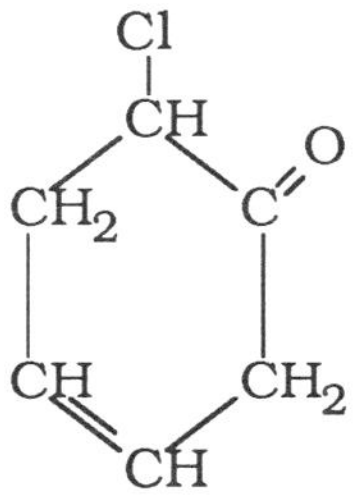

Set B

7.   A and C; B and D

8.   A 1,1-dimethylcyclohexane

E 1,3-dimethylcyclohexane   (NOT 1,5-dimethylcyclohexane from counting around the ring in the wrong way. You count from position one, picked as the location of one CH_3 group, along the shortest path around the ring to the next group.)

9.   (a)   ethyl   (Actually, ethyl group, but we may omit "group.")

(b)   isobutyl   (Any four-carbon alkyl group must be one of the four butyl groups. The two derived from isobutane are either the isobutyl or the *t*-butyl group.)

(c)   isopropyl   (Any three-carbon alkyl group must be one of the two propyl groups—propyl or isopropyl.)

(d)   butyl   (Straighten out the chain and you have a group based on butane, not isobutane. The two butyl groups based on butane are the butyl and the *sec*-butyl groups.)

(e)   isobutyl   (Compare b and e)

(f)   *t*-butyl   (It's the only butyl group where the unused bond is at a 3 ° carbon.)

(g)   *t*-butyl   (This is simply f rotated through part of a circle.)

(h)   isopropyl   (A C_3-alkyl group; it must be either propyl or isopropyl.)

(i)   isobutyl   (Compare b, e, and i.)

(j)   isopropyl   (Compare c, h, and j.)

(k)   *sec*-butyl   (It's the only butyl group where the unused bond is at a 2 ° carbon.)

(l)   *sec*-butyl   (Compare with k.)

(m)   *sec*-butyl   (Carefully compare k, l, and m. After chain straightening, the chain is straight in all three and the free bond comes from a 2 ° carbon.)

(n)   propyl

(o)   *sec*-butyl   (Compare with m and the accompanying note.)

(p)   butyl

(q)   *t*-butyl   (Just f tipped upside down. They have to be the same. Do you become someone else if you stand on your head?)

(r)   *sec*-butyl

10.   (a)   butane   (d)   isobutane or 2-methylpropane

      (b)   propane   (e)   propane

      (c)   butane

ANSWERS TO SELF–TESTING QUESTIONS

Completion

1.   molecular; carbon

2. $CH_3CHCH_2CH_3$ 3. CH_3
             | |
             CH_3 $CH_3CH_2CHCH_3$

(Your answers to 2 and 3 may be correct but still not look exactly like those given here. All that matters is that they show the correct nucleus-to-nucleus sequence while obeying the accepted conventions for condensing. One of these conventions is that as much of the structure as possible is written out on one line. Should you have questions about your answers, consult your instructor or one of the assistants.)

4.   identical

5.   (a)
```
     H  H
     |  |
  H—C—N—H
     |
     H
```
(b)
```
     H
     |
 Br—C—Br
     |
     H
```
(c)
```
     H  H  H
     |  |  |
  H—C—C—C—H
     |  |  |
     H  H  H
```
(d)
```
     H
     |
  H—C—C≡C—H
     |
     H
```
or
```
     H        H
     |        |
  H—C=C=C—H
              
```

6. There are two isomers: $CH_3-O-CH_2CH_3$ and $CH_3\underset{\underset{\textstyle OH}{|}}{C}HCH_3$.

7. butane

8. $Cl-CH_2\underset{\underset{\textstyle CH_3}{|}}{\overset{\overset{\textstyle CH_3}{|}}{C}}HCHCH_3$ 9. $CH_3\underset{\underset{\textstyle CH_3}{|}}{C}HCH_3$

10. (a) $CH_3\underset{\underset{\textstyle Cl}{|}}{C}HCH_3$ (b) $CH_3\underset{\underset{\textstyle CH_3}{|}}{\overset{\overset{\textstyle CH_3}{|}}{C}}-Br$ (c) $CH_3\underset{\underset{\textstyle }{}}{\overset{\overset{\textstyle CH_3}{|}}{C}}HCH_2-Cl$

   (d) $CH_3\underset{\underset{\textstyle Br}{|}}{C}HCH_2CH_3$ (e) $CH_3CH_2CH_2CH_2-I$ (f) $CH_3CH_2CH_2-Cl$

11. (a) isobutane   (b) butane   (c) isobutane
12. $2C_6H_{14} + 19O_2 \rightarrow 12CO_2 + 14H_2O$

13. $CH_3CH_2CH_2CH_2CH_2Br$ $CH_3CH_2CH_2\underset{\underset{\textstyle Br}{|}}{C}HCH_3$ $CH_3CH_2\underset{\underset{\textstyle Br}{|}}{C}HCH_2CH_3$

   1-bromopentane   2-bromopentane   3-bromopentane

14. $CH_3-\underset{\underset{\textstyle CH_3}{|}}{\overset{\overset{\textstyle CH_3}{|}}{C}}-CH_3$ 15.

Multiple–Choice

1. c and d	6. b and d	11. b
2. d	7. b	12. d
3. a, b, c, and d	8. b	13. b
4. a and b	9. d	14. c
5. a (or b), c, and d are isomers	10. b	15. b, c, d

2

UNSATURATED HYDROCARBONS

The two topics in this chapter that are most important in our development of the molecular basis of life are:

1. The addition reactions of the carbon-carbon double bond, particularly those with hydrogen and water. (Double bonds occur in all fats and oils. They also occur in the intermediate stages of the chemical breakdown of sugars, fatty acids, and amino acids as well as in the enzymes for these reactions.)

2. The fact that benzene rings in aromatic hydrocarbons do not behave as alkenes, in spite of being highly unsaturated. These rings take part in substitution reactions rather than in addition reactions. (The benzene ring occurs in a few amino acids and in many drugs.)

OBJECTIVES

The specific objectives for this chapter are the following. Return to these to test yourself after you have completed studying all of the chapter, including the Practice Exercises and Review Exercises.

1. Predict whether a given alkene can exist as cis and trans isomers.
2. Write the structures of alkenes from their IUPAC names.
3. Write the IUPAC names of alkenes.
4. Name cycloalkenes.
5. Write equations for the reactions of a given alkene with:
    (a) hydrogen; (b) a halogen: (c) a hydrogen halide; (d) sulfuric acid; and (e) water (in the presence of acid).

6.   Use Markovnikov's Rule to predict the correct products in the addition reactions of alkenes.
7.   Explain why Markovnikov's Rule works.
8.   Predict if a hydrocarbon will be oxidized by ozone, permanganate, and dichromate.
9.   Correctly use terms associated with polymerizations, like polymer, monomer, polymerization, and promoter.
10.   Describe the structure of benzene in molecular orbital terms.
11.   Explain why benzene takes part in substitution reactions rather than in addition reactions.
12.   Write equations for the reactions of benzene that result in halogenation, nitration, and sulfonation.
13.   Write the common names of all of the monosubstituted benzenes given in this chapter.
14.   Devise names for derivatives of benzene having two or more substituents.
15.   Define each of the terms in the Glossary and give examples or illustrations where applicable.

GLOSSARY

Addition Reaction. Any reaction in which two parts of a reactant molecule add to a double or a triple bond.

Alkene. A hydrocarbon whose molecules have double bonds.

Alkene Group. The carbon-carbon double bond.

Aliphatic Compound. Any organic compound whose molecules lack a benzene ring or a similar structural feature.

Alkyne. A hydrocarbon whose molecules have triple bonds.

Aromatic Compound. Any organic compound whose molecules have a benzene ring (or a feature very similar to this).

Carbocation. Any cation in which a carbon atom has just six outer level electrons; a carbonium ion.

Geometric Isomerism. Isomerism caused by restricted rotations that give different geometries to the same structural organization; cis-trans isomerism.

Geometric Isomers. Isomers with identical atomic organizations but different geometries; cis-trans isomers.

Macromolecule. A molecule with a very large formula mass, upwards of several thousand.

Markovnikov's Rule. In the addition of an unsymmetrical reactant to an unsymmetrical double bond of a simple alkene, the positive part of the reactant (usually H^+) goes to the carbon with the greater number of hydrogen atoms and the negative part goes to the other carbon of the double bond.

Monomer. Any compound that can be used to make a polymer.

Polymer. Any substance with a very high formula mass whose molecules have a repeating structural unit.

Polymerization. A chemical reaction that makes a polymer from a monomer.

KEY MOLECULAR "MAP SIGNS"

Key Molecular "Map Signs" In Organic Molecules	The Associated Chemical and Physical Properties
If the molecule is largely (or, of course, entirely) hydrocarbon-like	Expect the substance to be relatively insoluble in water and relatively soluble in organic solvents. Expect the substance to be less dense than water.
Alkanelike portions of all molecules	In these portions, expect no chemical changes involving water, acids, alkalis, or chemical oxidizing agents or reducing agents.
An alkene double bond Markovnikov's Rule applies	Adds H—H; becomes saturated Adds X—X; forms di-halo compounds (X = Cl, Br) Adds HX; forms alkyl halides (X = F, Cl, Br, I) Adds H_2O; forms alcohols Adds H_2SO_4; forms alkyl hydrogen sulfates Adds its own kind; polymers form
Benzene ring	Not like an alkene—gives substitution, not addition reactions. Ring can nitrate, sulfonate, and halogenate [with iron(III) salts as catalysts].

DRILL EXERCISES

The chapter in the text has a number of drills on the reactions of alkenes and benzene. Do these Exercises first.

EXERCISE I

To review condensed structures, write out the following as full structures.

1.   CH₃

2.

3.   NH₂

4.   H

EXERCISE II

As drill in several things—names, cis-trans isomers, and the recognition of nonequivalent positions—write the structures and the IUPAC names of all of the monochloro derivatives of all of the alkenes through the C$_4$ alkenes. Be sure to show cis and trans forms with correct geometry; for example, the structure of *cis*-2-chloro-2-butene is:

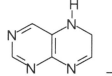

There are a total of 16 compounds.

EXERCISE III

Memorize the names and structures of the monosubstituted benzenes given in the text. Then study how to use the designations o-, m- and p- for disubstituted benzenes. When you have done this study, test yourself with these. Beneath each structure write the correct name.

1.   OH   2.   Cl   3.   CH₃   4.   NH₂

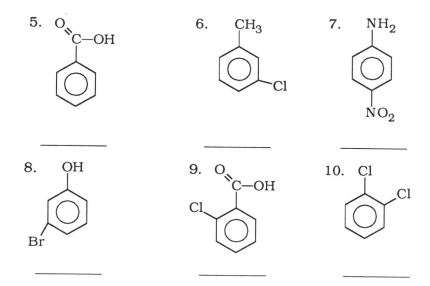

5. $\underset{\text{C}-\text{OH}}{\overset{\text{O}}{\|}}$

6. CH₃ ... Cl

7. NH₂ ... NO₂

8. OH ... Br

9. $\underset{\text{C}-\text{OH}}{\overset{\text{O}}{\|}}$ Cl

10. Cl ... Cl

How to Work Problems in Organic Reactions

The *overall goal* of the chapters on organic chemistry in the text is to learn the chemical properties of functional groups. *You know you've reached this goal when you can write the products made by the reaction of a given set of starting materials.* In other words, it is not enough to be able to recognize the names and structural features of functional groups—although you can't do anything else without starting here.

It's also not enough to memorize a sentence that summarizes a chemical fact, such as "alkenes react with hydrogen in the presence of a metal catalyst, heat, and pressure to give alkanes"—although memorizing such sentences is also absolutely essential to the goal. Such sentences are fundamental facts about the world in which we live. What you have to be able to do is apply such knowledge to a specific set of reactants.

A typical problem, for example, is "What forms, if anything, in the following reaction:

$$CH_3-CH=CH_2 \;+\; H_2 \quad \xrightarrow[\text{heat, pressure}]{\text{Ni}} \quad ?"$$

Here's the strategy to follow in all of the reactions of organic compounds that we will study.

1. Figure out the family or families to which the organic starting materials belong(s). You might even want to write the names of these families beneath the specific structures. (In our example, the family is "alkene.")

2. Review your "memory list" of chemical facts about all alkenes. In your mental "storage," you should have the chemical-fact sentences that you will prepare in the next Exercise.

   FOR EVERY FUNCTIONAL GROUP THAT WE STUDY YOU MUST PREPARE A MEMORIZED LIST OF CHEMICAL–FACT SENTENCES THAT SUMMARIZE THE CHEMICAL PROPERTIES THAT WE STUDY.

3. If you go through the list and find no match to the stated problem, then assume that no reaction takes place and write "no reaction" as the answer.

4. When you find the match between the specific problem and one of the listed chemical properties, stop and construct the structure of the answer. The many worked examples in the text develop the patterns for doing this, and study them thoroughly first.

EXERCISE IV. SUMMARIZING THE CHEMICAL PROPERTIES OF ALKENES

When you have completed this exercise, you will have a complete sentence summarizing each kind of chemical reaction of the carbon-carbon double bond that we have studied. These statements are the chemical properties of this double bond that must be memorized, but they have to be learned in such a way that you can apply them to specific situations. Hence, following this Exercise there are several drill problems to give you practice.

As you work each specific exercise among these drills, repeat to yourself the sentence statement that summarizes the property being illustrated. This kind of repeated reinforcement will soon give you a surprisingly good working knowledge of these organic reactions, and you will be able to apply what you have learned to much more complicated situations with ease.

1. The alkene double bond reacts with hydrogen (in the presence of a metal catalyst and heat) to give _____ .
2. The alkene double bond reacts with chlorine to give _____ .
3. The alkene double bond reacts with bromine to give _____ .
4. The alkene double bond reacts with hydrogen chloride to give _____ .
5. The alkene double bond reacts with hydrogen bromide to give _____ .
6. The alkene double bond reacts with concentrated sulfuric acid to give _____ .
7. The alkene double bond reacts with water (in the presence of an acid catalyst) to give

    _____ .
8. The complete combustion of any alkene produces _____ and _____ .

Another way to organize chemical facts about a functional group such as the carbon-carbon double bond is by means of a 5 x 8" note card. An example is given below, but for the remaining functional groups it is vitally important that you prepare the cards yourself. Part of the learning process is in this preparation, and having someone else do it for you robs you of that benefit.

(a) Markovnikov's rule applies

Notice that the functional group is put in the center and its chemical properties are arranged about it. All of the arrows point outward. Later on, it will be useful to prepare cards summarizing key reactants, like H_2O or H_2 on which you'll list all of the reactions studied involving these substances.

EXERCISE V. DRILL ON THE ADDITION OF HYDROGEN TO ALKENES

Write the structures of the products of the following reactions. If no reaction occurs, write "no reaction." (See Example 2.3 in the text.) In each, assume that the system is heated.

1.
$$CH_2{=}\overset{\overset{\displaystyle CH_3}{|}}{C}CH_2CH_3 \quad + \; H_2 \xrightarrow{\;\;Ni\;\;} \underline{\hspace{3cm}}$$

2.
$$CH_3CH{=}CHCH_3 \;\; + \; H_2 \xrightarrow{\;\;Ni\;\;} \underline{\hspace{3cm}}$$

3. 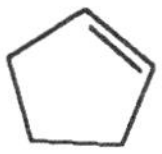
$\quad + \; H_2 \xrightarrow{\;\;Ni\;\;} \underline{\hspace{3cm}}$

4.
$$CH_2{=}\overset{\overset{\displaystyle CH_3}{|}}{C}CH{=}CH_2 \; + \; 2H_2 \xrightarrow{\;\;Ni\;\;} \underline{\hspace{3cm}}$$

5. 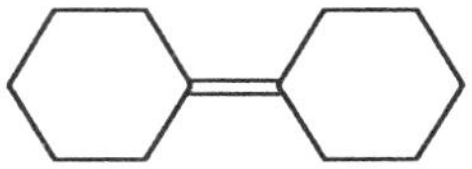
$+ \; H_2 \xrightarrow{\;\;Ni\;\;} \underline{\hspace{3cm}}$

EXERCISE VI. DRILL ON THE ADDITION OF CHLORINE OR BROMINE TO ALKENES

Write the structures of the products of the following reactions. If no reaction occurs, write "no reaction." (See Example 2.4 in the text.)

1. $CH_2{=}CHCH_3 \; + \; Cl_2 \quad \longrightarrow \underline{\hspace{3cm}}$

2.
$$CH_2{=}\overset{\overset{\displaystyle CH_3}{|}}{C}CH_2CH_2CH_3 \; + \; Br_2 \quad \longrightarrow \underline{\hspace{3cm}}$$

3. 
$+ \; Cl_2 \quad \longrightarrow \underline{\hspace{3cm}}$

4. $CH_3CH{=}CHCH{=}CH_2 \; + \; 2Br_2 \quad \longrightarrow \underline{\hspace{3cm}}$

5.
$CH_2{=}$ $+ \; Cl_2 \quad \longrightarrow \underline{\hspace{3cm}}$

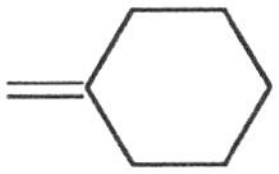

EXERCISE VII. DRILL ON THE ADDITION OF H—Cl(*g*) OR H—Br(*g*) TO ALKENES

Write the structures of the products of each reaction. If no reaction occurs, write "no reaction."

1. $CH_3CH=CHCH_3$ + H—Cl $\longrightarrow$ _____

2. ⬠ + HBr $\longrightarrow$ _____

3.
$$\overset{CH_3}{\underset{|}{CH_2=CCH_2CH_2CH_3}}$$ + H—Cl $\longrightarrow$ _____

4. ⬡—CH_3 + HBr $\longrightarrow$ _____

5. $CH_2=CHCH_2CH_2CH_2CH=CH_2$ + 2HBr $\longrightarrow$ _____

EXERCISE VIII. DRILL ON THE ADDITION OF H₂O TO ALKENES

Write the structures of the products of each reaction. If no reaction occurs, write "no reaction." Note: H⁺ represents an acid catalyst.

1. $CH_2=CHCH_2CH_3$ + H_2O $\xrightarrow{H^+}$ _____

2. $CH_3CH=CHCH_3$ + H_2O $\xrightarrow{H^+}$ _____

3.
$$\overset{CH_3}{\underset{|}{CH_2=CCH_2CH_3}}$$ + H_2O $\xrightarrow{H^+}$ _____

4. ⬠ + H_2O $\xrightarrow{H^+}$ _____

5. CH_3CH_2—⬡ + H_2O $\xrightarrow{H^+}$ _____

SELF-TESTING QUESTIONS

Use the self-testing questions as a final examination for the chapter after you have worked the exercises in the text and those given above.

COMPLETION

1. In a molecule of ethylene, the number of atoms whose nuclei are all in the same plane equals
    _____.

2. The structures of the geometric isomers of 2-pentene are:

    _____   _____

3. In a cis isomer, two reference groups lie on _____ side(s) of the double bond.

4. Write the overall equation for the reaction of 1,4-dimethylbenzene with each of the following reagents, assuming that only monosubstitution occurs.

    (a) Br_2 ($FeBr_3$ or Fe catalyst)   _____

    (b) HNO_3 (in concd H_2SO_4)   _____

    (d) H_2SO_4, concd   _____

5. When hydrogen chloride adds to isobutylene, the intermediate organic cation is _____ and has the structure:

    $(1°, 2°, or 3°)$

6. The other organic cation that, at least on paper, could also form from isobutylene is _____ and has the structure:

    $(1°, 2°, or 3°)$

7. Arrange the following organic cations in the order of their increasing stability by arranging the letters that identify them in the correct order on this line:

    _____ < _____ < _____

    _____

    least stable   most stable

    A. $CH_3CH_2^+$

    B. $CH_3CH_2\overset{\displaystyle CH_3}{\underset{\displaystyle CH_3}{\overset{\textstyle |}{\underset{\textstyle |}{C}}}}{}^+$

    C. $CH_3CH_2\overset{+}{C}HCH_3$

8. Complete these equations by writing the structure(s) of the product(s).

    (a)

    $CH_3\underset{\underset{\textstyle CH_3}{\textstyle |}}{C}{=}CHCH_3 + HCl(gas) \longrightarrow$ _____

    (b)

    $CH_2{=}\underset{\underset{\textstyle CH_3}{\textstyle |}}{C}CH_2CH_3 + H_2O \xrightarrow{\ H^+\ }$ _____

(c)

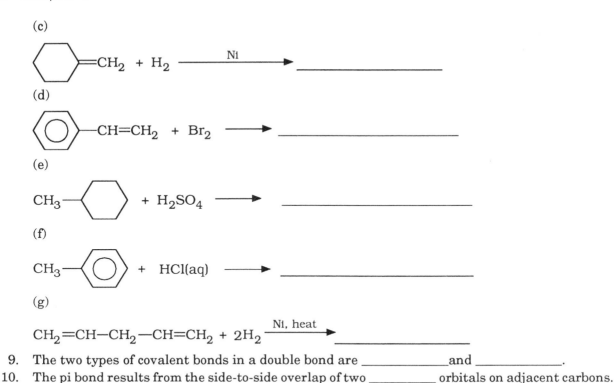

$$\text{(ring)}=CH_2 + H_2 \xrightarrow{\text{Ni}} \underline{\hspace{4cm}}$$

(d)

$$\text{(benzene)}-CH=CH_2 + Br_2 \longrightarrow \underline{\hspace{4cm}}$$

(e)

$$CH_3-\text{(ring)} + H_2SO_4 \longrightarrow \underline{\hspace{4cm}}$$

(f)

$$CH_3-\text{(benzene)} + HCl(aq) \longrightarrow \underline{\hspace{4cm}}$$

(g)

$$CH_2=CH-CH_2-CH=CH_2 + 2H_2 \xrightarrow{\text{Ni, heat}} \underline{\hspace{3cm}}$$

9.   The two types of covalent bonds in a double bond are _____ and _____.

10.   The pi bond results from the side-to-side overlap of two _____ orbitals on adjacent carbons.

MULTIPLE–CHOICE

1.   A family of organic compounds containing only carbon and hydrogen and having only single bonds are the
(a) alkenes   (c) alkynes
(b) alkanes   (d) cycloalkanes

2.   A family of organic compounds whose molecules will add water (under an acid catalysis) and change into alcohols are the
(a) alkenes   (c) aromatic hydrocarbons
(b) alkanes   (d) cycloalkenes

3.   A chemist, handed a sample of an organic compound, was told that it was either

$$CH_2=CHCH_2CH_3 \quad \text{or} \quad CH_3CH_2CH_2CH_3$$

Which one it was could be decided by determining if the sample would react with
(a)   sodium hydroxide
(b)   hydrogen (with nickel present and heated)
(c)   sodium chloride
(d)   water (in the presence of an acid catalyst)

4.   An aromatic hydrocarbon can be expected to undergo
(a)   substitution reactions   (c)   addition reactions
(b)   reaction with water   (d)   no reactions

5.   The combustion of 1-butene will produce
(a)   butylene and water   (c)   an alcohol
(b)   carbon dioxide and water   (d)   cyclobutane

6. An isomer of $\begin{array}{cc} CH_3 & CH_3 \\ | & / \\ CH=CH \end{array}$ is

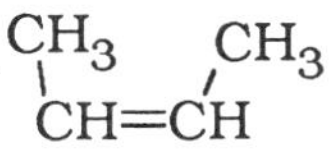

(a) 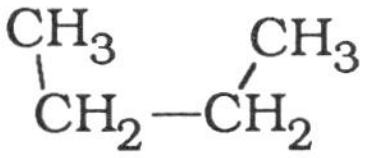 $\begin{array}{cc} CH_3 & CH_3 \\ | & / \\ CH_2-CH_2 \end{array}$

(b) 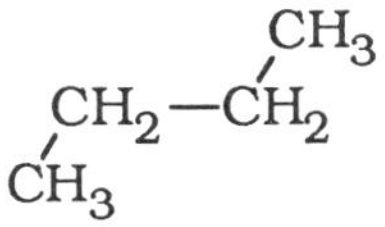 $\begin{array}{c} CH_3 \\ / \\ CH_2-CH_2 \\ / \\ CH_3 \end{array}$

(c) $\begin{array}{c} CH_3 \\ / \\ CH=CH \\ | \\ CH_3 \end{array}$

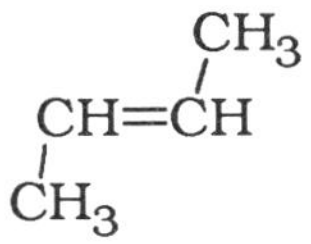

(d) 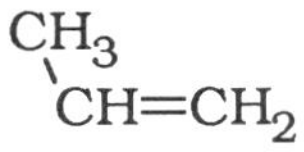 $\begin{array}{c} CH_3 \\ \backslash \\ CH=CH_2 \end{array}$

7. $CH_3{}^+$ is the
   (a) methyl radical
   (b) methyl group
   (c) methyl anion
   (d) methyl cation

8. The substance $CH_3-\overset{\displaystyle CH_3}{\underset{\displaystyle CH_3}{\overset{|}{\underset{|}{C}}}}-OH$ could be made by addition of water to

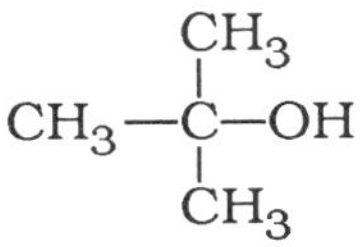

(a) $CH_3CH=CHCH_3$

(b) 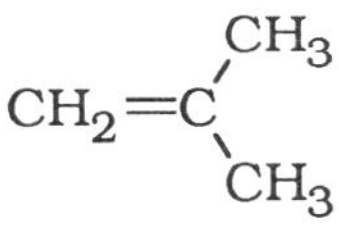 $\begin{array}{c} CH_3 \\ / \\ CH_2=C \\ \backslash \\ CH_3 \end{array}$

(c) $CH_2=CHCH_2CH_3$

(d) none of these

9. The substance $Cl-CH_2-\overset{\displaystyle CH_3}{\overset{|}{CH}}-CH_3$ could be made by the addition of HCl to

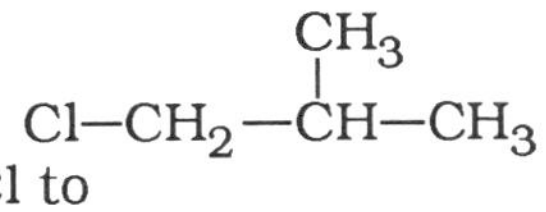

(a) $CH_3CH=CHCH_3$

(b) 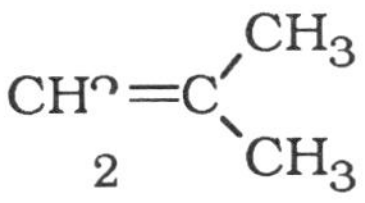 $\begin{array}{c} CH_3 \\ / \\ CH_2=C \\ \backslash \\ CH_3 \end{array}$

(c) $CH_2=CHCH_2CH_3$

(d) none of these

10. The substance ⬡ is

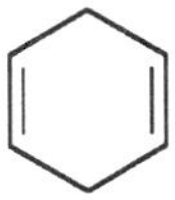

(a) benzene
(b) aromatic
(c) a diene
(d) none of these

11. Action of aqueous alkali on CH₃—⟨benzene ring⟩ could be expected to produce

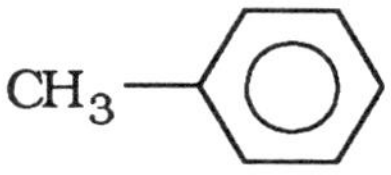

(a) 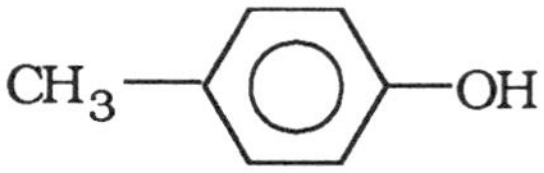 CH₃—⟨benzene ring⟩—OH

(b) 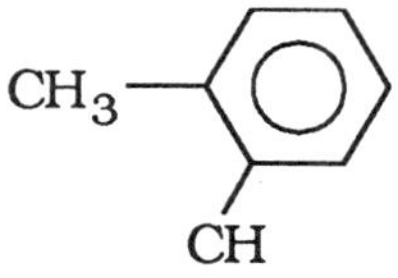 CH₃—⟨benzene ring⟩
CH

(c) 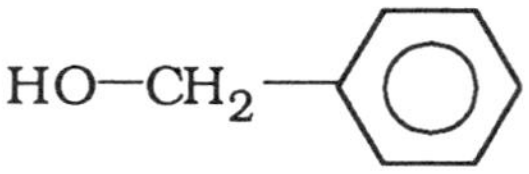 HO—CH₂—⟨benzene ring⟩

(d) no reaction

12. Action of aqueous hydrochloric acid on CH₃—⟨benzene ring⟩ could be expected to produce

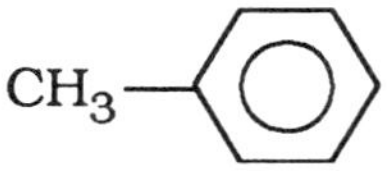

(a) 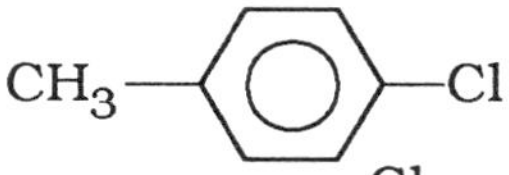 CH₃—⟨benzene ring⟩—Cl

(b) 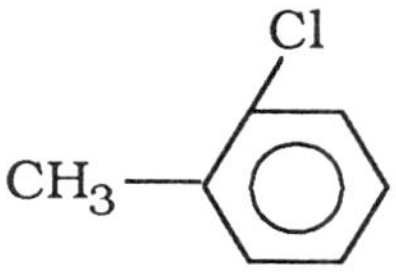 CH₃—⟨benzene ring⟩
Cl

(c) 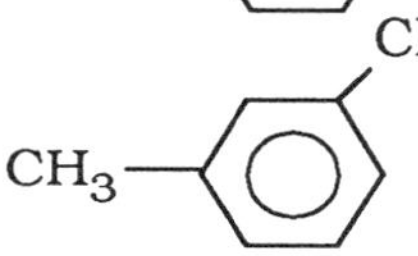 CH₃—⟨benzene ring⟩
Cl

(d) no reaction

13. Action of aqueous potassium permanganate on benzene could be expected to produce

(a) 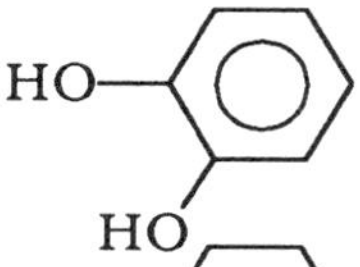 HO—⟨benzene ring⟩
HO

(b) 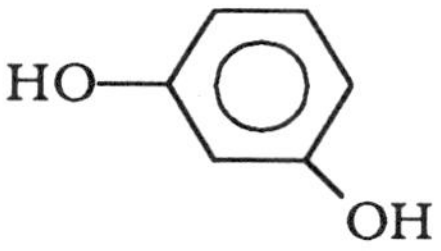 HO—⟨benzene ring⟩
OH

(c) 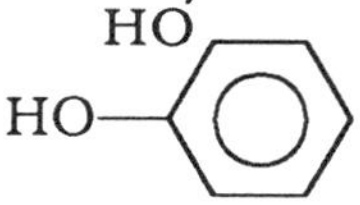 HO—⟨benzene ring⟩

(d) no reaction

14. The trans isomer of ⟨cyclopentane with CH₃ groups⟩ is

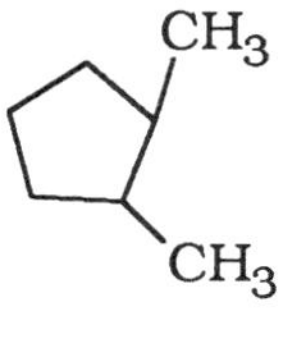

(a) 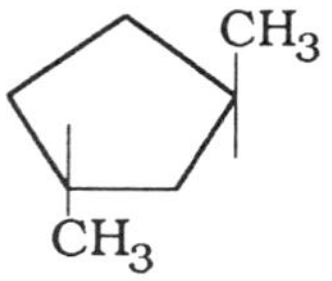 ⟨cyclopentane with CH₃ and CH₃⟩

(b) 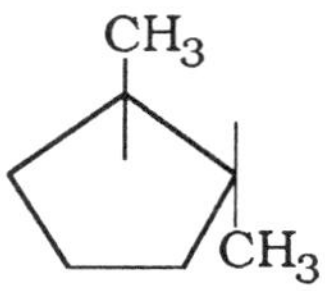 ⟨cyclopentane with CH₃ and CH₃⟩

(c) 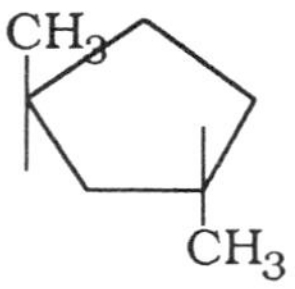 ⟨cyclopentane with CH₃ and CH₃⟩

(d) 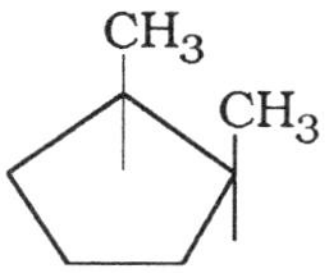 ⟨cyclopentane with CH₃ and CH₃⟩

ANSWERS

ANSWERS TO DRILL EXERCISES

Exercise I

1.

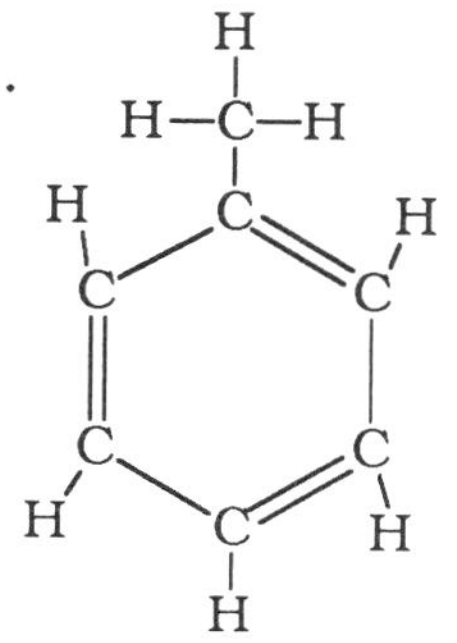

2.

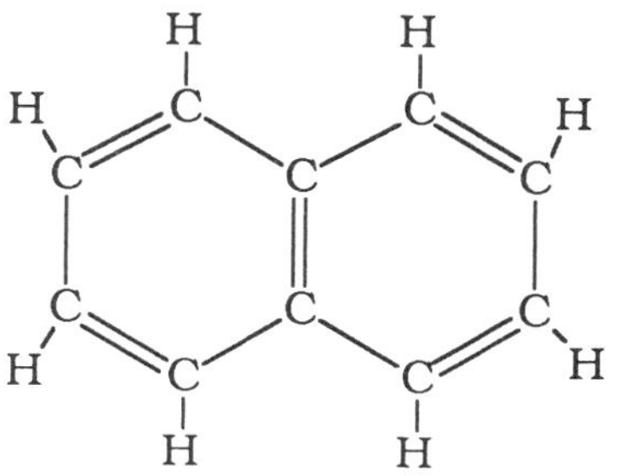

3.

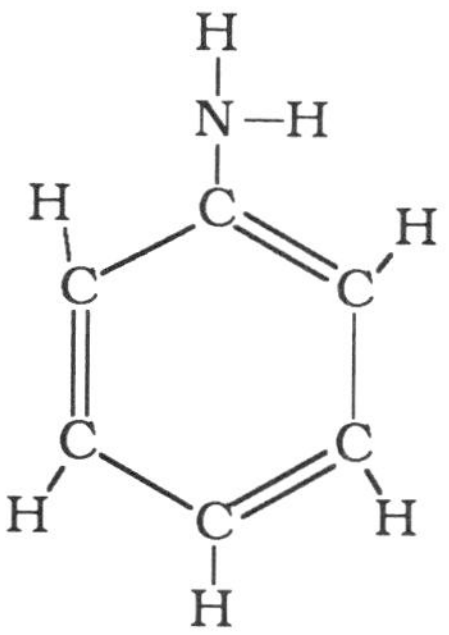

4.

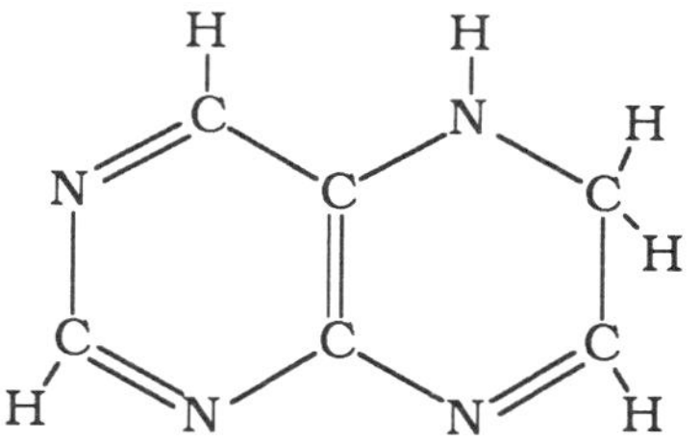

Exercise II

C_2 CH_2=CH–Cl   chloroethane

C_3 Cl–CH_2–CH=CH_2 3-chloro-1-propene

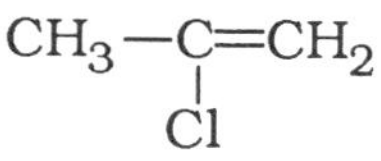

CH_3–C=CH_2
        |
        Cl

2-chloro-1-propene (in this case 2-chloropropene would be correct, too.)

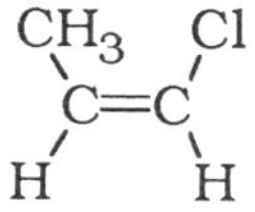

CH_3 Cl
   \   /
    C=C
   /   \
  H   H

cis-1-chloropropene

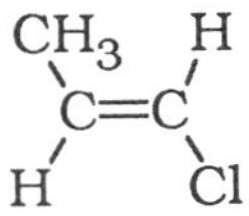

C_4

$$\underset{H}{\overset{CH_3}{\diagdown}} C = C \underset{Cl}{\overset{H}{\diagup}}$$

trans-1-chloropropene

$$\underset{H}{\overset{Cl}{\diagdown}} C = C \underset{H}{\overset{CH_2CH_3}{\diagup}}$$

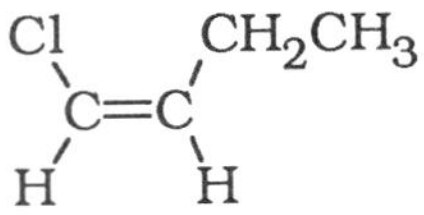

cis-1-chloro-1-butene

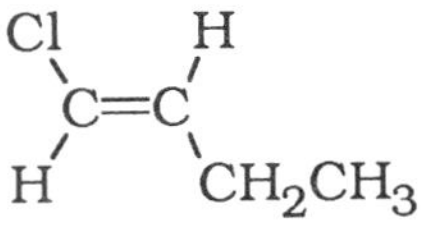

$$\underset{H}{\overset{Cl}{\diagdown}} C = C \underset{CH_2CH_3}{\overset{H}{\diagup}}$$

trans-1-chloro-1-butene

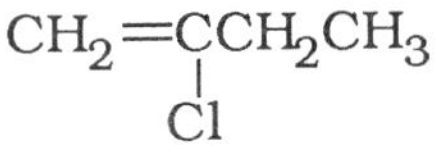

$$CH_2{=}\underset{Cl}{\overset{|}{C}}CH_2CH_3$$

2-chloro-1-butene

$$CH_2{=}CH\underset{Cl}{\overset{|}{C}}HCH_3$$

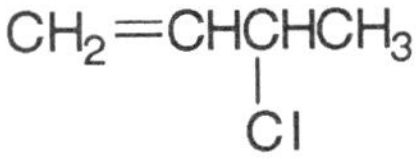

3-chloro-1-butene

$$CH_2{=}CHCH_2CH_2Cl$$

4-chloro-1-butene

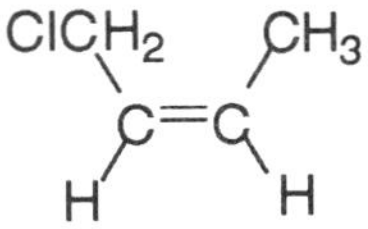

$$\underset{H}{\overset{ClCH_2}{\diagdown}} C = C \underset{H}{\overset{CH_3}{\diagup}}$$

cis-1-chloro-2-butene

$$\underset{Cl}{\overset{CH_3}{\diagdown}} C = C \underset{H}{\overset{CH_3}{\diagup}}$$

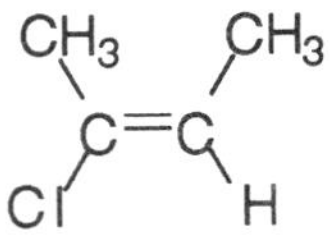

cis-2-chloro-2-butene (The CH_3 groups are cis.)

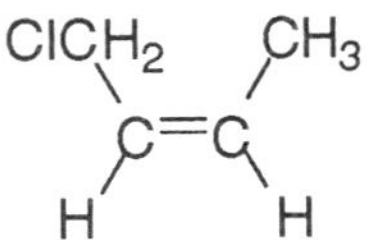

$$\underset{H}{\overset{ClCH_2}{\diagdown}} C = C \underset{H}{\overset{CH_3}{\diagup}}$$

trans-1-chloro-2-butene

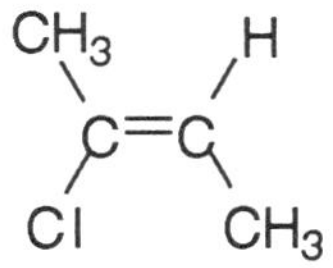

$$CH_3 \quad\quad H$$
$$C=C$$
$$Cl \quad\quad CH_3$$

trans-2-chloro-2-butene

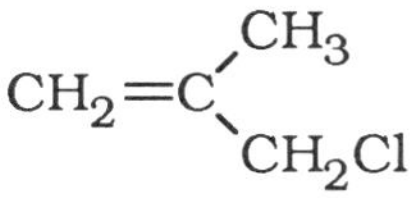

$$CH_2=C \begin{array}{c} CH_3 \\ CH_2Cl \end{array}$$

1-chloro-2-methylpropene
(-1-propene is unnecessary.)

$$Cl-CH=C \begin{array}{c} CH_3 \\ CH_3 \end{array}$$

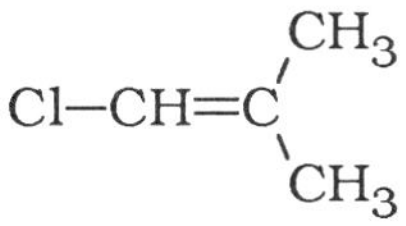

3-chloro-2-methylpropene
(-1-propene is unnecessary.)

Exercise III

1. phenol
2. chlorobenzene
3. toluene
4. aniline
5. benzoic acid

6. *m*-chlorotoluene
7. *p*-nitroaniline
8. *m*-bromophenol
9. *o*-chlorobenzoic acid
10. *o*-dichlorobenzene

Exercise IV

1. an alkane (or more generally, a saturated site)
2. a 1,2-dichloro compound (where 1 and 2 refer to relative locations)
3. a 1,2-dibromo compound
4. an alkyl chloride
5. an alkyl bromide
6. an alkyl hydrogen sulfate
7. an alcohol
8. carbon dioxide and water

Exercise V

1.  $$CH_3$$
$$CH_3CHCH_2CH_3$$

2. $CH_3CH_2CH_2CH_3$

3.

4. 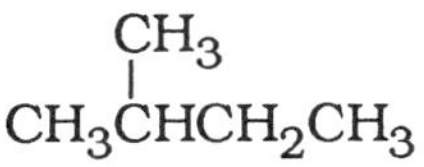
$$\underset{\displaystyle CH_3\overset{\displaystyle \overset{CH_3}{|}}{C}HCH_2CH_3}{}$$

5.

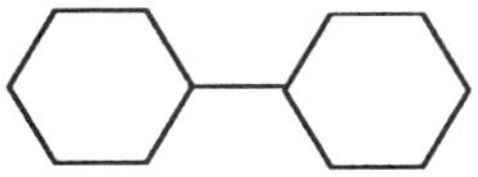

Exercise VI

1. 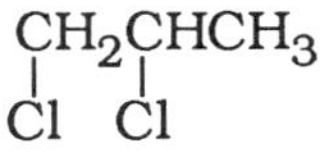
$$\underset{Cl\quad Cl}{CH_2CHCH_3}$$

2. 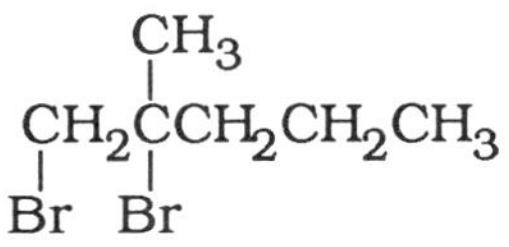
$$CH_2\overset{\overset{CH_3}{|}}{\underset{\underset{Br}{|}}{C}}CH_2CH_2CH_3 \qquad \underset{Br}{}$$

3.

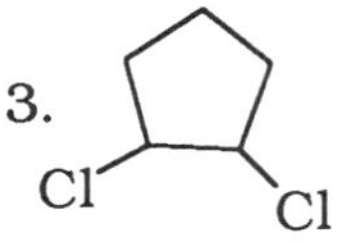

4. 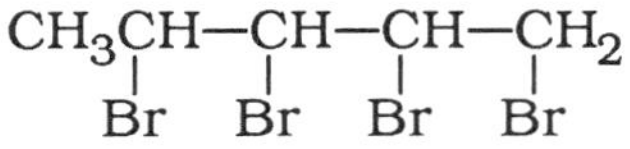
$$\underset{Br\quad Br\quad Br\quad Br}{CH_3CH-CH-CH-CH_2}$$

5. 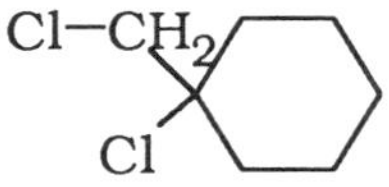
$$Cl-CH_2$$

Exercise VII

1. 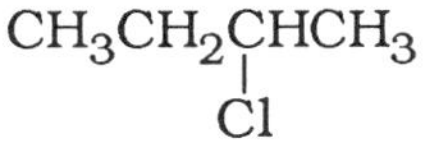
$$\underset{Cl}{CH_3CH_2CHCH_3}$$

2. 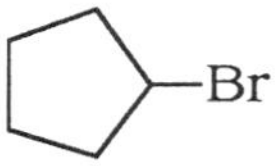
—Br

3. $CH_3\overset{\overset{CH_3}{|}}{\underset{\underset{Cl}{|}}{C}}CH_2CH_2CH_3$

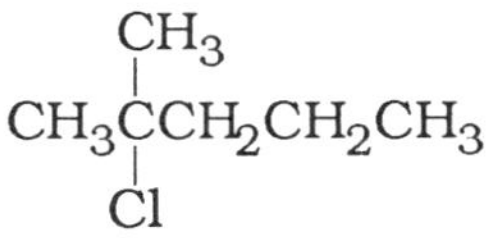

4.

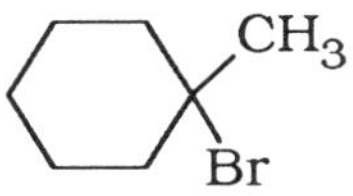

5. 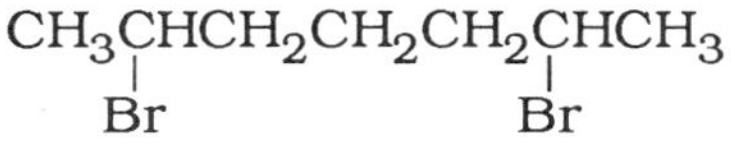
$$\underset{Br\qquad\qquad\qquad Br}{CH_3CHCH_2CH_2CH_2CHCH_3}$$

Exercise VIII

1. 
$$\underset{OH}{CH_3CHCH_2CH_3}$$

2. 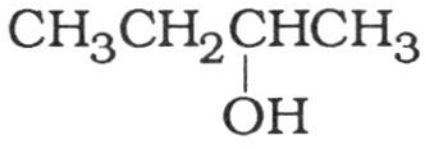
$$\underset{OH}{CH_3CH_2CHCH_3}$$

3.
$$CH_3\overset{\overset{CH_3}{|}}{\underset{\underset{OH}{|}}{C}}CH_2CH_3$$

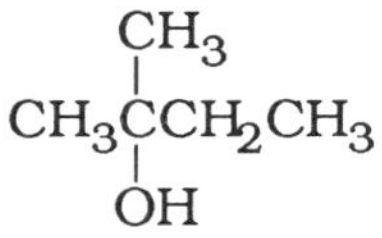

4. 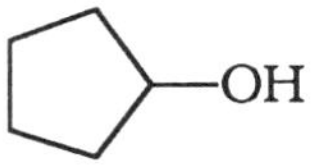
—OH

5. 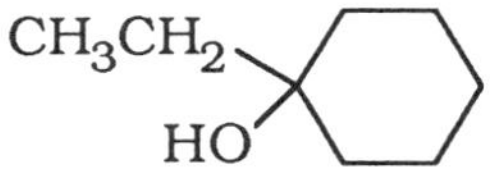
$$\underset{HO}{CH_3CH_2}$$

ANSWERS TO SELF–TESTING QUESTIONS

Completion

1. six

2.

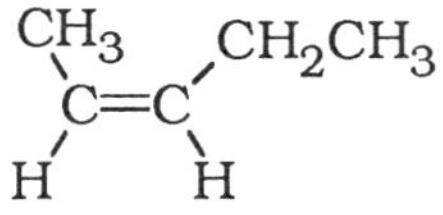

cis 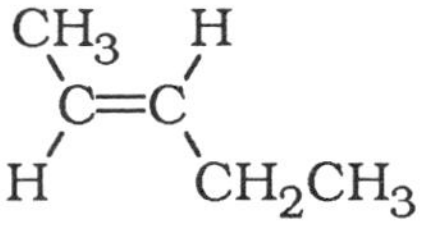   trans

3. the same

4. (a)

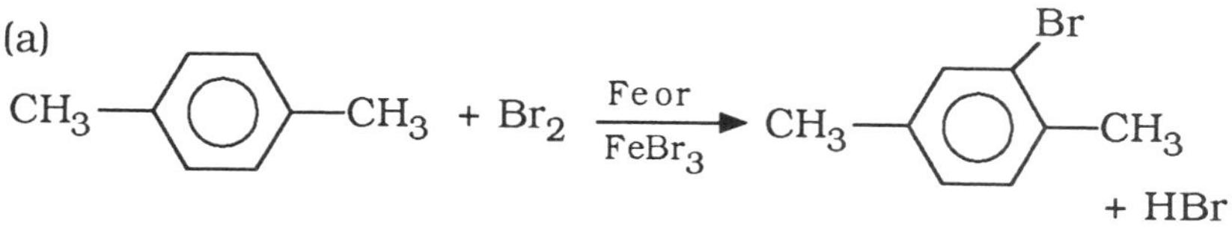

$$CH_3-\langle\bigcirc\rangle-CH_3 + Br_2 \xrightarrow[FeBr_3]{Fe or} CH_3-\langle\bigcirc\rangle\overset{Br}{-}CH_3$$

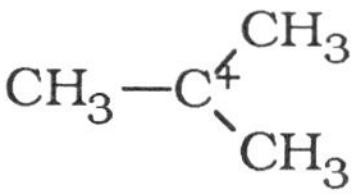

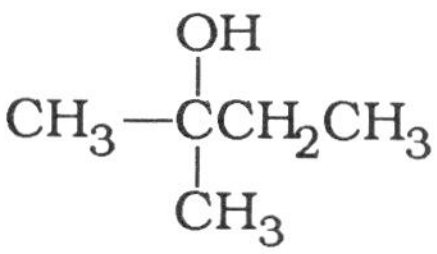

$$+ HBr$$

(b)

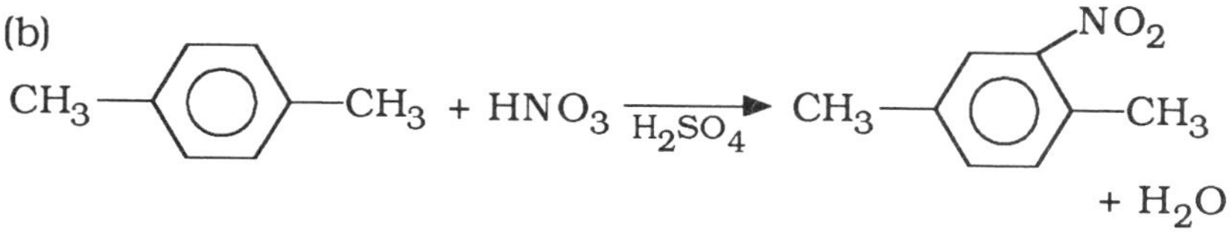

$$CH_3-\langle\bigcirc\rangle-CH_3 + HNO_3 \xrightarrow{H_2SO_4} CH_3-\langle\bigcirc\rangle\overset{NO_2}{-}CH_3$$

$$+ H_2O$$

(c)

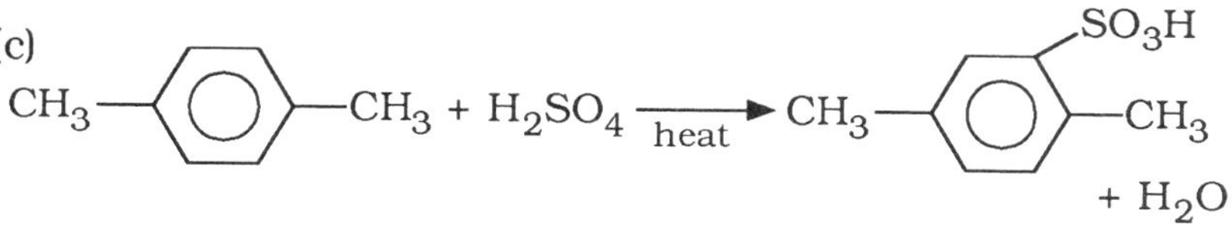

$$CH_3-\langle\bigcirc\rangle-CH_3 + H_2SO_4 \xrightarrow{heat} CH_3-\langle\bigcirc\rangle\overset{SO_3H}{-}CH_3$$

$$+ H_2O$$

5. 3°, $CH_3-\overset{\overset{CH_3}{|}}{\underset{\underset{CH_3}{|}}{C}}$

6. 1°, $^+CH_2-\overset{\overset{CH_3}{|}}{\underset{\underset{CH_3}{|}}{CH}}$

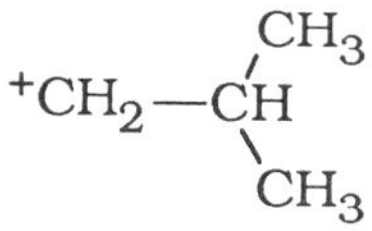

7. A < C < B

8. (a) 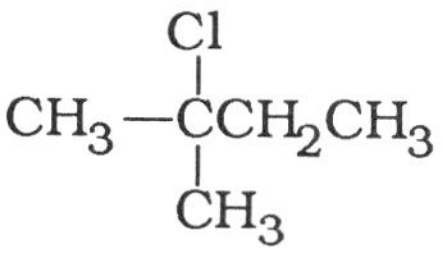 $CH_3-\overset{\overset{Cl}{|}}{\underset{\underset{CH_3}{|}}{C}}CH_2CH_3$ (b) $CH_3-\overset{\overset{OH}{|}}{\underset{\underset{CH_3}{|}}{C}}CH_2CH_3$ (c) $\langle\bigcirc\rangle-CH_3$

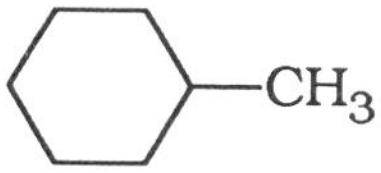

(d) 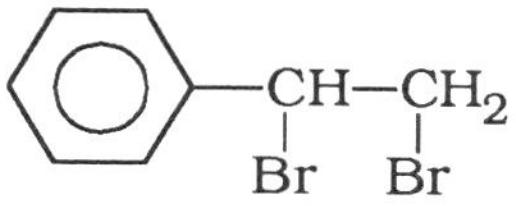 $\langle\bigcirc\rangle-\overset{}{\underset{\underset{Br}{|}}{CH}}-\overset{}{\underset{\underset{Br}{|}}{CH_2}}$ (Br$_2$ needs Fe or FeBr$_3$ to attack the ring.)

(e)   no reaction (Sulfuric acid does not attack alkanes or cycloalkanes.)
(f)   no reaction (Aqueous acids do not attack the benzene ring or alkyl groups.)
(g) $CH_3CH_2CH_2CH_2CH_3$

9. sigma, pi
10. p

Multiple–Choice

1. b and d
2. a and d
3. b and d
4. a
5. b
6. c

7. d
8. b
9. d (Markovnikov's Rule prevents use of any of the other choices.)
10. c
11. d (Benzene rings and alkyl groups are stable to aqueous alkali.)
12. d (Benzene rings and alkyl groups are stable to aqueous acids.)
13. d (Benzene is exceptionally stable toward oxidizing agents.)
14. b

3

ALCOHOLS, PHENOLS, ETHERS, AND THIOALCOHOLS

The three most important topics in this chapter relating to the molecular basis of life are:

1. Hydrogen bonds—which occur in nearly all of the biologically important molecules, such as carbohydrates, proteins, and nucleic acids.
2. Electron-transfer events (oxidations and reductions)—which occur at one or more of the stages in the breakdown of all substances used in the body's cells; and
3. Dehydrations of alcohols—which also occur several times in the reactions of metabolism.

We shall need the nomenclature of alcohols, ethers, and phenols if only to be able to talk and write about them.

OBJECTIVES

In greater detail, you should be able to do the following specific objectives after you have carefully studied this chapter and worked the Practice Exercises and Review Exercises in it.

1. Recognize the alcohol, phenol, ether, thioalcohol, and disulfide groups in structures.
2. Write the common names for alcohols (up through C_4).
3. Write the IUPAC names for alcohols.
4. Classify alcohols as 1°, 2°, or 3° and as mono-, di-, or polyhydric.
5. Write structures that illustrate how alcohol molecules form hydrogen bonds between each other or with water molecules in aqueous solutions.
6. Explain how alcohols have much higher boiling points and solubilities in water than do hydrocarbons of comparable formula mass.

7. Contrast phenols and alcohols in their acidities and in their abilities to neutralize strong bases (like OH⁻).

8. Give one reaction characteristic of phenols.

9. Given the structure of an alcohol, write the structure of the alkene or the ether that could be made from it by acid-catalyzed dehydration.

10. Given the structure of an alcohol, write the structure of the aldehyde and carboxylic acid that could be made from it (if it is a primary alcohol) or the ketone that could be made from it (if it is a secondary alcohol) by oxidation.

11. Write the structure of the disulfide that could be made by the oxidation of a thioalcohol.

12. Write the structure of the thioalcohol(s) that could be made by the reduction of a disulfide.

13. Write sentences that summarize each of the chemical reactions of alcohols, thioalcohols, disulfides, phenols, and ethers that are studied in this chapter.

14. Prepare 5 × 8" cards for each functional group studied on which are summarized the chemical properties they give.

15. Make lists of the functional groups we have so far studied that are chemically affected by (a) oxidizing agents, (b) reducing agents, (c) hot acid catalysts, (d) neutralizing agents (at room temperature) such as aqueous acids and aqueous hydroxides.

16. Define each term in the Glossary. Where appropriate, illustrate a definition with a structure or a reaction.

GLOSSARY

Alcohol. Any organic compound whose molecules have the OH group attached to a saturated carbon; R—OH.

Alcohol Group. The OH group when joined to a saturated carbon.

Dihydric Alcohol. An alcohol with two OH groups; a glycol.

Disulfide. A compound whose molecules have the sulfur-sulfur unit, S—S as in R—S—S—R.

Ether. An organic compound whose molecules have an oxygen attached by single bonds to separate carbon atoms neither of which is a carbonyl carbon atom; R—O—R'.

Glycol. A dihydric alcohol.

Mercaptan. A thioalcohol; R—S—H.

Monohydric Alcohol. An alcohol with one OH group per molecule.

Phenol. Any organic compound whose molecules have an OH group attached to a benzene ring.

Primary Alcohol. An alcohol in whose molecules an OH group is attached to a primary carbon, as in RCH_2OH.

Secondary Alcohol. An alcohol in whose molecules an OH group is attached to a secondary carbon atom; R_2CH—OH.

Tertiary Alcohol. An alcohol in whose molecules an OH group is held by a carbon from which three bonds extend to other carbon atoms; R_3C—OH.

Thioalcohol. A compound whose molecules have the SH group attached to a saturated carbon atom; a mercaptan.

Trihydric Alcohol. An alcohol with three OH groups per molecule.

KEY MOLECULAR "MAP SIGNS"

We have likened the functional groups to "map signs" that enable us to "read" the structural formulas of complicated systems. The principal map signs in this chapter are the following.

Key Molecular "Map Signs" In Organic Molecules	What to Expect When This Functional Group is Present
—C—O—H Alcohol group	*Influence on physical properties:* • The OH group is a good hydrogen-bond donor and hydrogen-bond acceptor. • Molecules with the OH group tend to be more polar (e.g., they have higher boiling points) and more soluble in water than are molecules without any group that can participate in hydrogen bonding (formula masses being about the same). *Influence on chemical properties:* A molecule with the alcohol group is vulnerable to • dehydrating agents (acids + heat); either double bonds are introduced or ethers are made. • oxidizing agents (those that can pull out the pieces of the element hydrogen): 1° alcohols → aldehydes → acids 2° alcohols → ketones 3° alcohols → (no reaction) (The list for alcohols will be completed in later chapters.)
OH Phenol system	A phenolic OH group is a weak acid, but strong enough to neutralize the hydroxide ion.
—C—O—C— Ether group (To be a simple ether, the carbons shown here must hold either Hs or Rs.)	*Influence on physical properties:* The oxygen of an ether molecule can accept hydrogen bonds; ethers therefore are slightly more soluble in water than alkanes; they are also slightly more polar. *Influence on chemical properties:* None. (As far as our study goes, the simple ether group is not attacked by water, dilute acids or bases, or oxidizing or reducing agents. We have to be able to recognize it, but then we can ignore it.)

—S—H Thioalcohol (sulfhydryl) group	Easy oxidation to S—S; this property, which we need to understand in its relation to proteins, is the only property that concerns us.
—S—S— Disulfide	Easy reduction to 2 SH groups; this property, which we also need to understand in its relation to proteins, is the only property that concerns us.

DRILL EXERCISES

I.   EXERCISES IN HYDROGEN BONDS

We define an H-bond donor as a molecule with a δ+ on a hydrogen attached to oxygen or nitrogen (O—H or N—H).

An H-bond acceptor is a molecule with a δ– on an oxygen or a trisubstituted nitrogen in any functional group.

H-bond donors can establish H-bonds not only to their own kind but also to water molecules and to any other H-bond acceptors. H-bond donors are invariably H-bond acceptors.

Some molecules, such as ethers, ketones, aldehydes, and esters, can only be H-bond acceptors. They have oxygens (or nitrogens, as in examples not yet studied), but they do not have OH or NH (i.e., a hydrogen with a δ+ and held by oxygen or nitrogen).

Examine these structures and answer the questions that follow.

A. $CH_3CH_2O—H$

B. $CH_3CH_2—O—CH_2CH_3$

C. ⟨O⟩—O—H

D. $CH_3—\overset{\overset{\textstyle O}{\|}}{C}—O—CH_3$

E. $CH_3—\overset{\overset{\textstyle O}{\|}}{C}—O—H$

F. $CH_3CH_2CH_2CH_3$

1.   Which are H-bond donors? _____
2.   Which are H-bond acceptors? _____
3.   Which are H-bond acceptors only? _____
4.   Which would be completely insoluble in water? _____
5.   Which would be more soluble in water, A or B? _____
6.   Which would be more soluble in water, C or E? _____

II.   DRILL IN WRITING THE PRODUCTS OF THE DEHYDRATION OF ALCOHOLS

After you have studied Example 3.2 in the text and have tried Practice Exercise 5 of Chapter 3, you might feel the need for further drill. Write the products of the dehydration of the following alcohols.

1. $CH_3\overset{\overset{\displaystyle CH_3}{|}}{C}HCH_2OH$ $\xrightarrow[\text{heat}]{H^+}$ _____

2. ⬡$-\overset{\underset{\displaystyle OH}{|}}{C}HCH_3$ $\xrightarrow[\text{heat}]{H^+}$ _____

3. ⬠$-OH$ $\xrightarrow[\text{heat}]{H^+}$ _____

4. ⬡$-\overset{\overset{\displaystyle CH_3}{|}}{\underset{\underset{\displaystyle OH}{|}}{C}}CH_3$ $\xrightarrow{H^+}$ _____

5. CH_3-⬡$-OH$ $\xrightarrow{H^+}$ _____

III. DRILL IN WRITING THE PRODUCTS OF THE OXIDATION OF ALCOHOLS

Examples 3.3 and 3.4 plus Practice Exercises 6, 7, and 8 of Chapter 3 are the places to begin this study. For more drill, write the products of the oxidation of the following alcohols. If oxidation in the sense being studied cannot occur, then write "no reaction." If the initial product can be oxidized further, then write the next product, too.

1. $CH_3CH_2CH_2CH_2OH\ +\ (O)\ \longrightarrow$ _____

2. $Cl-$⬡$-CH_2OH\ +\ (O)\ \longrightarrow$ _____

3. $CH_3\overset{\overset{\displaystyle CH_3}{|}}{\underset{\underset{\displaystyle CH_3}{|}}{C}}CH_2CH_3\ +\ (O)\ \longrightarrow$ _____

4. ⬡$-\overset{\overset{\displaystyle OH}{|}}{C}HCH_3\ +\ (O)\ \longrightarrow$ _____

5. $Cl-$⬡$-\overset{\overset{\displaystyle CH_3}{|}}{C}HOH\ +\ (O)\ \longrightarrow$ _____

6. $CH_3\overset{\overset{\displaystyle OH}{|}}{\underset{\underset{\displaystyle CH_3}{|}}{C}H}CHCH_3\ +\ (O)\ \longrightarrow$ _____

$$\text{7.} \quad CH_3CH_2\overset{\overset{\displaystyle CH_3}{|}}{C}HCH_2OH \ + \ (O) \longrightarrow \underline{\hspace{4cm}}$$

$$\text{8.} \quad \langle\bigcirc\rangle\!\!-\!\!\overset{\overset{\displaystyle CH_3}{|}}{\underset{\underset{\displaystyle CH_3}{|}}{C}}OH \ + \ (O) \longrightarrow \underline{\hspace{3cm}}$$

IV. DRILL IN WRITING THE STRUCTURES OF ETHERS THAT CAN BE MADE FROM ALCOHOLS

Example 3.5 shows how this is done. In order to make this kind of exercise more helpful for applications in the next chapter, we will include in the drill examples in which you'll construct a structure of an unsymmetrical ether, like R—O—R', from two different given alcohols. We will also make this a review of the names of alcohols. If only one alcohol is named, then the question is what symmetrical ether can be made from this alcohol? If two alcohols are named, then the question is what unsymmetrical ether can be made from the two?

1. isobutyl alcohol _____

2. *t*-butyl alcohol _____

3. 4-methylcyclohexanol _____

4. isopropyl alcohol and methyl alcohol _____

5. cyclopentanol and *sec*-butyl alcohol _____

V. DRILL IN WRITING THE PRODUCT OF THE OXIDATION OF A THIOALCOHOL

Example 3.6 in the text shows how to do this kind of exercise. For practice, write the products of the oxidation of the following thioalcohols.

$$\text{1.} \ CH_3SH \ + \ (O) \longrightarrow \underline{\hspace{3cm}}$$

$$\text{2.} \ CH_3CH_2CH_2CH_2SH \ + \ (O) \longrightarrow \underline{\hspace{3cm}}$$

$$\text{3.} \ \langle\bigcirc\rangle\!\!-\!\!SH \ + \ (O) \longrightarrow \underline{\hspace{3cm}}$$

$$\text{4.} \ \langle\bigcirc\rangle\!\!-\!\!CH_2SH \ + \ (O) \longrightarrow \underline{\hspace{3cm}}$$

$$\text{5.} \ CH_3\overset{\overset{\displaystyle CH_3}{|}}{C}HCH_2SH \ + \ (O) \longrightarrow \underline{\hspace{3cm}}$$

VI. DRILL IN WRITING THE PRODUCT OF THE REDUCTION OF A DISULFIDE

See page 95 in the text on how to do this. Remember that disulfides aren't always symmetrical, like R—S—S—R. Those that we'll encounter in biochemistry usually are not; they're like R—S—S—R'. The reduction of this kind gives two different thioalcohols, so both of them have to be written.

1. $CH_3-S-S-\overset{\overset{\displaystyle CH_3}{|}}{C}HCH_3$ + 2(H) ⟶ _____

2. ⟨○⟩—$CH_2-S-S-CH_2CH_3$ + 2(H) → _____

3. $CH_3CH_2OCH_2CH_2-S-SCH_3$ + 2(H) → _____

4. $\begin{array}{c} S-S \\ CH_2 \qquad CH_2 \\ H_2C-CH_2 \end{array}$ + 2(H) ⟶

_____

5. ⟨○⟩—S—S—⟨○⟩ + 2(H)

_____

SELF-TESTING QUESTIONS

Treat the following questions as a final examination for the chapter. Work them only after you have done the drills with the in-chapter and end-of-chapter Review Exercises. As a review for these tests, go back to the chapter objectives, one by one, and see if you can do them.

COMPLETION

Test your knowledge of chemical properties by writing the structure(s) of the principal organic product(s) that would be expected to form in each of the following. If no reaction will occur, write "none." Reread Review Exercise 3.43 in the text, for stipulations concerning problems of this type. Notice particularly that whenever the dehydration of any given alcohol is intended to go to the corresponding ether, then the coefficient "2" will appear before the structure of that alcohol. If the "2" is absent, then the intended reaction is the internal dehydration of the alcohol to an alkene (if possible). (Normally, coefficients are not put in until after all structures of reactants and products have been written; "balancing comes last.") Where the oxidation of a 1° alcohol is indicated, show both the aldehyde and the carboxylic acid that are formed.

1. $CH_3CH_2CH_3 \xrightarrow[\text{heat}]{H_2SO_4}$ _____

2. $CH_3\overset{\overset{\displaystyle CH_3}{|}}{C}HCH_2OH$ + (O) $\longrightarrow$ _____

3. $2CH_3CH_2CH_2OH$ $\xrightarrow[\text{heat}]{H_2SO_4}$ _____

4. $CH_3\overset{\overset{\displaystyle CH_3}{|}}{\underset{\underset{\displaystyle CH_3}{|}}{C}}-OH$ $\xrightarrow[\text{heat}]{H_2SO_4}$ _____

5. $CH_3CH_2CH_2OH$ + (O) $\longrightarrow$ _____

6. $CH_3\overset{\overset{\displaystyle CH_3}{|}}{C}HCH_2SH$ + (O) $\longrightarrow$ _____

7. 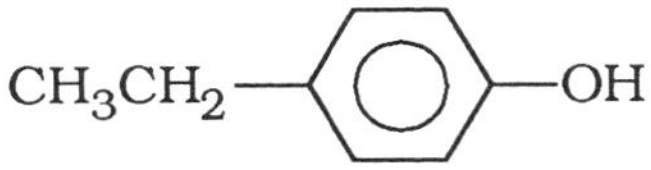 CH_3CH_2—⬡—OH + NaOH(aq) $\longrightarrow$ _____

8. $CH_3-\overset{\overset{\displaystyle CH_3}{|}}{\underset{\underset{\displaystyle CH_3}{|}}{C}}-CH_2-S-S-CH_3$ + 2(H) $\longrightarrow$

_____

9. 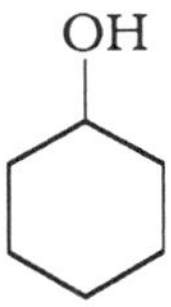 ⬡OH + (O) $\longrightarrow$

_____

10. $CH_3CH_2\overset{\overset{\displaystyle OH}{|}}{C}HCH_2CH_3$ + (O) $\longrightarrow$ _____

11. 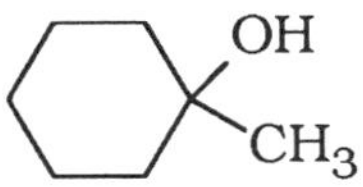 ⬡$\overset{OH}{\underset{CH_3}{\big<}}$ + (O) $\longrightarrow$

_____

12. $CH_3CH_2CH_2CH_3$ + H_2O $\xrightarrow[\text{heat}]{H^+}$ _____

13. CH_3OH + (O) $\longrightarrow$ _____

14. $HOCH_2CH_2CH_3$ $\xrightarrow[\text{heat}]{H_2SO_4}$ _____

15. $CH_2{=}CHCH_2OH + H_2 \xrightarrow{\text{Ni, heat}}$ _____

16. $CH_3CH_2{-}O{-}CH_2{-}CH_2{-}\overset{\overset{\displaystyle OH}{|}}{C}HCH_3 + (O) \longrightarrow$ _____

MULTIPLE–CHOICE

1. The oxidation of $CH_3{-}\overset{\overset{\displaystyle OH}{|}}{C}H{-}CH_2{-}CH_3$ could be made to
    to produce

    (a) $CH_3{-}\overset{\overset{\displaystyle OH}{|}}{C}H{-}O{-}CH_2{-}CH_3$

    (b) $H{-}\overset{\overset{\displaystyle O}{||}}{C}{-}CH_2CH_2CH_3$

    (c) $CH_3{-}CH_2{-}\overset{\overset{\displaystyle O}{||}}{C}{-}CH_3$

    (d) $HO{-}\overset{\overset{\displaystyle O}{||}}{C}CH_2CH_2CH_3$

2. What is the best explanation for the solubility of glycerol in water?

    $HO{-}CH_2{-}\overset{\overset{\displaystyle}{|}}{\underset{\underset{\displaystyle OH}{|}}{C}}H{-}CH_2{-}OH$
    Glycerol

    (a) Glycerol is a small molecule.
    (b) Glycerol molecules are polar.
    (c) Glycerol molecules can donate and accept hydrogen bonds to and from water molecules in the solution.
    (d) Glycerol's ions are well-solvated by water.

3. The substance whose structure is $CH_3{-}OH$ is known as
    (a) wood alcohol
    (b) grain alcohol
    (c) methanol
    (d) potable alcohol

4. The substance whose structure is $CH_3CH_2OCH_2CH_3$ is
    (a) a common fuel
    (b) a common antifreeze
    (c) an anesthetic
    (d) an eye irritant in smog

5. The dehydration of 3-hexanol would produce
    (a) 2-hexene only
    (b) 3-hexene only
    (c) a mixture of 2-hexene and 3-hexene
    (d) cyclohexene

6. The oxidation of  —OH would

    (a) destroy the ring   (c) produce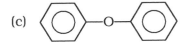

    (b) produce hydroquinone   (d) cyclohexene

7. The reaction of 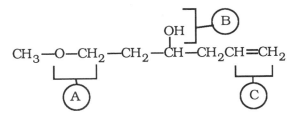 —OH with NaOH(aq) gives

    (a)   (c)

    (b) $-O^-Na^+ + H_2O$ (d) none of these

8. The common name of CH_3CHCH_3 is
                                           OH

    (a) isopropanol   (c) 2-propanol
    (b) 2-propyl alcohol   (d) isopropyl alcohol

Multiple-choice questions 9 through 11 refer to this structure:

$$CH_3-O-CH_2-CH_2-CH-CH_2CH{=}CH_2$$

with labels OH—B, A, C

9. The group labeled Ⓐ is
    (a) an easily hydrolyzed group
    (b) an easily oxidized group
    (c) an easily reduced group
    (d) a generally unreactive group

10. The group labeled Ⓑ could be
    (a) oxidized to a ketone
    (b) oxidized to an aldehyde
    (c) involved in an acid-catalyzed dehydration
    (d) reduced to a ketone

11.   The group labeled ⓒ could be
   (a)   made to react with dilute sodium hydroxide
   (b)   made to add a water molecule (if an acid catalyst were available)
   (c)   reduced to a carbon-carbon single bond by hydrogen (with a catalyst and heat)
   (d)   involved in hydrogen bonding

12.   A substance that can neutralize aqueous sodium hydroxide is

(a)    (c)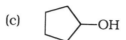

(b) 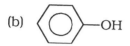   (d)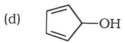

13.   A mild reducing agent will react with
   (a) CH_3CH_2SH (c) $CH_3—S—S—CH_3$
   (b) CH_3CH_2OH (d) $CH_3CH_2—O—CH_2CH_3$

ANSWERS

ANSWERS TO DRILL EXERCISES

I.   Exercises in Hydrogen Bonds

   1.   A, C, and E

   2.   A, B, C, D, and E

   3.   B and D

   4.   F   (an alkane)

   5.   A   (It's of lower formula mass, and it can both donate and accept H—bonds to water, the solvent. B can only accept H—bonds from water.)

   6.   E   (It's less hydrocarbon like, of lower-formula mass, and has two H— bond accepting sites—the two oxygens—whereas structure C has only one H—bond accepting site.)

II.   Drill in Writing the Products of the Dehydration of Alcohols

1. $CH_3\overset{\overset{\displaystyle CH_3}{|}}{C}=CH_2$ 2.   ⟨⟩—$CH=CH_2$ 3. ⬠

4.   ⟨⟩—$\overset{\overset{\displaystyle CH_3}{|}}{C}=CH_2$ 5.   no reaction

III.   Drill in Writing the Products of the Oxidation of Alcohols

1. $CH_3CH_2CH_2CHO \xrightarrow{\text{more (O)}} CH_3CH_2CH_2CO_2H$

2. 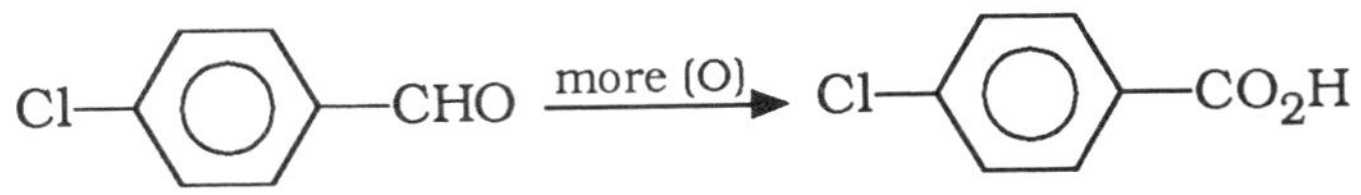   Cl—⬡—CHO $\xrightarrow{\text{more (O)}}$ Cl—⬡—CO_2H

3.   no reaction

4. 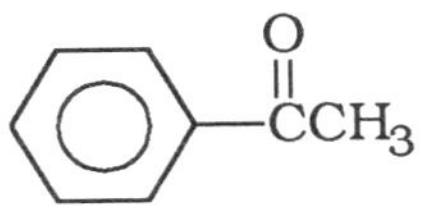   ⬡—$\overset{\overset{\displaystyle O}{\|}}{C}CH_3$

5. 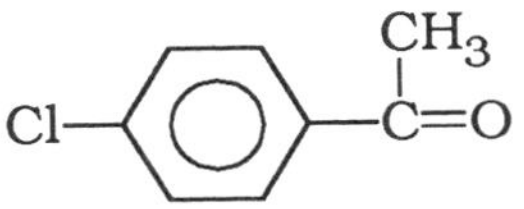   Cl—⬡—$\overset{\overset{\displaystyle CH_3}{|}}{C}{=}O$

6. 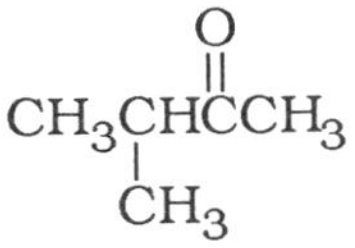 $CH_3\overset{\overset{\displaystyle O}{\|}}{C}\underset{\underset{\displaystyle CH_3}{|}}{C}HCH_3$

7. 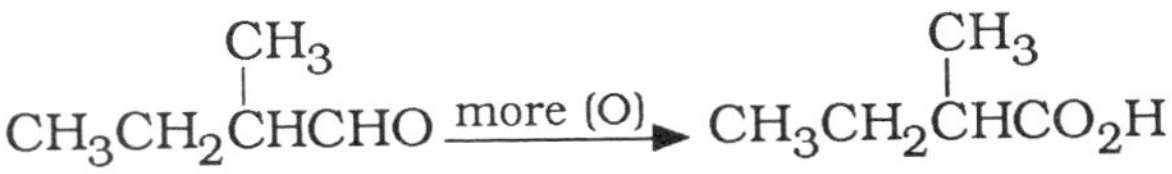 $CH_3CH_2\overset{\overset{\displaystyle CH_3}{|}}{C}HCHO \xrightarrow{\text{more (O)}} CH_3CH_2\overset{\overset{\displaystyle CH_3}{|}}{C}HCO_2H$

8.   no reaction

IV.   Drill in Writing the Structures of Ethers That Can Be Made from Alcohols

1.  $CH_2\overset{\overset{\displaystyle CH_3}{|}}{C}HCH_2OCH_2\overset{\overset{\displaystyle CH_3}{|}}{C}HCH_3$

2. $CH_3\overset{\overset{\displaystyle CH_3}{|}}{\underset{\underset{\displaystyle CH_3}{|}}{C}}-O-\overset{\overset{\displaystyle CH_3}{|}}{\underset{\underset{\displaystyle CH_3}{|}}{C}}CH_3$

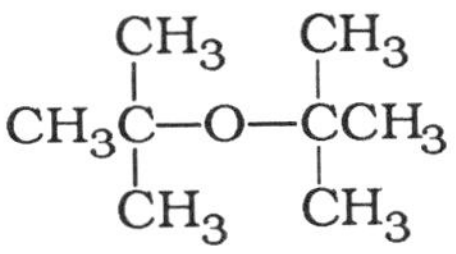

3. CH_3—⬡—O—⬡—CH_3

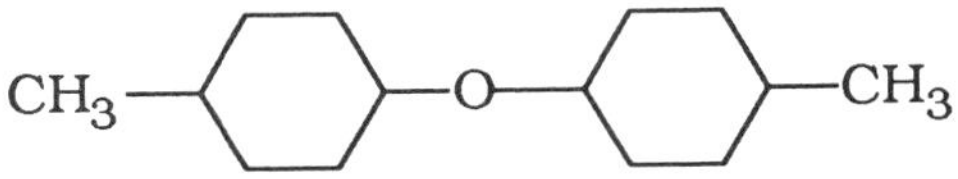

4. 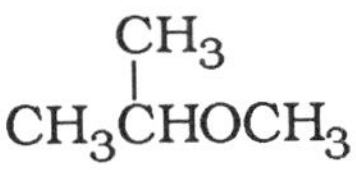 $CH_3\overset{\overset{\displaystyle CH_3}{|}}{C}HOCH_3$

5. 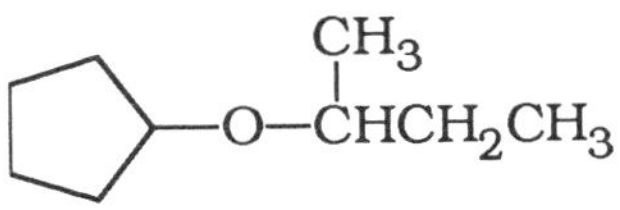   ⬠—O—$\overset{\overset{\displaystyle CH_3}{|}}{C}HCH_2CH_3$

V.   Drill in Writing the Product of the Oxidation of a Thioalcohol

1. $CH_3{-}S{-}S{-}CH_3$

2. $CH_3CH_2CH_2CH_2{-}S{-}S{-}CH_2CH_2CH_2CH_3$

3. 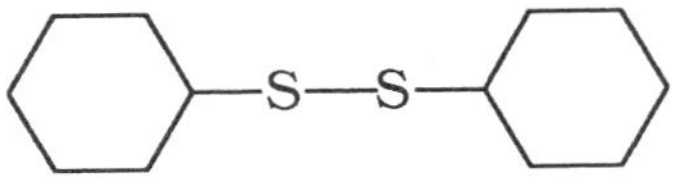   ⬡—S—S—⬡

4. 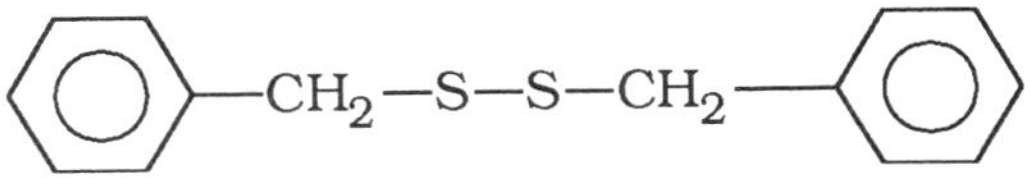   ⬡—$CH_2{-}S{-}S{-}CH_2$—⬡

5. $\underset{\underset{CH_3}{|}}{CH_3CHCH_2}-S-S-\underset{\underset{CH_3}{|}}{CH_2CHCH_3}$

VI.   Drill in Writing the Product of the Reduction of a Disulfide

1. $CH_3SH \; + \; \underset{\underset{CH_3}{|}}{HSCHCH_3}$ 2. 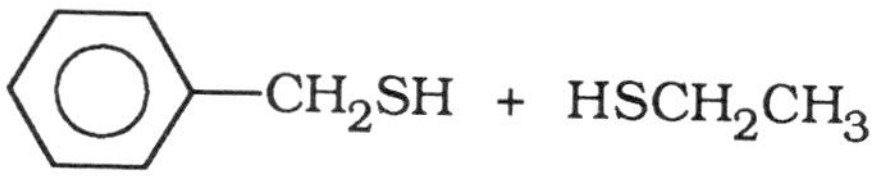   ⬡$-CH_2SH \; + \; HSCH_2CH_3$

3. $CH_3CH_2OCH_2CH_2SH + HSCH_3$ 4. $HSCH_2CH_2CH_2CH_2SH$

5. 2 ⬡$-SH$

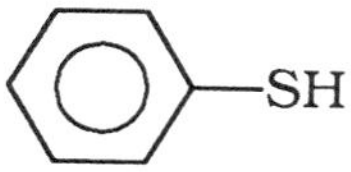

ANSWERS TO SELF–TESTING QUESTIONS

Completion

1. none

2. $\underset{\underset{CH_3}{|}}{CH_3CH}-\overset{\overset{O}{||}}{C}-H \xrightarrow{(O)} \underset{\underset{CH_3}{|}}{CH_3CH}-\overset{\overset{O}{||}}{C}-OH$

3. $CH_3CH_2CH_2-O-CH_2CH_2CH_3 + H_2O$ (Note the coefficient "2," our signal that the dehydration is intended to produce an ether, not an alkene.)

4. $CH_2{=}C{\overset{CH_3}{\underset{CH_3}{\big\langle}}} \; + \; H_2O$

5. $CH_3CH_2\overset{\overset{O}{||}}{C}-H \xrightarrow{(O)} CH_3CH_2\overset{\overset{O}{||}}{C}-OH$

6. $\underset{\underset{CH_3}{|}}{CH_3CHCH_2}-S-S-\underset{\underset{CH_3}{|}}{CH_2CHCH_3} \; + \; H_2O$

7. 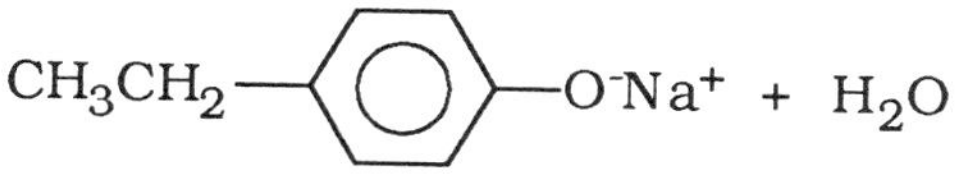 CH_3CH_2-⬡$-O^-Na^+ \; + \; H_2O$ (The electrical charges can be omitted.)

8. $\underset{\underset{CH_3}{|}}{\overset{\overset{CH_3}{|}}{CH_3CCH_2}}SH \; + \; HSCH_3$ 9. 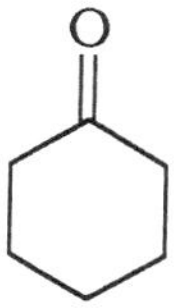 [cyclohexanone structure]   10. $CH_3CH_2\overset{\overset{O}{||}}{C}CH_2CH_3$

11. none (3° alcohols do not oxidize.)

12.   none (Reactant is an alkane.)

13. $$H-\overset{\overset{O}{\|}}{C}-H \xrightarrow{(O)} H-\overset{\overset{O}{\|}}{C}-OH$$
      (This will also continue to oxidize to
      $HO-\overset{\overset{O}{\|}}{C}-OH$, which is carbonic acid.)

14. $CH_2{=}CHCH_3 + H_2O$

      (Note the absence of the coefficient "2," a signal that internal dehydration to an alkene is intended.)

15. $CH_3CH_2CH_2OH$
      (Only the double bond responds to hydrogenation. We studied no reaction of the alcohol group with hydrogen—and none, in fact, exists; therefore, we can assume that this group is unaffected by the hydrogen and all the conditions present for hydrogenation.)

16. $CH_3CH_2-O-CH_2CH_2\overset{\overset{O}{\|}}{C}CH_3 + H_2O$

      (Only the 2° alcohol is affected by an oxidizing agent. We studied no reaction of an ether, the other functional group, with oxygen. Therefore, we can assume that it is untouched by the oxidizing agent, which is, in fact, true.)

Multiple–Choice

1.   c
2.   c
3.   a and c
4.   c
5.   c (Water can split out from the carbon holding the OH group in two directions.)
6.   c (The reactant is cyclohexanol, not phenol.)
7.   b (The reactant is phenol.)
8.   d (Choices a and b mix the common system and the IUPAC system—a definite no-no in organic chemistry. Choice c is the IUPAC name.)
9.   d
10.   a and c
11.   b and c
12.   b
13.   c

4

ALDEHYDES AND KETONES

This chapter will prepare you for the study of carbohydrates as well as for those pathways of metabolism that involve aldehyde and ketone groups.

OBJECTIVES

After studying this chapter and working the exercises in it, you should be able to do the following. Pay particular attention to objectives 1, 5, and 8 through 13. They are very important to our future needs.

1. Recognize the following families in the structures of compounds: aldehydes, ketones, carboxylic acids, hemiacetals, hemiketals, acetals, and ketals.
2. Give the common and the IUPAC names and structures for aldehydes, ketones, and acids through C_4.
3. Compare the physical properties of aldehydes and ketones with those of alcohols or hydrocarbons of comparable formula masses.
4. Write specific examples of reactions for the synthesis of simple aldehydes and ketones from alcohols.
5. Write specific examples of reactions for the oxidation of aldehydes.
6. Describe what one does and sees in a positive Tollens' Test.
7. Describe what one does and sees in a positive Benedict's Test.
8. Illustrate specific reactions for the reduction of aldehydes and ketones.
9. Give specific equations for the addition of an alcohol to an aldehyde (or ketone) to form a hemiacetal (or a hemiketal).
10. Write the structures of the original alcohol and aldehyde (or ketone) from the structure of a hemiacetal (or hemiketal).

11.   Give specific examples of reactions forming acetals or ketals.
12.   Give specific examples of reactions showing the hydrolysis of acetals or ketals.
13.   Give definitions of the terms in the Glossary and provide illustrations where applicable.

   Be sure to prepare and learn the one-sentence statements of chemical properties of aldehydes, ketones, hemiacetals, hemiketals, acetals, and ketals. Also, be sure to prepare the 5 × 8" reaction summary cards, and the cards that list the functional groups affected by various kinds of inorganic reactants.

GLOSSARY

Acetal. Any organic compound in which two ether linkages extend from one CH unit, as in:

$$RCH(OR')_2$$

Aldehyde. An organic compound that has a carbonyl group joined to H on one side and C on the other, R—CH=O

Aldehyde Group.   —CH=O

Benedict's Reagent. A solution of copper(II) sulfate, sodium citrate, and sodium carbonate that is used in the Benedict's test.

Benedict's Test. The use of Benedict's reagent to detect the presence of any compound whose molecules have easily oxidized functional groups—α-hydroxyaldehydes and α-hydroxyketones—such as those present in monosaccharides. In a positive test the intensely blue color of the reagent disappears and a reddish precipitate of copper(I) oxide separates.

Carbonyl Group.   C=O

Complex. (See *Complex Ion.*)

Complex Ion. A combination of a metal ion and one or more electron-rich molecules or anions.

Hemiacetal. Any compound whose molecules have both an OH group and an ether linkage coming to a —CH— unit:

$$\underset{\displaystyle |}{\overset{\displaystyle OH}{—CH—O—C—}} \qquad \text{as in} \qquad \underset{\displaystyle |}{\overset{\displaystyle OH}{R—CH—O—R'}}$$

Hemiketal. Any compound whose molecules have both an OH group and an ether linkage coming to a carbon that otherwise bears no H atoms:

$$\overset{\displaystyle OH}{—C—O—C—} \qquad \text{as in} \qquad \overset{\displaystyle OH}{\underset{\displaystyle R}{R—C—O—R'}}$$

Ketal. A substance whose molecules have two ether linkages joined to a carbon that also holds two hydrocarbon groups as in $R_2C(OR')_2$.

Keto Group. The carbonyl group when it is joined on each side to carbon atoms.

Ketone. Any compound with a carbonyl group attached to two carbon atoms, as in $R_2C{=}O$.

Tollens' Reagent. A slightly alkaline solution of the diammine complex of the silver ion, $Ag(NH_3)_2^+$, in water.

Tollens' Test. The use of Tollens' reagent to detect an easily oxidized group such as the aldehyde group.

KEY MOLECULAR "MAP SIGNS"

Key Molecular "Map Signs" In Organic Molecules	What to Expect When This Functional Group is Present
O ‖ —C—H Aldehyde group	*Influence on physical properties*: The aldehyde group is moderately polar; it can accept H-bonds. *Influence on chemical properties*: • One of the most easily oxidized groups: Changes into a carboxyl group Gives Tollens' test • Can be reduced to a 1° alcohol group • Adds an alcohol molecule to form a hemiacetal • Can be converted into an acetal system
O ‖ R—C—R Ketone	*Influence on physical properties*: Same as the aldehyde group *Influence on chemical properties*: • Strongly resists oxidation (unlike the aldehydes) • Can be reduced to a 2° alcohol group • Adds an alcohol to form a hemiketal (although not as readily as an aldehyde gives the hemiacetal) • Can be converted into a ketal system
OH | R—C—H | O—R' Hemiacetal OH | R—C—R | O—R' Hemiketal	*Two properties of importance*: • Both the hemiacetal and the hemiketal systems are unstable; they exist in equilibrium with the aldehyde or ketone and the alcohol (R'OH) that formed them. • Both systems can be changed to the acetal or ketal system by a reaction with another alcohol molecule.

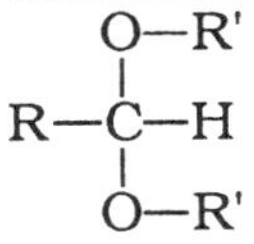

$$
\begin{array}{c}
\text{O--R'} \\
| \\
\text{R--C--H} \\
| \\
\text{O--R'} \\
\text{Acetal}
\end{array}
$$

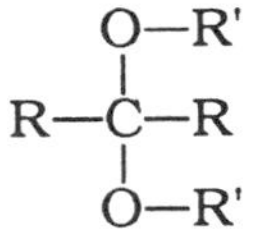

$$
\begin{array}{c}
\text{O--R'} \\
| \\
\text{R--C--R} \\
| \\
\text{O--R'} \\
\text{Ketal}
\end{array}
$$

One important property:
• Both the acetal and the ketal systems react with water when an acid catalyst is present, but they do not react in the presence of a basic catalyst; in so doing, they revert back to the original aldehyde

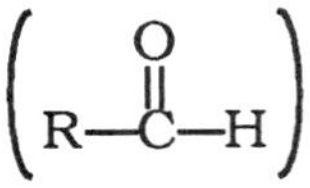

$$
\left(\underset{R-C-H}{\overset{O}{\parallel}} \right) \text{ or ketone } \left(\underset{R-C-R}{\overset{O}{\parallel}} \right)
$$

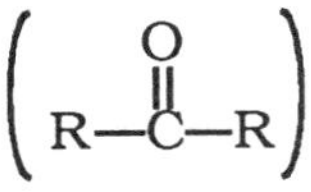

and alcohol (2R'OH)

SELF-TESTING QUESTIONS

COMPLETION

1. Write the full structure of each compound in the space above or to the side of its condensed structure.

(a)

CH₃CHO

(b)

$$
\underset{\text{OH}}{\overset{}{\text{CH}_3\text{CHOCH}_3}}
$$

(c)

CH₃CH₂OH

(d)

$$
\underset{\text{OCH}_3}{\overset{}{\text{CH}_3\text{CHOCH}_3}}
$$

2. What functional groups have we studied that will be attacked by oxidizing agents? (Give the name of the product, too.)

   (a) In this chapter   _____

   _____

   (b) In previous chapters   _____

   _____

   _____

   _____

3. What functional groups have we studied that will be attacked by reducing agents? (Give the name of the product, too.)

   (a) In this chapter   _____

   _____

   (b) In previous chapters   _____

   _____

   _____

   _____

4. By means of short statements, summarize the chemical reactions of alcohols (from both this and earlier chapters).

   Example: *Alcohols can be dehydrated to form alkenes.* _____

   _____

   _____

   _____

   _____

   _____

   _____

   _____

   _____

   _____

5. Examine the structures shown in question 1, parts a through d. Answer the following questions by writing the condensed structures of both the reactants and the products.

   (a) Which of the structures in question 1 will give a positive test with Tollens' Reagent?

   _____

   (b) Which can be hydrolyzed? _____

   (c) Which can be oxidized under basic conditions? _____

   (d) Which can be reduced? _____

6. Write the structure(s) of the principal organic product(s) that could be expected to form in each case. If no reaction occurs, write "none."

   (a) $CH_3CH=O$ $\xrightarrow{(O)}$ _____

   (b) $CH_3CH=O$ + $HOCH_3$ $\rightleftharpoons$ _____

   (c) 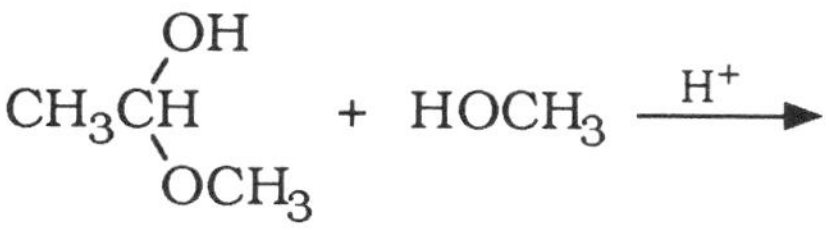 $CH_3\overset{OH}{\underset{OCH_3}{CH}}$ + $HOCH_3$ $\xrightarrow{H^+}$ _____

   (d) 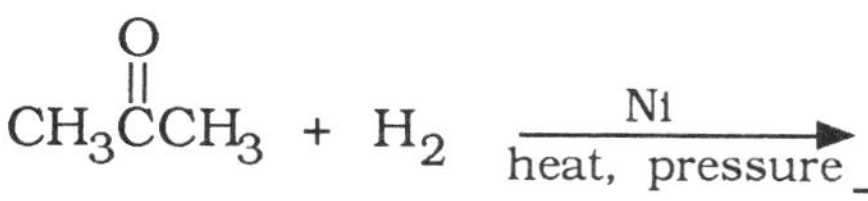 $CH_3\overset{O}{\overset{||}{C}}CH_3$ + H_2 $\xrightarrow[\text{heat, pressure}]{Ni}$ _____

   (e) 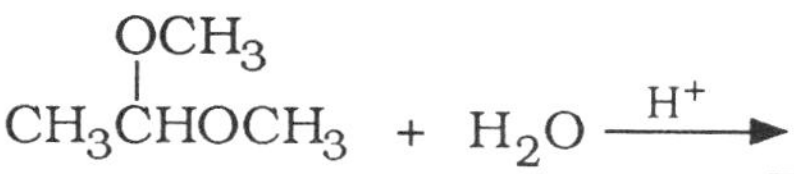 $CH_3\overset{OCH_3}{\underset{|}{CH}}OCH_3$ + H_2O $\xrightarrow{H^+}$ _____ + _____

(f) $O=CHCH_2CH_3$ $\xrightarrow{(O)}$ _____

(g) $CH_3-S-S-CH_2CO_2H$ $\xrightarrow[\text{agent}]{\text{mild reducing}}$ _____ + _____

(h) $HOCH_2CH_2\overset{\overset{\displaystyle O}{||}}{C}CH_3$ $\xrightarrow[\text{heat}]{H_2SO_4}$ _____

7. Write the common and the IUPAC name of each of the following.

	Common	IUPAC

(a) $CH_3CH_2\overset{\overset{\displaystyle O}{||}}{C}H$ _____ _____

(b) $CH_3\overset{\overset{\displaystyle O}{||}}{C}CH_3$ _____ _____

(c) $CH_3\overset{\overset{\displaystyle O}{||}}{C}H$ _____ _____

(d) $CH_3CH_2\overset{\overset{\displaystyle O}{||}}{C}CH_2CH_3$ _____ _____

MULTIPLE–CHOICE

1. Which of these compounds could be hydrolyzed the most easily (assuming acid catalysis)?
   (a) $CH_3CH_2OCH_2CH_3$
   (b) $CH_3OCH_2OCH_3$
   (c) $CH_3OCH_2CH_2OCH_3$
   (d) $CH_3CH_2CH_2CH_2OH$

2. Which of these compounds could be the most easily oxidized under mild conditions:

   (a) $CH_3CH_2OCH_2CH_3$

   (b) $CH_3CH_2CH_2CH_2\overset{\overset{\displaystyle O}{||}}{C}H$

   (c) $CH_3CH_2\overset{\overset{\displaystyle O}{||}}{C}CH_2CH_3$

   (d) $CH_3CH_2O\overset{\overset{\displaystyle CH_3}{|}}{C}HOCH_2CH_3$

3. Which of these compounds would be most soluble in water:
   (a) $CH_3CH_2CH_2CH_2OH$
   (c) $CH_3CH_2CH_2CH_2CH_3$

   (b) $CH_3CH_2CH_2CH_2\overset{\overset{\displaystyle O}{||}}{C}H$

   (d) $CH_3CH_2CH_2OCH_2CH_3$

4. The substance that precipitates in a positive Benedict's Test is
    (a) CuO   (b) Ag   (c) Cu_2O (d) Ag^+

5. The acetal that could be hydrolyzed to CH_3CH_2OH and CH_3CHO is

    (a) $CH_3CH_2\overset{\displaystyle OH}{\underset{\displaystyle O-CH_3}{CH}}$

    (b) $CH_3CH_2O\underset{\displaystyle OCH_2CH_3}{CHCH_3}$

    (c) $CH_3\overset{\displaystyle OH}{\underset{\displaystyle O-CH_2CH_3}{CH}}$

    (d) none of these

6. The alcohol that could be oxidized to $CH_3\overset{\displaystyle CH_3}{CH}-\overset{\displaystyle O}{\overset{\displaystyle \|}{CH}}$ is

    (a) $CH_3\overset{\displaystyle CH_3}{CH}-\overset{\displaystyle O}{\overset{\displaystyle \|}{C}}-OH$

    (b) $CH_3\overset{\displaystyle CH_3}{CH}-OH$

    (c) $CH_3\overset{\displaystyle CH_3}{CH}CH_2CH_2OH$

    (d) CH_3CH-CH_2OH with CH_3 below

7. Isopropyl alcohol could be made by the catalytic reduction of
    (a) acetone   (c) methyl ethyl ketone
    (b) propionaldehyde   (d) acetaldehyde

8. The oxidation of $CH_3CH_2CH_2\overset{\displaystyle O}{\overset{\displaystyle \|}{CH}}$ would give

    (a) isobutyraldehyde   (c) butyraldehyde
    (b) butyric acid   (d) 1-butanol

9. Which choice contains the best description of the functional group(s) present in this structure?

    (a) a hemiacetal   (c) a hemiketal
    (b) an acetal   (d) a ketal

10. If a hydride donor were used, which of the following systems would be able to accept it?
    (a) $CH_3CH_2CH_2CH_3$ (c) $CH_3OCH_2CH_3$

    (b) $CH_3CH_2CH_2OH$ (d) $CH_3CH_2CH_2\overset{\displaystyle O}{\overset{\displaystyle \|}{CH}}$

ANSWERS

ANSWERS TO SELF–TESTING QUESTIONS

Completion

1. (a)
$$\begin{array}{ccc} & H & O \\ & | & \| \\ H- & C-C & -H \\ & | \\ & H \end{array}$$

(c)
$$\begin{array}{ccc} & H & H \\ & | & | \\ H- & C-C & -O-H \\ & | & | \\ & H & H \end{array}$$

(b)
$$\begin{array}{cccc} & H & H & H \\ & | & | & | \\ H- & C-C & -O-C & -H \\ & | & | & | \\ & H & O-H & H \end{array}$$

(d)
$$\begin{array}{cccc} & H & H & H \\ & | & | & | \\ H- & C-C & -O-C & -H \\ & | & | & | \\ & H & O & H \\ & & | \\ & & H-C-H \\ & & | \\ & & H \end{array}$$

2. (a) Aldehydes are oxidized to carboxylic acids.
    (b) Alkenes are oxidized to various products (which we did not study in any detail).
    1° alcohols are oxidized to aldehydes or to carboxylic acids.
    2° alcohols are oxidized to ketones.
    Mercaptans are oxidized to disulfides.

3. (a) Aldehydes are hydrogenated to 1° alcohols.
    Ketones are hydrogenated to 2° alcohols.
    (b) Alkenes are hydrogenated to alkanes.
    Alkynes are hydrogenated to alkenes and thence to alkanes.
    Disulfides are reduced to mercaptans.

4. Alcohols can be dehydrated to form ethers.
    1° Alcohols can be oxidized to aldehydes, and then to carboxylic acids.
    2° Alcohols can be oxidized to ketones.
    Alcohols will add to aldehydes or ketones to form hemiacetals or hemiketals (both of which are generally too unstable to be isolated).
    Alcohols will react with hemiacetals or hemiketals to form acetals or ketals.

5. (a)
$$CH_3CHO \longrightarrow CH_3\overset{\overset{\displaystyle O}{\|}}{C}-OH$$

The hemiacetal, b, will also give a positive test because, in solution, it exists in equilibrium with acetaldehyde and methyl alcohol; acetaldehyde will react with Tollens' Reagent.

(b) "To be hydrolyzed" means to undergo a chemical reaction with water. Only the acetal, d, reacts with water:

$$\underset{\underset{OCH_3}{|}}{CH_3CHOCH_3} + H_2O \xrightarrow{H^+} CH_3CHO + 2CH_3OH$$

(c) The aldehyde, a, the hemiacetal, b, and the alcohol, c, can be oxidized under basic conditions. The alcohol will give acetaldehyde or acetic acid, depending on other conditions. The oxidation of a and b was discussed in the answer to part a of this question. The acetal is stable in a base and will not hydrolyze; therefore, it cannot break down into oxidizable compounds.

(d) The aldehyde, a, can be reduced to CH_3CH_2OH. The hemiacetal, b, will also react because it is present in equilibrium with its parent aldehyde (CH_3CHO) and alcohol (CH_3OH). As with a, the aldehyde will be taken out by the reducing agent and changed to ethyl alcohol.

6. (a) CH_3CO_2H (b) $CH_3\overset{\overset{\displaystyle OH}{|}}{C}HOCH_3$ (c) $CH_3\overset{\overset{\displaystyle OCH_3}{\diagup}}{\underset{\underset{\displaystyle OCH_3}{\diagdown}}{C}}H$ (+ H_2O)

(d) $CH_3\overset{\overset{\displaystyle OH}{|}}{C}HCH_3$ (e) $CH_3CH{=}O + 2CH_3OH$ (f) $CH_3CH_2CO_2H$

(g) $CH_3SH + HSCH_2CO_2H$ (h) $CH_2{=}CH\overset{\overset{\displaystyle O}{||}}{C}CH_3$

7.   *Common*   *IUPAC*
    (a) propionaldehyde   propanal
    (b) acetone   propanone (2-propanone is all right, but the number 2 is
                                  unnecessary here.)
    (c) acetaldehyde   ethanal
    (d) diethyl ketone   3-pentanone

Multiple–Choice

1. b (an acetal)

    (a is an ether; c is a di-ether and not an acetal; and d is an alcohol)

2. b (an aldehyde)

    (a is an ether; c is a ketone; and d is an acetal. The oxidizing reagent is neutral or basic, as in Tollens' Test or Benedict's Test; therefore, the acetal will hold together. However, if the reagent is acidic, the acetal will hydrolyze to produce some acetaldehyde (and ethyl alcohol) and the aldehyde will be oxidized.)

3. a (an alcohol)

    (It is the only one that can both accept and donate hydrogen bonds.)

4. c
5. b
6. d
7. a
8. b
9. a
10. d

5

CARBOXYLIC ACIDS AND ESTERS

The study of this chapter prepares you to study both lipids (e.g., fats and oils) and proteins as well as the metabolic pathways that involve the carboxyl group, its anionic form, and the ester group. Also introduced here are organophosphates, and their esters and anhydride system, which are systems of great importance at the molecular level of life.

OBJECTIVES

After you have studied this chapter and worked the Practice Exercises and Review Exercises in it, you should be able to do the following. Objectives 1 through 5 are very important for understanding the applications of this chapter in biochemistry and in our study of the molecular basis of life.

1. Recognize as structural "map signs" the carboxylic acid group, the carboxylate ion group, the ester group and ester bond, the acid chloride group, and the anhydride group.
2. By examining a structure of a substance, determine the probable ability of the substance to neutralize either a base or an acid or to be hydrolyzed, saponified, or esterified.
3. Write equations that are specific examples of the formation of:
   (a) an acid from an alcohol or an aldehyde
   (b) a carboxylic acid salt from the acid
   (c) a carboxylic acid from its salt
   (d) an ester from an alcohol and
      (1) a carboxylic acid
      (2) a carboxylic acid chloride
      (3) a carboxylic acid anhydride

(e) an alcohol and an acid from an ester

(f) an alcohol and the salt of a carboxylic acid from an ester

4. Give a general explanation for the relative acidity of an acid over that of an alcohol.

5. Explain through equations and discussion how the carboxyl group may be used as a "solubility switch."

6. Write the structural features common to phosphate, diphosphate, and triphosphate esters.

7. Define the terms in the Glossary, and give illustrations where applicable.

Reaction summary cards should be made for acids, their salts, acid chlorides, acid anhydrides, and esters. The partly completed card for alcohols can now be completed. A list of sentences that summarize chemical facts should be prepared and learned. The lists of reactions organized by key reagents (e.g., acids, bases, water and so forth) should be brought up to date.

GLOSSARY

Acid Anhydride. In organic chemistry, a compound formed by splitting water out between two OH groups of the acid function of an organic acid. The structural features are:

Carboxylic acid anhydride system   Phosphoric acid anhydride system

Acid Chloride. A derivative of an acid in which the OH group of the acid has been replaced by Cl.

Acid Derivative. Any organic compound that can be made from an organic acid or that can be changed back to the acid by hydrolysis. (Examples are acid chlorides, acid anhydrides, esters, and amides.)

Acyl Group.

Acyl Group Transfer Reaction. Any reaction in which an acyl group transfers from a donor to an acceptor.

Amide. Any organic compound whose molecules have a carbonyl-nitrogen unit,

Carboxylic Acid. A compound whose molecules have the carboxyl group, CO_2H.

Ester. A derivative of an acid and an alcohol that can be hydrolyzed to these parent compounds. Esters of carboxylic acids and phosphoric acid occur in living systems.

$$-\overset{|}{\underset{|}{C}}-O-\overset{O}{\overset{||}{C}}-$$

System in an ester of
a carboxylic acid

$$-\overset{|}{\underset{|}{C}}-O-\overset{O}{\overset{||}{\underset{\underset{OH}{|}}{P}}}-OH$$

System in an ester of
phosphoric acid

Esterification. The formation of an ester.

Fatty Acid. Any carboxylic acid that can be obtained by the hydrolysis of animal fats or vegetable oils.

Saponification. The reaction of an ester with sodium or potassium hydroxide to give an alcohol and the salt of an acid.

KEY MOLECULAR "MAP SIGNS"

Key Molecular "Map Signs" In Organic Molecules	What to Expect When This Functional Group is Present
$$-\overset{O}{\overset{||}{C}}-O-H$$ Carboxyl group [Often written as CO_2H or COOH]	*Influence on physical properties:* The carboxyl group is a very polar group; it can both donate and accept H-bonds *Influence on chemical properties:* • Can neutralize OH^- (or HCO_3^- or CO_3^{2-}); in so doing, CO_2H becomes CO_2^- • Can be changed into an ester by reacting with an alcohol when an acid catalyst is present.
$$-\overset{O}{\overset{||}{C}}-O^-$$ Carboxylate ion	*Influence on physical properties:* • One of the most effective groups at bringing long hydrocarbon chains into solution. • All salts of the carboxylic acids are solids at room temperature. *Influence on chemical properties:* • Aqueous solutions will test slightly basic. • Can neutralize H^+
$$-\overset{O}{\overset{||}{C}}-O-\overset{|}{\underset{|}{C}}-$$ Ester	*Influence on physical properties:* The ester group is a moderately polar group; it can accept H-bonds but cannot donate them. *Influence on chemical properties:* • Can be hydrolyzed (to the carboxylic acid and alcohol). • Can be saponified (to the carboxylate ion and alcohol).

$$\overset{\overset{\displaystyle O}{\|}}{—C—Cl}$$

Acid chloride

$$\overset{\overset{\displaystyle O}{\|}}{—C}—O—\overset{\overset{\displaystyle O}{\|}}{C—}$$

Acid anhydride

Properties of importance:
Acid chlorides and acid anhydrides are
• easily hydrolyzed (to the carboxylic acid).
• easily react with alcohols to give esters.

$$R—O—\overset{\overset{\displaystyle O}{\|}}{\underset{\underset{\displaystyle OH}{|}}{P}}—O—\overset{\overset{\displaystyle O}{\|}}{\underset{\underset{\displaystyle \uparrow OH}{|}}{P}}—O—H$$

Influence on physical properties: Both esters exist at the pH of body fluids as negatively charged ions; therefore they are very soluble in water.

$$R—O—\overset{\overset{\displaystyle O}{\|}}{\underset{\underset{\displaystyle OH}{|}}{P}}—O—\overset{\overset{\displaystyle O}{\|}}{\underset{\underset{\displaystyle \uparrow OH}{|}}{P}}—O—\overset{\overset{\displaystyle O}{\|}}{\underset{\underset{\displaystyle \uparrow OH}{|}}{P}}—O—H$$

Di- and triphosphate esters

Influence on chemical properties:
• Can neutralize OH⁻ (to the extent they begin with OH groups on phosphorus and have not been neutralized).
• In absence of a catalyst, they only react very slowly with water at the pH of body fluids.
• In the presence of the appropriate enzymes, they react rapidly with the alcohol (and amino) groups of biochemicals and suffer breakage of the bonds indicated by the arrows (↑).

Additional Reactions of Functional Groups Introduced in Earlier Chapters

$$—\overset{\overset{\displaystyle |}{}}{\underset{\underset{\displaystyle |}{}}{C}}—O—H$$

Alcohol (or phenol)

• Forms esters with carboxylic acids, either by reacting with the carboxylic acid (when a mineral acid catalyst is present) or by reacting with an acid chloride or acid anhydride.

DRILL EXERCISES

I. EXERCISES IN STRUCTURES AND NAMES

1. To make sure that you understand the condensed structures that are often used with carbonyl compounds, write out the following condensed structures as full structures.

(a) CH_3CO_2H

(b) $HO_2CCH_2CH_3$

(c) CH_3CHO

(d) CH_3CH_2OH

(e) $CH_3CO_2CH_3$

(f) CH_3CONH_2

(g) $HOOCCH_3$

2. It is important to be able to recognize quickly the presence of functional groups in structures;

otherwise, the chemical and physical properties of such structures cannot be "read" (in the sense of "reading" a map with knowledge of "map signs"). Study each of the following structures and assign them to their correct families. Some will have more than one functional group; name them all.

Examples:

$$\underset{\text{CH}_3\overset{\displaystyle O}{\overset{\|}{\text{C}}}\text{OH}}{} \quad \text{Carboxylic acid} \qquad \text{CH}_3\overset{\displaystyle O}{\overset{\|}{\text{C}}}\text{CH}_2\overset{\displaystyle O}{\overset{\|}{\text{C}}}\text{CH}_3 \quad \text{Two keto groups}$$

$$\text{C}_6\text{H}_5\overset{\displaystyle O}{\overset{\|}{\text{C}}}\text{H} \quad \text{Aldehyde} \qquad \text{CH}_3\overset{\displaystyle O}{\overset{\|}{\text{C}}}\text{O}\overset{\displaystyle O}{\overset{\|}{\text{C}}}\text{CH}_3 \quad \text{Anhydride}$$

$$\underset{\underset{\text{OH}}{|}}{\text{CH}_3\text{CH}\overset{\displaystyle O}{\overset{\|}{\text{C}}}\text{OH}} \quad \begin{array}{l}\text{Carboxylic acid} \\ \text{and alcohol}\end{array} \qquad \text{H}\overset{\displaystyle O}{\overset{\|}{\text{C}}}\text{OH} \quad \begin{array}{l}\text{Carboxylic acid} \\ \text{(This acid is not in the} \\ \text{aldehyde class, but it} \\ \text{does have some aldehyde} \\ \text{properties.)}\end{array}$$

(a) $\text{HO}\overset{\displaystyle O}{\overset{\|}{\text{C}}}\text{CH}_3$

(b) $\text{H}\overset{\displaystyle O}{\overset{\|}{\text{C}}}\text{C}_6\text{H}_5$

(c) $\text{CH}_3\overset{\displaystyle O}{\overset{\|}{\text{C}}}\text{Cl}$

(d) $\text{H}\overset{\displaystyle O}{\overset{\|}{\text{C}}}-\text{OCH}_3$

(e) $\text{CH}_3\text{O}\overset{\displaystyle O}{\overset{\|}{\text{C}}}\text{CH}_3$

(f) $\text{CH}_3\overset{\displaystyle O}{\overset{\|}{\text{C}}}\text{CH}_3$

(g) $\text{CH}_3\text{CH}_2\text{OH}$

(h) $\text{CH}_3\overset{\displaystyle O}{\overset{\|}{\text{C}}}\text{ONa}$

(i) $\text{H}\overset{\displaystyle O}{\overset{\|}{\text{C}}}\text{H}$

(j) 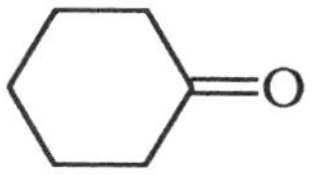 $\text{cyclohexanone} = \text{O}$

(k) $\text{C}_6\text{H}_5\overset{\displaystyle O}{\overset{\|}{\text{C}}}\text{OH}$

(l) $\text{C}_6\text{H}_5\text{O}\overset{\displaystyle O}{\overset{\|}{\text{C}}}\text{H}$

(m) $\text{HOCH}_2\text{CH}_2\overset{\displaystyle O}{\overset{\|}{\text{C}}}\text{CH}_3$

(n) $\underset{\underset{\text{OH}}{|}}{\text{CH}_3\text{CH}\overset{\displaystyle O}{\overset{\|}{\text{C}}}\text{CH}_3}$

(o) $(\text{CH}_3)_2\text{CHCOOH}$

(p) $\text{H}\overset{\displaystyle O}{\overset{\|}{\text{C}}}\text{CH}_2\text{CH}_2\overset{\displaystyle O}{\overset{\|}{\text{C}}}\text{CH}_3$

(q) $CH_3O\overset{O}{\overset{\|}{C}}CH_2CH_3$

(r) $CH_3OCH_2\overset{O}{\overset{\|}{C}}CH_3$

(s) $CH_3O\overset{O}{\overset{\|}{C}}CH_2\overset{O}{\overset{\|}{C}}H$

(t) $CH_3\overset{O}{\overset{\|}{C}}O\overset{O}{\overset{\|}{C}}CH_2CH_3$

(u) $CH_3CH_2CO_2H$

(v) $CH_3CH_2OCH\overset{O}{\overset{\|}{C}}CH_3$ with CH_3 branch

(w) $CH_3O\overset{O}{\overset{\|}{C}}CH_2\overset{O}{\overset{\|}{C}}OCH_3$

(x) $C_6H_5\overset{O}{\overset{\|}{C}}O\overset{O}{\overset{\|}{C}}C_6H_5$

(y) $C_6H_5\overset{O}{\overset{\|}{C}}Cl$

(z) $HO-\langle\bigcirc\rangle-\overset{O}{\overset{\|}{C}}H$

(aa) $CH_2{=}CHO\overset{O}{\overset{\|}{C}}CH_3$

(bb) $CH_3CH_2O\overset{O}{\overset{\|}{C}}CH{=}CH_2$

(cc) $CH_3\overset{O}{\overset{\|}{C}}OCH_2CH_2O\overset{O}{\overset{\|}{C}}CH_3$

(dd) $CH_3\overset{O}{\overset{\|}{C}}CH_2CH_2\overset{O}{\overset{\|}{C}}OH$

3. If after studying the ways to name esters and salts discussed in the text, you continue to have trouble, try the following. Prefixes in the common names for esters and acid salts (and aldehydes) relate to the acid portion of the structure. You may need some drill simply in recognizing what the acid portion is. It is that part of the structure that contains the carbonyl group plus whatever else consists solely of carbon and hydrogen. In other words, the acid portion is the part of the structure of the acid derivative that came from the parent acid.

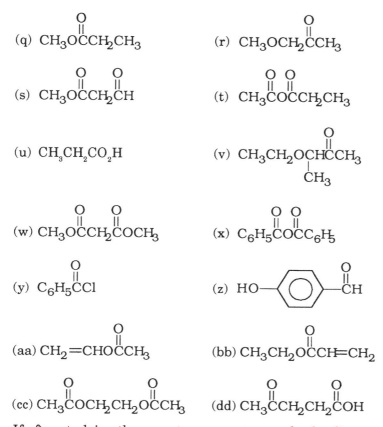

For the following structures, circle the acid portion in each structure and then write the prefix associated with its common name.

Examples:

$CH_3O\boxed{\overset{O}{\overset{\|}{C}}CH_3}$

Acet

$\boxed{CH_3CH_2CH_2\overset{O}{\overset{\|}{C}}}OCH_2CH_2CH_2CH_3$

Butyr

Note that the oxygen that has only single bonds and the R-group attached to it are not included.

(a) $\overset{\text{O}}{\overset{\|}{\text{HCOCH}_3}}$

(b) $\overset{\text{O}}{\overset{\|}{\text{CH}_3\text{CH}_2\text{CO}^-\text{Na}^+}}$

(c) $\overset{\text{O}}{\overset{\|}{\text{CH}_3\text{CH}_2\text{OCH}}}$

(d) $\overset{\text{O}}{\overset{\|}{\text{CH}_3\text{CH}_2\text{OCCH}_3}}$

(e) $\overset{\text{O}}{\overset{\|}{\text{CH}_3\text{CH}_2\text{COCH}_3}}$

(f) $\overset{\text{O}}{\overset{\|}{\text{CH}_3\text{OCCH}_2\text{CH}_2\text{CH}_3}}$

4. Complete the following table according to the example given.

STRUCTURE	FAMILY	COMMON NAME
$\overset{\text{O}}{\overset{\|}{\text{CH}_3\text{COCH}_3}}$	<u>ester</u>	<u>methyl acetate</u>

(a) $(\text{CH}_3)_2\text{CHOCCH}_3$ (with $\overset{\text{O}}{\|}$ on C)   _____   _____

(b) $\overset{\text{O}}{\overset{\|}{\text{CH}_3\text{CH}_2\text{CO}^-\text{Na}^+}}$ _____   _____

(c) $\overset{\text{O}}{\overset{\|}{\text{CH}_3\text{OCCH}_2\text{CH}_2\text{CH}_3}}$ _____   _____

(d) $\overset{\text{O}}{\overset{\|}{(\text{CH}_3)_2\text{CHCH}_2\text{OCCH}_2\text{CH}_2\text{CH}_3}}$ _____   _____

(e) $\overset{\text{O}}{\overset{\|}{\text{H}-\text{CCH}_2\text{CH}_3}}$ _____   _____

5. Write the IUPAC names for compounds (a) – (e) of the previous exercise.
   (a) _____
   (b) _____
   (c) _____
   (d) _____
   (e) _____

6. Write the condensed structural formulas of the following.

   (a) methyl ethanoate

   (b) 2,3-dimethylbutanal

(c) propanoic anhydride

(d) *sec*-butyl ethanoate

(e) 2-methylbutanoyl chloride

(f) sodium 3-chloropropanoate

II. DRILL ON THE REACTIONS OF CARBOXYLIC ACIDS WITH STRONG BASES

Write the structures of the salts that form in the following situations. Assume in every case that the aqueous base is being used at room temperature. You may write the structures showing the electrical charges (e.g., $CH_3CO_2^-Na^+$) or not (e.g., CH_3CO_2Na), but you should never draw a line between the two parts of the salts (e.g., CH_3CO_2—Na) because a line means a *covalent* bond, which is not present here.

1. $CH_3CH_2CH_2CO_2H$ + KOH(*aq*) → _____

2. $CH_3(CH_2)_6CO_2H$ + NaOH(*aq*) → _____

3. CH_3—O—CH_2CO_2H + NaOH(*aq*) → _____

4. $HOCH_2CH_2CO_2H$ + KOH(*aq*) → _____

5. $HO_2CCH_2CH_2CO_2H$ + 2NaOH(*aq*) → _____

III. DRILL ON THE REACTIONS OF CARBOXYLIC ACID SALTS WITH STRONG ACIDS

Write the structures of the acids that form in the following situations. Assume in every case that the aqueous acid is being used at room temperature.

1. $HCO_2^-Na^+$ + HCl(*aq*) → _____

2. $K^+O_2CCH_2CH_2CH_3$ + HCl(*aq*) → _____

3. $CH_3(CH_2)_8CO_2^-Na^+$ + HCl(*aq*) → _____

4. CH_3—⬡—$CO_2^-K^+$ + HCl(*aq*) → _____

5. $CH_3CH_2OCH_2CH_2CO_2^-Na^+$ + HCl(*aq*) → _____

IV. DRILL ON WRITING THE STRUCTURES OF ESTERS THAT CAN FORM FROM GIVEN ACIDS AND ALCOHOLS

Example 5.2 in the text provides the pattern. If after doing Review Exercises 6 and 7 you feel the need for more practice, try the following. Write the structures of the esters that can form between the following pairs of compounds. This will also serve as a review of the names of acids and alcohols. On the lines above the names write the structures of the reactants, and then form the structures of the products.

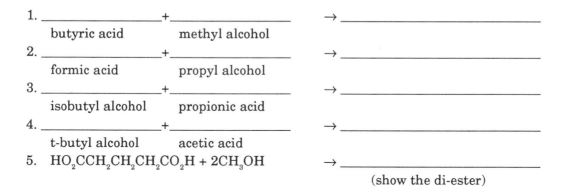

1. _____+_____   → _____
     butyric acid   methyl alcohol

2. _____+_____   → _____
     formic acid   propyl alcohol

3. _____+_____   → _____
     isobutyl alcohol   propionic acid

4. _____+_____   → _____
     t-butyl alcohol   acetic acid

5. $HO_2CCH_2CH_2CH_2CO_2H + 2CH_3OH$ → _____
                                              (show the di-ester)

V.   DRILL ON WRITING THE PRODUCTS OF THE HYDROLYSIS OF ESTERS

1. $CH_3CH_2\overset{\overset{\displaystyle O}{\|}}{C}OCH_3$ + $H_2O \xrightarrow[\text{heat}]{H^+}$ _____

2. $CH_3O\overset{\overset{\displaystyle O}{\|}}{C}CH_3$ + $H_2O \xrightarrow[\text{heat}]{H^+}$ _____

3. CH_3—⟨O⟩—$\overset{\overset{\displaystyle O}{\|}}{C}O\overset{\overset{\displaystyle CH_3}{|}}{C}HCH_3$ + $H_2O \xrightarrow[\text{heat}]{H^+}$ _____

4. $CH_3CH_2\overset{\overset{\displaystyle CH_3}{|}}{C}H-O-\overset{\overset{\displaystyle O}{\|}}{C}\underset{\underset{\displaystyle CH_3}{|}}{C}HCH_3$ + $H_2O \xrightarrow[\text{heat}]{H^+}$ _____

5. $CH_3CH_2O\overset{\overset{\displaystyle O}{\|}}{C}CH_2CH_2CH_2\overset{\overset{\displaystyle O}{\|}}{C}OCH_3$ + $2H_2O \xrightarrow[\text{heat}]{H^+}$ _____

VI.   DRILL IN WRITING THE PRODUCTS OF THE SAPONIFICATION OF ESTERS

Example 5.5 in the text describes how to do this kind of exercise. For additional drill, write the structures of the products of the complete saponification of the esters of the preceding drill exercise. Use NaOH(aq) as the saponifying agent.

1. _____

2. _____

3. _____

4. _____

5. _____

SELF-TESTING QUESTIONS

Use the following questions as your own final examination for the chapter. As a review, go back to the chapter objectives and find out if you can do them.

COMPLETION

1. Write the structure(s) of the organic product(s) that would form in each reaction. If no reaction occurs, write "none." Some of the reactions will involve a review of earlier chapters.

(a) CH_3CO_2H + NaOH $\xrightarrow{\text{water}}$ _____

(b) CH_3CO_2H + CH_3OH $\xrightarrow[\text{heat}]{H^+}$ _____

(c) $CH_3CH_2CH_2CH_2CO_2^-Na^+$ + $HCl(aq)$ $\longrightarrow$ _____

(d) 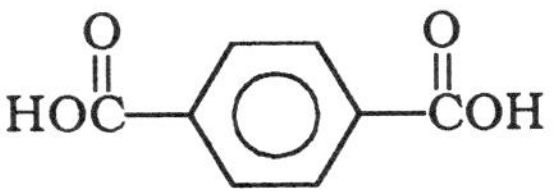 $HO\overset{O}{\overset{\|}{C}}$—⬡—$\overset{O}{\overset{\|}{C}}OH$ + $2CH_3CH_2OH$ $\xrightarrow[\text{heat}]{H^+}$ _____

(e) $CH_3O\overset{O}{\overset{\|}{C}}CH_3$ + H_2O $\xrightarrow[\text{heat}]{H^+}$ _____ + _____

(f) $CH_3\underset{\underset{OH}{|}}{C}HCO_2H$ $\xrightarrow{(O)}$ _____

(g) $CH_3O\overset{O}{\overset{\|}{C}}CH_2CH_2\overset{O}{\overset{\|}{C}}OCH_3$ $\xrightarrow[\text{water}]{H_2O,H^+}$ _____ + _____

(h) $CH_3OCH_2CH_2\overset{O}{\overset{\|}{C}}OCH_3$ $\xrightarrow[\text{water}]{H_2O,H^+}$ _____ + _____

(i) $CH_3CH_2CH_2\overset{\overset{\displaystyle O}{\|}}{C}-Cl$ + CH_3CH_2OH ⟶ _____

(j) $CH_3S-S-CH_2-\overset{\overset{\displaystyle O}{\|}}{C}-O^-Na^+$ + HCl $\xrightarrow[water]{}$ _____

(k) $CH_3\overset{\overset{\displaystyle O}{\|}}{O}CCH_3$ + NaOH $\xrightarrow[heat]{}$ _____ + _____

2. Which of the reactions in question 1 illustrate
   (a) esterification _____
   (b) saponification _____

MULTIPLE–CHOICE

1. The compound whose structure is

   $CH_3-O-CH_2-\overset{\overset{\displaystyle O}{\|}}{C}-O-CH_3$ has which functional group(s)?

   (a) two ether groups
   (b) an ether and an ester group
   (c) two ester groups
   (d) two ether and one ketone group

2. The structure of sodium butyrate is

   (a) $Na^+O\overset{\overset{\displaystyle O}{\|}}{C}OCH_2CH_2CH_3$ (c) $CH_3CH_2CH_2-\overset{\overset{\displaystyle O}{\|}}{C}-O-Na$

   (b) $Na^+\overset{\overset{\displaystyle O}{\|}}{C}-O-CH_2CH_2CH_2CH_3$ (d) $CH_3CH_2CH_2\overset{\overset{\displaystyle O}{\|}}{C}O_2Na^+$

3. The name of $CH_3\overset{\overset{\displaystyle CH_3}{|}}{C}HCO_2CH_2\overset{\overset{\displaystyle CH_3}{|}}{C}HCH_3$ is

   (a) isopropyl isobutyrate
   (b) isobutyl 2-methylpropanoate
   (c) isobutyl butyrate
   (d) isopropyl 3-methylbutanoate

4. The acid derivative that is most reactive toward water is
   (a) the ester   (c) the acid chloride
   (b) the acid salt   (d) both (b) and (c)

5. The action of aqueous hydrochloric acid on $CH_3CH_2CH_2CH_2CO_2^-K^+$ gives
   (a) $HO_2CCH_2CH_2CH_2CH_3$ + KCl
   (b) $CH_3CH_2CH_2CH_3$ + CO_2 + KCl

(c) $CH_3CH_2CH_2CH_2COCl + KOH$

(d) none of these

6. The action of aqueous potassium hydroxide at room temperature on CH_3CH_2—O—$CH_2CH_2CO_2H$ gives

(a) $CH_3CH_2H + HOCH_2CH_2CO_2^-K^+$

(b) $CH_3CH_2OCH_2CH_2CO_2^-K^+$

(c) $K^+OCH_2CH_2OCH_2CH_2CO_2H + H_2O$

(d) no reaction

7. In the following structure, the numbered arrows point toward functional groups. What are the numbers of the arrows pointing toward groups readily attacked by relatively mild oxidizing agents?

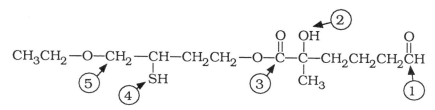

(a) 1   (b) 2   (c) 3   (d) 4   (e) 5

8. In the structure of question 7, which groups, if any, are subject to acid-catalyzed hydrolysis?

(a) 1   (b) 2   (c) 3   (d) 4   (e) 5   (f) none

9. In the structure of question 7, which groups, if any, will neutralize aqueous sodium hydroxide at room temperature?

(a) 1   (b) 2   (c) 3   (d) 5   (e) none of these

10. The action of one mole of water (containing a trace of acid catalyst) on one mole of compound Y produces one mole of CH_3CO_2H and one mole of CH_3CH_2OH. The structure of Y is

(a) $CH_3CH_2OCH_2CH_3$

(c) $CH_3CH-O-CH_2CH_3$
           |
          OH

(b) $CH_3CH_2O\overset{\overset{\displaystyle O}{\|}}{C}CH_3$

(d) $CH_3\overset{\overset{\displaystyle O}{\|}}{C}-O-\overset{\overset{\displaystyle O}{\|}}{C}CH_2CH_3$

11. The esterification of propionic acid by ethyl alcohol gives

(a) $CH_3CH_2CH_2O\overset{\overset{\displaystyle O}{\|}}{C}CH_3$

(c) $CH_3CH_2\overset{\overset{\displaystyle O}{\|}}{C}CH_2CH_3$

(b) $CH_3CH_2\overset{\overset{\displaystyle O}{\|}}{C}OCH_2CH_3$

(d) $CH_3CH_2\overset{\overset{\displaystyle OH}{|}}{\underset{\underset{\displaystyle H}{|}}{C}}OCH_2CH_3$

12. The saponification of $CH_3CH_2CH_2O-CH_2CH_2\overset{\overset{\displaystyle O}{\|}}{C}-O-CH_3$ by sodium hydroxide would produce

(a) $CH_3CH_2CH_2OH$ + $HOCH_2CH_2\overset{\overset{O}{\|}}{C}O^-Na^+$ + $HOCH_3$

(b) $CH_3CH_2CH_2$—O—$CH_2CH_2\overset{\overset{O}{\|}}{C}$—$O^-Na^+$ + $HOCH_3$

(c) $CH_3CH_2CH_2O^-Na^+$ + $HOCH_2CH_2\overset{\overset{O}{\|}}{C}OCH_3$

(d) $CH_3CH_2CH_2OCH_2CH_2\overset{\overset{O}{\|}}{C}OH$ + Na^+OCH_3

ANSWERS

ANSWERS TO DRILL EXERCISES

I. Exercises in Structures and Names

1. (a) 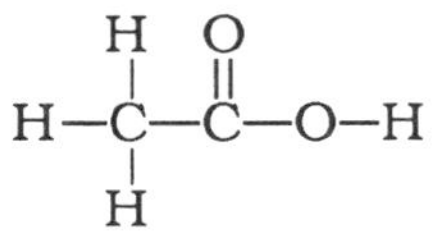
```
    H  O
    |  ||
H—C—C—O—H
    |
    H
```
(b)
```
 O  H  H
 ||  |  |
H—O—C—C—C—H
     |  |
     H  H
```

(c)
```
    H  O
    |  ||
H—C—C—H
    |
    H
```
(d) 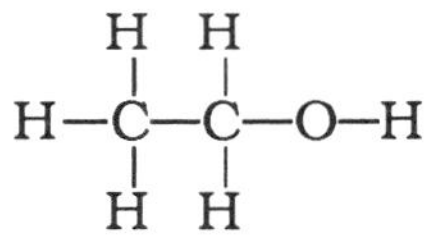
```
 H  H
 |  |
H—C—C—O—H
 |  |
 H  H
```

(e) 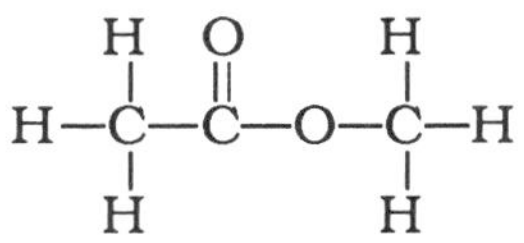
```
    H  O     H
    |  ||     |
H—C—C—O—C—H
    |         |
    H         H
```
(f) 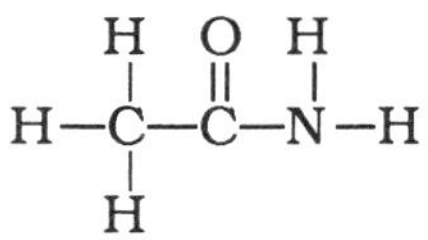
```
 H  O  H
 |  ||  |
H—C—C—N—H
 |
 H
```

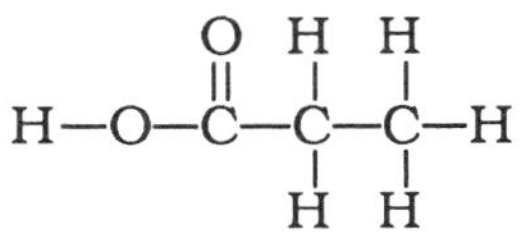

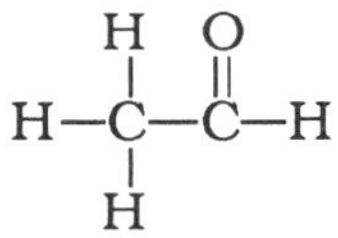

(g) 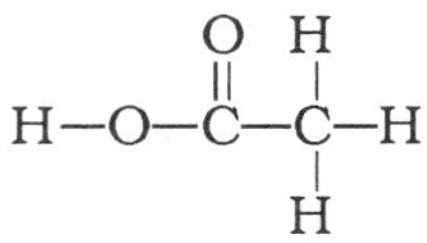
```
    O  H
    ||  |
H—O—C—C—H
         |
         H
```

2. (a) acid   (b) aldehyde

(c) acid chloride   (d) ester

(e) ester   (f) ketone

(g) alcohol   (h) acid salt

(i) aldehyde   (j) ketone

(k) acid   (l) ester

(m) alcohol, ketone   (n) alcohol, ketone

(o) acid   (p) aldehyde, ketone

(q) ester   (r) ether, ketone

(s) ester, aldehyde   (t) anhydride

(u) acid   (v) ether, ketone

(w) diester   (x) anhydride

(y) acid chloride   (z) phenol, aldehyde

(aa) alkene, ester   (bb) ester, alkene

(cc) diester   (dd) ketone, acid

3. (a)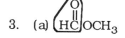

Form

(b)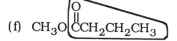

Propion

(c) CH_3CH_2O

Form

(d) CH_3CH_2O 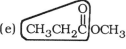 CCH_3

Acet

(e) CH_3CH_2C OCH_3

Propion

(f) CH_3O $CCH_2CH_2CH_3$

Butyr

4. (a) ester   isopropyl acetate

(b) salt   sodium propionate

(c) ester   methyl butyrate

(d) ester   isobutyl butyrate

(e) aldehyde   propionaldehyde

5. (a) isopropyl ethanoate (d) isobutyl butanoate

(b) sodium propanoate (e) propanal

(c) methyl butanoate

6. (a) $CH_3CO_2CH_3$

(b) $CH_3\overset{\overset{\displaystyle CH_3}{|}}{C}H\overset{\overset{}{|}}{C}HCHO$ with CH_3 below

(c) $CH_3CH_2\overset{\overset{\displaystyle O}{||}}{C}O\overset{\overset{\displaystyle O}{||}}{C}CH_2CH_3$

(d) $CH_3CO_2\overset{\overset{\displaystyle CH_3}{|}}{C}HCH_2CH_3$

(e) $CH_3CH_2\overset{\overset{\displaystyle CH_3}{|}}{C}HCOCl$

(f) $ClCH_2CH_2CO_2{}^-Na^+$

II. Drill on the Reactions of Carboxylic Acids with Strong Base

1. $CH_3CH_2CH_2CO_2{}^-K^+$
2. $CH_3(CH_2)_6CO_2{}^-Na^+$
3. $CH_3{-}O{-}CH_2CO_2{}^-Na^+$
4. $HOCH_2CH_2CO_2{}^-K^+$
5. $Na^+O_2CCH_2CH_2CO_2{}^-Na^+$

III. Drill on the Reactions of Carboxylic Acid Salts with Strong Acids

1. HCO_2H
2. $HO_2CCH_2CH_2CH_3$

3. $CH_3(CH_2)_8CO_2H$
4. 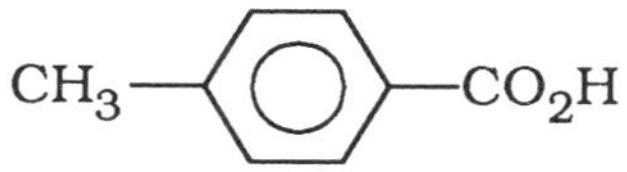 $CH_3{-}\bigcirc{-}CO_2H$

5. $CH_3CH_2OCH_2CH_2CO_2H$

IV. Drill on Writing the Structures of Esters That Can Form From Given Acids and Alcohols

1. $CH_3CH_2CH_2CO_2CH_3$
2. $HCO_2CH_2CH_2CH_3$

3. $CH_3CH_2CO_2CH_2\overset{\overset{\displaystyle CH_3}{|}}{C}HCH_3$
4. $CH_3CO_2\overset{\overset{\displaystyle CH_3}{|}}{\underset{\underset{\displaystyle CH_3}{|}}{C}}CH_3$

5. $CH_3O_2CCH_2CH_2CH_2CO_2CH_3$

V. Drill on Writing the Products of the Hydrolysis of Esters

1. $CH_3CH_2CO_2H + HOCH_3$
2. $CH_3OH + HO_2CCH_3$

3. CH_3—⬡—CO_2H + $HOCHCH_3$ (with CH_3 group above the CH)

CH₃ above:
3. CH_3—⬡—CO_2H + $HO\overset{\underset{|}{CH_3}}{C}HCH_3$

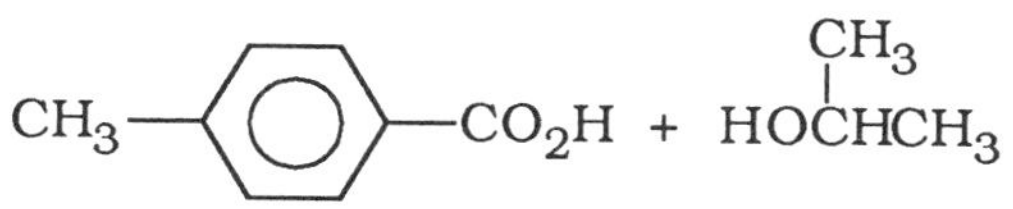

4. $CH_3CH_2\overset{\underset{|}{CH_3}}{C}HOH$ + $HO_2C\overset{\underset{|}{CH_3}}{C}HCH_3$

5. $CH_3CH_2OH + HO_2CCH_2CH_2CH_2CO_2H + HOCH_3$

VI.   Drill in Writing the Products of the Saponification of Esters

1. $CH_3CH_2CO_2Na + HOCH_3$

2. $CH_3OH + CH_3CO_2Na$

3.  CH_3—⬡—CO_2Na + $HOCH(CH_3)_2$

4. $CH_3CH_2\overset{\underset{|}{CH_3}}{C}HOH$ + $CH_3\overset{\underset{|}{CH_3}}{C}HCO_2Na$

5. $CH_3CH_2OH + NaO_2CCH_2CH_2CH_2CO_2Na + HOCH_3$

ANSWERS TO SELF–TESTING QUESTIONS

Completion

1. (a) $CH_3CO_2^-Na^+ (+ H_2O)$
   (b) $CH_3CO_2CH_3 (+ H_2O)$
   (c) $CH_3CH_2CH_2CH_2CO_2H (+ NaCl)$

   (d) 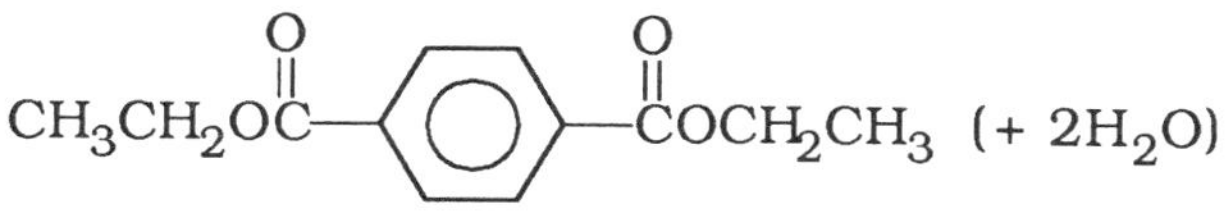 $CH_3CH_2O\overset{\overset{\textstyle O}{\|}}{C}$—⬡—$\overset{\overset{\textstyle O}{\|}}{C}OCH_2CH_3$ $(+ 2H_2O)$

   (e) $CH_3OH + CH_3CO_2H$

   (f) $CH_3\overset{\overset{\textstyle O}{\|}}{C}CO_2H$

   (g) $2CH_3OH + HO_2CCH_2CH_2CO_2H$

   (h) $CH_3OCH_2CH_2CO_2H + HOCH_3$

$$\text{(i)} \quad CH_3CH_2CH_2\overset{\displaystyle O}{\overset{\displaystyle \|}{C}}\!-\!O\!-\!CH_2CH_3 \quad (+\ HCl)$$

(j) $CH_3S\!-\!SCH_2CO_2H$ (+ NaCl)

(k) $CH_3OH + CH_3CO_2^-\ Na^+$

2. (a) b, d, and i   (b) k

Multiple–Choice

1.	b	2.	d	3.	b
4.	c	5.	a	6.	b
7.	a and d	8.	c	9.	e
10.	b	11.	b	12.	b

6

AMINES AND AMIDES

The chemistry of the amino group and the amide function is essential to our study of proteins and nucleic acids as well as those pathways of metabolism that involve the amine and amide functional groups.

OBJECTIVES

After you have studied Chapter 6 in the text and worked its Practice Exercises and Review Exercises, you should be able to do the following.

1. Identify the amine and amide functions in given structures, whether open-chain or heterocyclic.
2. Write the names of simple amines and amides.
3. Write the structures of simple amines and amides from their names.
4. Describe hydrogen bonding as it occurs among amines and amides and discuss its effects on physical properties.
5. Write equations for the reactions of amines with strong, aqueous acids.
6. Write equations for the reactions of protonated amines with strong, aqueous bases.
7. Write the structures of amides that can be formed (directly or indirectly) from given carboxylic acids and NH_3, RNH_2, or R_2NH.
8. Given the structure of an amide, write the products that form when it is hydrolyzed.
9. Define the terms in the Glossary.

GLOSSARY

Alkaloid. A physiologically active, heterocyclic amine isolated from plants.

Amide. Any organic compound whose molecules have a carbonyl-nitrogen unit, $-\overset{\overset{\displaystyle O}{\|}}{C}-\overset{|}{N}-$.

Amide Bond. The single bond that holds the carbonyl group to the nitrogen atom in an amide.

Amine. Any organic compound whose molecules have a trivalent nitrogen atom, as in $R-NH_2$, $R-NH-R$, or R_3N.

Amine Salt. Any organic compound whose molecules have a positively charged, tetravalent, protonated nitrogen atom, as in RNH_3^+, $R_2NH_2^+$, or R_3NH^+.

Base Ionization Constant (K_b). For the equilibrium (where B is a base), $B + H_2O \rightleftharpoons BH^+ + OH^-$,

$$K_b = \frac{\left[BH^+\right]\left[OH^-\right]}{B}$$

You should prepare the $5 \times 8"$ "reactions" cards for the amino group, the protonated amino group, and the amide group. Now is the time to bring up to date the lists of functional groups that react with aqueous acids, aqueous bases, and water.

KEY MOLECULAR "MAP SIGNS"

Key Molecular "Map Signs" In Organic Molecules	What to Expect When This Functional Group is Present
$-\overset{\mid}{\underset{\mid}{N}}:$ Amino group	*Influence on physical properties*: • If present as $R-NH_2$ or R_2NH, the amino group can both donate and accept hydrogen bonds to and from water, amines, or alcohols. • If present as R_3N, the amino group can only accept hydrogen bonds; it is as soluble in water as alcohols are. *Influence on chemical properties*: • The presence of an amino group makes the molecule a proton acceptor, a Brønsted base. • The easy room-temperature changes:

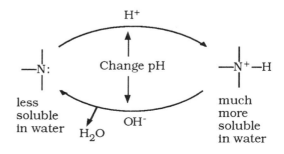

	Makes the amino group one of the major solubility switches in nature's biochemical molecules.
Amines (with at least one H on N)	• Form amides with acids; best done by a reaction of the amine with the acid chloride or anhydride.

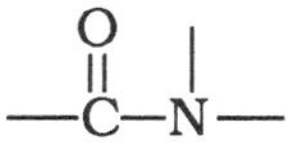

$$\overset{\overset{\textstyle O}{\|}}{-C}-\overset{\overset{\textstyle \|}{}}{N}-$$

Amide

Influence on physical properties:
Amides have a very polar group (particularly if at least one hydrogen is attached to the nitrogen); virtually all amides are solids at room temperature.

Influence on chemical properties:
• Amides are not basic; they are not proton acceptors in the way that makes amines basic.
• Can be hydrolyzed by aqueous acids or bases, if heated, to give the carboxylic acid and the amine (or ammonia).

DRILL EXERCISES

I.   DRILL IN WRITING PRODUCTS OF THE REACTIONS OF AMINES WITH STRONG ACIDS

Example 6.1 in the text describes how these products can be written. For additional practice, write the structures of the organic cations that form when each of the following amines reacts with something like hydrochloric acid (which is really a reaction with H_3O^+).

1. $CH_3CH_2CH_2NH_2$ _____

2. 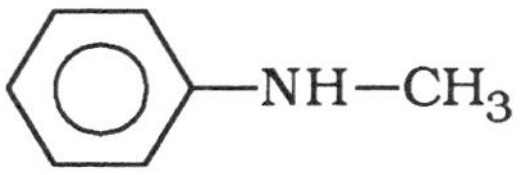   ⬡—NH—CH₃   _____

3. $CH_3CH_2\underset{\underset{\textstyle CH_3}{|}}{N}CH_2CH_3$ _____

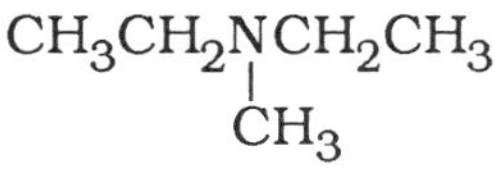

4.    ⬠NH   _____

5. $CH_3NHCH_2CH_2NHCH_3$ (and 2HCl)   _____

II. DRILL IN WRITING THE PRODUCTS OF THE REACTIONS OF PROTONATED AMINES WITH STRONG, AQUEOUS BASE

Example 6.2 in the text explains how to work this kind of problem. For further drill, write the products of the deprotona- tion of the following cations.

1. $CH_3CH_2NH_3^+$ _____

2. $CH_3CH_2\overset{+}{N}HCH_3$
    $\qquad\quad |$
    $\qquad\quad CH_3$ _____

3. ⬡—$\overset{CH_3}{\underset{}{\overset{|+}{N}}}HCH_3$ _____

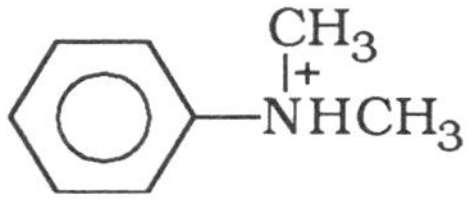

4. $\overset{+}{N}H_3CH_2CH_2\overset{O}{\overset{||}{C}}CH_3$ _____

5. $\overset{+}{N}H_3CHCO_2^-$
    $\qquad |$
    $\qquad CH_3$ _____

III. DRILL IN WRITING THE STRUCTURES OF THE AMIDES THAT CAN BE MADE (DIRECTLY OR INDIRECTLY) FROM GIVEN CARBOXYLIC ACIDS AND AMINES

See Example 6.3 for a discussion of how to work this kind of problem. The following will give extra opportunities to drill yourself. Write the structure of the amides that can be made from the given acids and amines (or ammonia).

1.  ⬡—CO_2H + NH_3 _____

2. $CH_3CH_2CO_2H + NH_3$ _____

3. $\overset{CH_3}{\overset{|}{CH_3CHCH_2CO_2H}}$ + NH_3 _____

4. 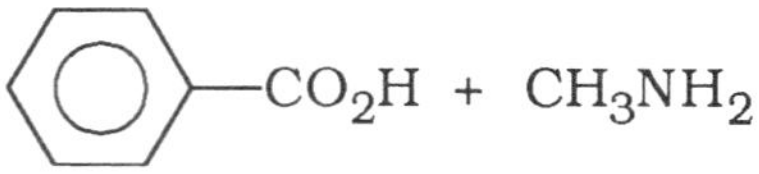 ⬡—CO_2H + CH_3NH_2 _____

5. $CH_3CO_2H + CH_3CH_2NH_2$ _____

IV. DRILL IN WRITING THE PRODUCTS OF THE HYDROLYSIS OF AMIDES

Example 6.4 in the text discusses how this kind of problem can be worked. For more practice, do the following. Write the structures of the products that can form by the hydrolysis of the following amides. Show the acids as acids, not as their salts. Similarly, show the amines as amines, not in protonated forms.

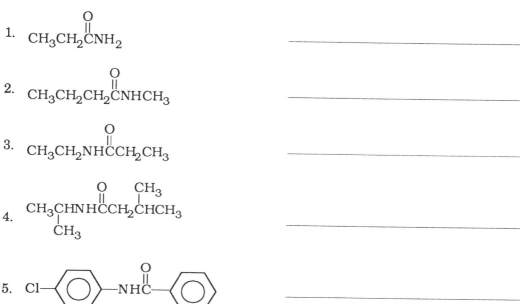

1. CH₃CH₂CNH₂ _____

2. CH₃CH₂CH₂CNHCH₃ _____

3. CH₃CH₂NHCCH₂CH₃ _____

4. CH₃CHNHCCH₂CHCH₃ _____

5. Cl—〈 〉—NHC—〈 〉 _____

V.   EXERCISES IN HYDROGEN BONDS

Amines can accept and donate H-bonds like alcohols, but the H-bonds in amines are weaker. (Amines with three groups on nitrogen and no N—H bond left can only accept hydrogen bonds.) The questions in this exercise refer to the following structures.

A. $CH_3CH_2CH_2$—$\overset{..}{\underset{..}{O}}$—H   B. $CH_3CH_2CH_2$—$\overset{H}{\underset{..}{N}}$—H

C. CH_3—$\overset{H}{\underset{..}{N}}$—H   D. CH_3—$\overset{CH_3}{\underset{CH_3}{N}}$:

E. $CH_3CH_2CH_2CH_2$—Br   F.

G. (benzene ring)—CH_2—$\overset{\overset{\displaystyle H}{|}}{N}$—H   H. (benzene ring)—$\overset{\overset{\displaystyle H}{|}}{\underset{\displaystyle ..}{N}}$—H   I. (ring)—$\overset{\overset{\displaystyle CH_3}{}}{N:}$

1. Which can donate H-bonds? _____
2. Which can accept H-bonds? _____
3. Which can only accept H bonds? _____
4. Which is totally insoluble in water? _____
5. Which has a higher boiling point, A or B? _____
6. Which is (are) the aromatic amine(s)? _____

A BRIEF SURVEY OF THE PRINCIPAL FUNCTIONAL GROUPS ATTACKED BY WATER, OXIDIZING AGENTS, REDUCING AGENTS OR NEUTRALIZING AGENTS

[**Note:** We summarize here only those reactions we have studied. In terms of the whole field of organic chemistry, the list is of course very incomplete. Omitted are the reactions of acid chlorides and anhydrides, including the phosphoric anhydrides.]

1. Groups that are split apart by water (hydrolysis) in the presence of acids or appropriate enzymes.

   Acetals and **ketals** are hydrolyzed to aldehydes and alcohols. For example, using the acetal,

   $$RC\overset{\overset{\displaystyle OR'}{/}}{\underset{\underset{\displaystyle OR'}{\backslash}}{H}} + H_2O \xrightarrow{H^+} RCHO + 2HOR'$$

   Esters are hydrolyzed to acids and alcohols.

   $$R\overset{\overset{\displaystyle O}{||}}{C}-OR' + H_2O \xrightarrow{H^+} RCO_2H + HOR'$$

   Amides are hydrolyzed to acids and amines (or ammonia).

   $$R\overset{\overset{\displaystyle O}{||}}{C}-\overset{\overset{\displaystyle |}{}}{N}- + H_2O \xrightarrow{H^+} RCO_2H + H-\overset{\overset{\displaystyle |}{}}{N}-$$

2. Groups that are affected by oxidizing agents.

   Alkenes are attacked by ozone and permanganate.

   Alcohol groups are converted to carbonyl groups.

   $$-\overset{\overset{\displaystyle |}{}}{\underset{\underset{\displaystyle H}{|}}{C}}-\overset{\displaystyle O}{\underset{\underset{\displaystyle H}{\backslash}}{}} + (O) \longrightarrow \overset{\displaystyle \backslash}{\underset{\displaystyle /}{C}}=O + H_2O$$

   Mercaptans are changed to disulfides.

   $$2R-S-H + (O) \longrightarrow R-S-S-R + H_2O$$

   Aldehydes are converted to carboxyl groups.

   $$RCHO + (O) \longrightarrow RCO_2H$$

3. Groups that are affected by reducing agents (e.g., H_2 or donors of $H{:}^-$).
   Carbon-carbon double bonds become saturated.

$$\text{C=C} + H_2 \xrightarrow[\text{heat}]{\text{Ni}} H-\overset{|}{\underset{|}{C}}-\overset{|}{\underset{|}{C}}-H$$

   Disulfides are changed to mercaptans.

$$R{-}S{-}S{-}R + 2(H) \longrightarrow 2RSH$$

   Aldehydes and **ketones** are changed to 1° and 2° alcohols.

$$\text{C=O} + H_2 \xrightarrow[\text{P, heat}]{\text{Ni}} \text{CHOH}$$

4. Groups that can neutralize strong acids at room temperature.

   Amines: $RNH_2 + H_3O^+ \longrightarrow R\overset{+}{N}H_3 + H_2O$

   Acid Salts: $R\overset{O}{\overset{||}{C}}O^- + H_3O^+ \longrightarrow R\overset{O}{\overset{||}{C}}OH + H_2O$

5. Groups that can neutralize strong bases at room temperature.

   Amine Salts: $R\overset{+}{N}H_3 + {}^-OH \longrightarrow RNH_2 + H_2O$

   Acids: $RCO_2H + {}^-OH \longrightarrow RCO_2^- + H_2O$

   Phenols: ⟨○⟩—OH + ${}^-$OH ⟶ ⟨○⟩—O$^-$ + H_2O

6. Other reactions involving water.
   Water is a *reactant* when carbon-carbon double bonds add water to form alcohols.
   Water is a *product* when carbon-carbon double bonds are introduced by the dehydration of an
      alcohol; acetals are formed from hemiacetals and alcohols; esters are formed from acids and
      alcohols; amides are formed from acids and amines.

SELF-TESTING QUESTIONS

COMPLETION

   Complete the following equations by writing the structures of the principal organic products. If no
reaction occurs, write "none." Several of the following involve a review of reactions studied in earlier
chapters.

1. $NH_2\overset{O}{\overset{||}{C}}CH_3 + H_2O \xrightarrow{\text{heat}}$ _____

2. $CH_3{-}S{-}H + HCl(aq) \longrightarrow$ _____

3. $CH_3CH_2CH_2NH_2$ + $HCl(aq)$ ⟶ _____

4. CH_3NH_2 + $NaOH(aq)$ ⟶ _____

5. $CH_3CH_2CH_2$—SH + (O) ⟶ _____

6. ⬡—S—S—⬡ + $2(H)$ ⟶ _____

7. ⬡ + OH^- ⟶ _____

8. $CH_3CH_2NHCH_2CH_2$ + $HCl(aq)$ ⟶ _____

9. $CH_3CH_2CH_2\overset{+}{N}H_3$ + OH^- ⟶ _____

10. ⬡$\overset{+}{N}\overset{H}{\underset{H}{}}$ + OH^- ⟶ _____

11. CH_3—O—CH_2CH_2—SH + (O) ⟶ _____

12. CH_2=$CHCH_2\overset{+}{N}H_3$ + OH^- ⟶ _____

13. CH_3CH_2—O—CH=$CHCH_2NH_2$ + H_2 $\xrightarrow[\text{heat}]{\text{Ni}}$ _____

14. ⬡—NH_2 + $HCl(aq)$ $\xrightarrow{\text{room temperature}}$ _____

15. ⬠—$\overset{+}{N}H_2CH_3$ + $NaOH(aq)$ ⟶ _____

16. $NH_2CH_2\overset{\overset{O}{\|}}{C}NHCH\overset{\overset{O}{\|}}{C}OH$ $\xrightarrow[\text{heat}]{H_2O}$ _____ + _____
    $\underset{CH_3}{|}$

17. $CH_3CH_2O\overset{\overset{O}{\|}}{C}CH_2CH_2CH_3$ + H_2O $\xrightarrow[\text{heat}]{H^+}$ _____

18. CH_3CO_2H + $NaOH(aq)$ ⟶ _____

19. $CH_3CH_2CH_2CO_2^-$ + H_3O^+ ⟶ _____

20. CH_3CHOCH_3 + H_2O $\xrightarrow[\text{heat}]{H^+}$ _____
         |
       OCH_3

MULTIPLE–CHOICE

1. Organic functional groups that are hydrolyzed by water (usually in the presence of an acid catalyst and heat) are
    (a) ethers   (d) disulfides
    (b) acetals   (e) esters
    (c) amides   (f) carboxylic acids

2. Organic functional groups that are rather easily oxidized are
    (a) alkanes   (d) ketones
    (b) aromatic hydrocarbons   (e) aldehydes
    (c) mercaptans   (f) amides

3. Organic functional groups that are good proton acceptors are
    (a) amino groups   (d) carboxylate ions
    (b) amides   (e) aromatic hydrocarbons
    (c) alkanes   (f) mercaptans

4. Organic functional groups that are good proton donors are
    (a) amino groups   (d) substituted ammonium ions
    (b) amides   (e) carboxylic acids
    (c) alkanes   (f) alcohols

5. Which compound is the most acidic?
    (a) CH_3CH_2OH

                        O
                        ||
    (c) CH_3C-NH_2

    (b) $CH_3CH=O$

    (d) CH_3CO_2H

6. Which compound is the most basic?
    (a) CH_3CH_2OH (c) $CH_3CH_2NH_3{}^+Cl^-$

                                            O
                                            ||
    (b) $CH_3CH_2NH_2$ (d) CH_3C-NH_2

                        O
                        ||
7. The compound $CH_3O-CH_2C-NH_2$ could be made from

              O   O
              || ||
    (a) CH_3OH and $HOCH_2CNH_2$ (c) CH_3OCH_2CH and NH_3

(b) CH_3OCH_2OH and $H-\overset{\overset{O}{\|}}{C}-NH_2$ (d) $CH_3OCH_2\overset{\overset{O}{\|}}{C}OH$ and NH_3

8. If the following compounds were arranged in the order of their increasing boiling points, that order would be

A. CH_3CH_3 B. CH_3OH C. CH_3NH_2
(a) A < B < C   (c) C < A < B
(b) A < C < B   (d) B < A < C

9. If the following compounds were arranged in the order of their increasing solubility in water, that order would be

A. $CH_3CH_2CH_2CH_2CH_2OH$ B. $CH_3CH_2CH_2CH_2CH_2CH_3$ C. $CH_3CH_2CH_2CH_2CH_2NH_2$
(a) A < B < C   (c) B < A < C
(b) C < B < A   (d) B < C < A

10. Which compound could neutralize aqueous sodium hydroxide?
(a) CH_3OH (c) $CH_3S-S-CH_3$
(b) CH_3NH_2 (d) $CH_3NH_3^+Cl^-$

11. The compound whose structure is $CH_3CH_2CH_2NH_2$ is called
(a) methylethylamine   (c) 1-aminopropane
(b) propylamine   (d) butylamine

12. The reaction of $CH_3-\underset{\underset{CH_3}{|}}{N}H_2^+$ with aqueous sodium hydroxide at room temperature will give

(a) CH_3NHCH_3 (c) $CH_3-\underset{\underset{CH_3}{|}}{N}{}^-$

(b) $CH_3\underset{\underset{CH_3}{|}}{N}-OH$ (d) $CH_3-\underset{\underset{CH_3}{|}}{N}-CH_3$

13. The presence of NH_2 group in an organic compound makes the compound
(a) a good proton donor   (c) more soluble in water
(b) less soluble in water   (d) a good proton acceptor

14. What is the best representation for hydrogen bonding in methylamine?

(a) $CH_3-\underset{\underset{H}{|}}{N}-H\cdots CH_3-\underset{\underset{H}{|}}{N}-H$ (c) $CH_3-N\overset{\overset{H\ H}{|}}{\underset{\underset{H\ H}{|}}{\cdots}}N-CH_3$

(b) $CH_3-\underset{\underset{H}{|}}{N}-H\cdots H-\underset{\underset{H}{|}}{N}-CH_3$ (d) $CH_3-\underset{\underset{H}{|}}{N}-H\cdots \underset{\underset{H}{|}}{N}-CH_3$

Questions fifteen through eighteen refer to this compound:

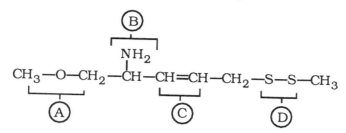

15. The group at (A) would
    (a) react with NaOH(aq)
    (b) react with HCl(aq)
    (c) be reduced by catalytic hydrogenation
    (d) be oxidizable by mild reagents
    (e) none of these

16. The group at (B) would
    (a) neutralize aqueous HCl   (c) react with H_2
    (b) accept H-bonds from water (d) none of these

17. The group at (C) would
    (a) add hydrogen chloride
    (b) donate hydrogen bonds
    (c) be attacked by OH⁻
    (d) none of these

18. The group at (D) would
    (a) be easily reduced   (c) neutralize NaOH(aq)
    (b) react with HCl(aq)   (d) react with water

ANSWERS

ANSWERS TO DRILL EXERCISES

I. **Drill in Writing Products of the Reactions of Amines with Strong Acids**

1. $CH_3CH_2CH_2NH_3^+$ 2. ⟨O⟩—$\overset{+}{N}H_2CH_3$ 3. $CH_3CH_2\overset{+}{N}HCH_2CH_3$
                                                                    $\overset{|}{C}H_3$

4. ⬠—$\overset{+}{N}H_2$ 5. $CH_3\overset{+}{N}H_2CH_2CH_2\overset{+}{N}H_2CH_3$

II.   Drill in Writing the Products of the Reactions of Protonated Amines with Strong, Aqueous Base

1. $CH_3CH_2NH_2$ 2. $CH_3CH_2\underset{\underset{\displaystyle CH_3}{|}}{N}CH_3$ 3.

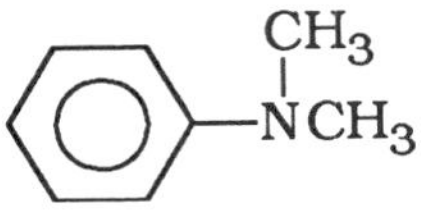

4. $NH_2CH_2CH_2\overset{\displaystyle O}{\overset{\displaystyle \|}{C}}CH_3$ 5. $NH_2\underset{\underset{\displaystyle CH_3}{|}}{C}HCO_2^-$

III.   Drill in Writing the Structures of the Amides That Can Be Made (Directly or Indirectly) from Given Carboxylic Acids and Amines

1. 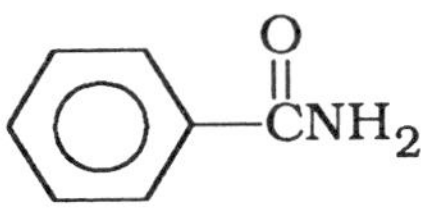   2. $CH_3CH_2\overset{\displaystyle O}{\overset{\displaystyle \|}{C}}NH_2$ 3. $CH_3\underset{\underset{\displaystyle CH_3}{|}}{C}HCH_2\overset{\displaystyle O}{\overset{\displaystyle \|}{C}}NH_2$

4. 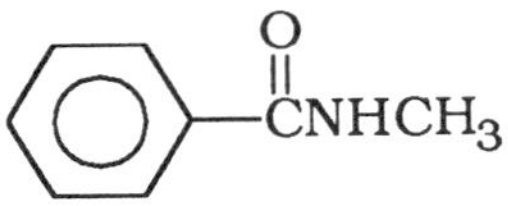   5. $CH_3\overset{\displaystyle O}{\overset{\displaystyle \|}{C}}NHCH_2CH_3$

IV.   Drill in Writing the Products of the Hydrolysis of Amides

1. $CH_3CH_2CO_2H + NH_3$ 2. $CH_3CH_2CH_2CO_2H + NH_2CH_3$

3. $CH_3CH_2NH_2 + HO_2CCH_2CH_3$

4. $CH_3\underset{\underset{\displaystyle CH_3}{|}}{C}HNH_2 + HO_2CCH_2\underset{\underset{\displaystyle CH_3}{|}}{C}HCH_3$

5.

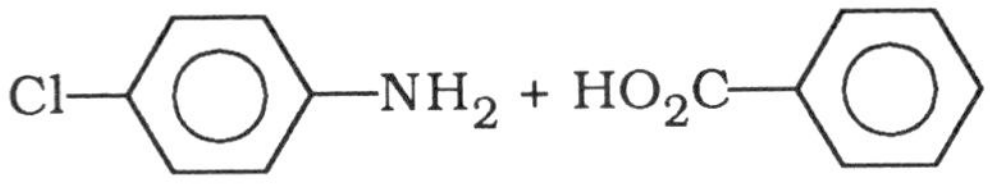

V.   Exercises in Hydrogen Bonds

1. A, B, C, F, G, and H

2. A, B, C, D, F, G, H, and I

3. D and I

4. E

5. A   (The amine, B, boils at a lower temperature because it forms weaker H-bonds than does the alcohol. The formula masses of both are essentially equal.)

6. H   (To be an aromatic amine, the amino group must be joined directly to the aromatic ring. While G is an aromatic compound, it is not an aromatic amine; instead, it is an aliphatic amine.)

ANSWERS TO SELF–TESTING QUESTIONS

Completion

1. $NH_3 + HO_2CCH_3$ (The hydrolysis of an amide.)

2. none   (The only reaction of mercaptans, as far as we are concerned, is their oxidation to disulfides.)

3. $CH_3CH_2CH_3NH_3Cl$

4. none (Amines do not react with aqueous bases.)

5. $CH_3CH_2CH_2$—S—S—$CH_2CH_2CH_3 + H_2O$

6. 2 [cyclohexane ring]—SH

7. none (Benzene has no reaction with bases.)

8. $CH_3CH_2-\overset{\overset{\displaystyle H}{|}+}{\underset{\underset{\displaystyle H}{|}}{N}}-CH_2CH_3Cl^-$

9. $CH_3CH_2CH_2NH_2 + H_2O$

10. [cyclohexane ring]$\overset{H}{N\diagup}$ + H_2O

11. CH_3—O—CH_2CH_2—S—S—CH_2CH_2—O—$CH_3 + H_2O$ (The ether group is unaffected; it remains intact during the reaction.)

12. CH_2=$CHCH_2NH_2 + H_2O$ (The double bond is not affected by an aqueous hydroxide ion.)

13. CH_3CH_2—O—$CH_2CH_2CH_2NH_2$ (Only the double bond is hydrogenated; neither the ether group nor the amino group is changed by hydrogenation.)

14. [cyclohexene ring]—$\overset{+}{N}H_3Cl^-$ (Aqueous hydrochloric acid does not affect the double bond at room temperature.)

15. [cyclopentane ring]—NH—CH_3 (+H_2O)

16. $NH_2CH_2CO_2H$ + $NH_2\underset{\underset{CH_3}{|}}{C}HCO_2H$ (Only the amide group is affected.)

17. CH_3CH_2OH + $HO_2CCH_2CH_2CH_3$ (The hydrolysis of an ester.)

18. $CH_3CO_2^-Na^+$ (+H_2O) 19. $CH_3CH_2CH_2CO_2H$ (+H_2O)

20. CH_3CHO + $2HOCH_3$ (The hydrolysis of an acetal.)

Multiple–Choice

1. b, c, e	2. c, e	3. a, d
4. d, e	5. d	6. b
7. d	8. b	9. d
10. d	11. b, c	12. a
13. c, d	14. d	15. e
16. a, b	17. a	18. a

7

OPTICAL ISOMERISM

Life is as dependent on molecular geometry as it is on other features of molecules. Nearly all biochemical substances consist of molecules that exhibit some type of *chirality*, or handedness. This is particularly true of all enzymes. Here we learn that the chemical properties of a substance whose molecules are chiral can depend, often dramatically, upon its *kind* of chirality. The chapter develops this point and gives some illustrations. You, in turn, should learn the definitions of the terms of optical isomerism given in the Glossary in such a way that you are able to illustrate them by examples.

OBJECTIVES

After studying this chapter and working the Practice and Review Exercises in it, you should be able to do all of the following.

1. Examine the structures of a set of isomers and pick out which are related as constitutional isomers and which are related as stereoisomers.
2. Examine a molecular structure and pick out any tetrahedral stereocenters.
3. Examine the tetrahedral stereocenters in a structure and pick out which are identical and which are different.
4. If all of the tetrahedral stereocenters in a structure are different, calculate the number of optical isomers possible.
5. Examine a structure and predict whether it can exist in two forms related as enantiomers.
6. Examine the structures of a set of stereoisomers and pick out which are related as enantiomers, which are related as diastereomers, and which are meso compounds. (From Special Topic 7.1.)
7. Write structures that illustrate the composition of a racemic mixture. (See Special Topic 7.1.)
8. Calculate the specific rotation of a substance from its observed rotation, its concentration, and the path length of light passing through it.

9. Calculate the concentration of an optically active substance from data on its observed optical rotation, its specific rotation, and the path length of light passing through it.

10. Define all of the terms in the Glossary and give illustrations where applicable.

GLOSSARY

Achiral. Not possessing chirality; that quality of a molecule (or other object) that allows it to be superimposed on its mirror image.

Chiral. Having handedness in a molecular structure. (See also *Chirality.*)

Chiral Carbon. (See *Tetrahedral Stereocenter.*)

Chirality. The quality of handedness that a molecular structure has that prevents this structure from being superimposable on its mirror image.

Constitutional Isomer. One of a set of isomers whose molecules differ in their atom-to-atom sequence.

Dextrorotatory. That property of an optically active substance by which it can cause the plane of plane-polarized light to rotate clockwise.

Enantiomer. One of a pair of stereoisomers that are related as an object is related to its mirror image but that cannot be superimposed one on the other.

Levorotatory. The property of an optically active substance that causes a counterclockwise rotation of the plane of plane-polarized light.

Optical Activity. The ability of a substance to rotate the plane of polarization of plane-polarized light.

Optical Isomer. One of a set of compounds whose molecules differ only in their chiralities.

Optical Rotation. The degrees of rotation of the plane of plane-polarized light caused by an optically active solution; the observed rotation of such a solution.

Plane-Polarized Light. Light whose electrical field vibrations are all in the same plane.

Polarimeter. An instrument for detecting and measuring optical activity.

Racemic Mixture. A 1:1 mixture of enantiomers which is therefore optically inactive.

Specific Rotation [α]. The optical rotation of a solution per unit of concentration per unit of path length.

$$[\alpha] = \frac{\alpha}{cl}$$

where α = observed rotation; c = concentration in g/mL;

and l = path length in decimeters.

Stereoisomer. One of a set of isomers whose molecules have the same atom-to-atom sequences but different geometric arrangements; a geometric (cis-trans) or optical isomer.

Substrate. The substance on which an enzyme performs its catalytic work.

Superimposition. An operation to see if one molecular model can be made to blend simultaneously at exactly every point with another model.

Tetrahedral Stereocenter. An atom in a molecule to which are attached four different atoms or groups.

SELF-TESTING QUESTIONS

COMPLETION

1. These two structures:

$$CH_3 \quad \text{and} \quad CH_3$$

(Cl, Cl on top carbon; H, H on bottom) and (H, Cl; Cl, H)

represent what kind of isomerism:_____

2. These two structures:

(H, Cl; H, Cl) and (Cl, H; H, Cl)

represent _____

3. These two structures:
$$CH_3\!-\!O\!-\!CH_2\!-\!CO_2H \text{ and } HO\!-\!CH_2CO_2CH_3$$
represent what kind of isomerism? _____

4. Complete the two perspective structures to represent the two enantiomers of 1-bromo-1-chloroethane:

$$\underset{\underset{Cl}{|}}{\overset{\overset{Br}{|}}{CH_3CH}}$$

5. Add two Cls and two Hs to each of these structures to show two diastereomers. (Cf. Special Topic 7.1)

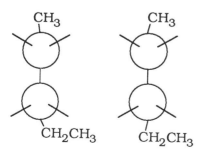

6. Let the surface depicted below represent the surface of an enzyme with the structural features A, B, and C that are involved in the enzyme's work as a catalyst. Examine structures 1, 2, and 3. If the reaction of this enzyme requires that A meet A, B meet B, and C meet C simultaneously as the molecule nestles to the surface of the enzyme, then which molecule(s) can interact with the enzyme? _____

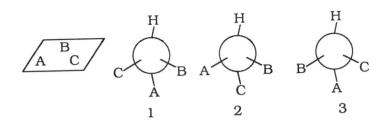

1

2

3

MULTIPLE–CHOICE

1. A substance that is optically active can
    (a) rotate polarized light
    (b) rotate the plane of light
    (c) polarize light
    (d) rotate the plane of plane-polarized light

2. If a substance has a specific rotation described as

    $[\alpha]_D^{20} = -15.6°$, the substance is

    (a) levorotatory   (c) optically active
    (b) dextrorotatory   (d) superimposable

3. A bottle labeled "(+)-glucose" contains a substance that is
    (a) achiral   (c) dextrorotatory
    (b) optically active   (d) positively charged

4. If a solution of 0.06 g/mL of compound X in a polarimeter tube 20 centimeters long has an optical rotation of 1.2°, then its specific rotation is
    (a) 23°   (b) 10°   (c) 100°   (d) 1°

5. If the solution of question 4 were put in a tube 1 dm long, the observed rotation would be
    (a) 5°   (b) 10°   (c) 50°   (d) 0.6°

6. How many tetrahedral stereocenters are present in the following structure?

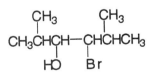

    (a) 1   (b) 2   (c) 3   (d) 4

7. The molecule in question 6 could exist as
    (a) one enantiomer   (c) two pairs of enantiomers
    (b) two enantiomers   (d) one pair of enantiomers

ANSWERS

ANSWERS TO SELF–TESTING QUESTIONS

Completion

1. stereoisomerism (or cis-trans isomerism or geometrical isomerism)
2. constitutional isomerism
3. constitutional isomerism

4.

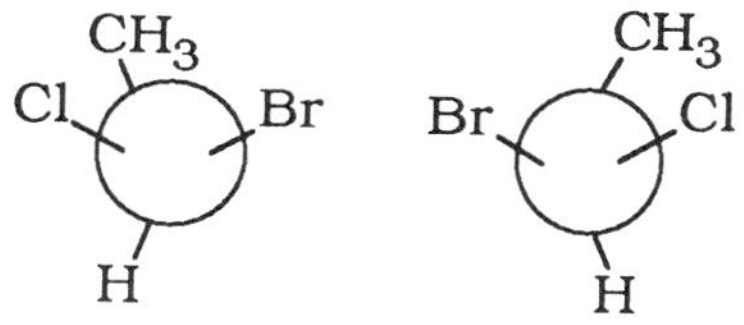

5.

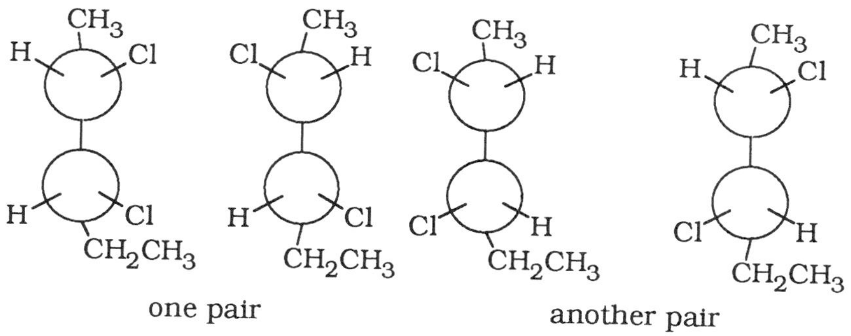

one pair   another pair

6. Structure 1 (Imagine that you rotate it counterclockwise 120° about the bond from the central atom to the H. The other two structures are enantiomers of structure 1.)

Multiple–Choice

1. d   5. d
2. a, c   6. b
3. b, c   7. c
4. b (20 cm = 2 dm)

8

CARBOHYDRATES

In a course of study such as this, *we cannot claim to know what carbohydrates are unless we know their molecular structures.* The emphasis in this chapter, therefore, is structure, particularly the structure of glucose (the most important monosaccharide), of maltose (a representative disaccharide), and of amylose (a representative polysaccharide and one of the two constituents of starch).

OBJECTIVES

After studying this chapter and working the Exercises in it, you should be able to do the following.

1. Describe carbohydrates by their structural features.
2. Name the three classes of carbohydrates we have studied.
3. Interpret such terms as "aldose," "hexose," and "aldohexose."
4. Name the three nutritionally important monosaccharides and give at least one source of each.
5. Name the three nutritionally important disaccharides, give a source of each, and name the products each gives when hydrolyzed.
6. Name three polysaccharides made entirely from glucose units and state where each is found in nature.
7. Write the structures for the α–, β–, and the open forms of glucose. (This would be a minimum goal as far as monosaccharide structures are concerned.)
8. Look at the cyclic structure of a given monosaccharide and point out its hemiacetal system.
9. Look at the cyclic structure of a given disaccharide and point out its acetal-oxygen bridge; tell if the disaccharide will give a positive result in Benedict's or Tollens' Test.
10. Explain what is meant by "deoxy–."

11. Write the structure of α– and β– maltose. (This is a minimum goal as far as disaccharide structures are concerned.)
12. Write the structure of the repeating unit in amylose (as a minimum goal for illustrating what polysaccharides are).
13. Describe an instance in which two polysaccharides differ only in the orientation (i.e., geometry) of their oxygen bridges.
14. Describe what one does and sees in the starch-iodine test.
15. Name the principal components of starch.
16. Describe the structural relations between amylopectin and amylose. (Do this in words if not by writing the structures.)
17. Write the plane projection structure of D-glucose and explain why it is in the D-family.
18. Tell from the plane projection structure of any aldose or ketose whether it is in the D- or L-family.
19. Define all of the terms in the Glossary.

GLOSSARY

Absolute Configuration. The arrangement in space about each tetrahedral stereocenter in a molecule.

Aldohexose. A monosaccharide whose molecules have six carbon atoms and an aldehyde group.

Aldose. A monosaccharide whose molecules have an aldehyde group.

Biochemistry. The study of the structures and properties of substances found in living systems.

Blood Sugar. The carbohydrates – mostly glucose – that are present in blood.

Carbohydrate. Any naturally occurring substance whose molecules are polyhydroxyaldehydes or polyhydroxyketones or can be hydrolyzed to such compounds.

D-Family; L-Family. The names of the two optically active families to which substances can belong when they are considered solely according to one kind of molecular chirality (molecular handedness) or the other.

Disaccharide. A carbohydrate that can be hydrolyzed into two monosaccharides.

Glycosides. Acetals of carbohydrates.

Glycosidic Link. The acetal oxygen bridge between monosaccharide units in di- and polysaccharides.

Iodine Test. The test for starch by which a drop of iodine reagent produces an intensely purple color if starch is present.

Ketohexose. A monosaccharide whose molecules contain six carbon atoms and have a keto group.

Ketose. A monosaccharide whose molecules have a ketone group.

Monosaccharide. A carbohydrate that cannot be hydrolyzed.

Mutarotation. The gradual change in the specific rotation of a substance in solution but without a permanent, irreversible chemical change occurring.

Photosynthesis. A series of reactions in plants, powered by solar energy absorbed by chlorophyll, by which CO_2, H_2O, and minerals are used to make complex molecules like glucose.

Polysaccharide. A carbohydrate whose molecules are polymers of monosaccharides.

Reducing Carbohydrate. A carbohydrate that gives a positive Benedict's Test.

Simple Sugar. Any monosaccharide.

DRILL EXERCISES

EXERCISES IN CARBOHYDRATE STRUCTURES AND SYMBOLS

In the simplified structures of the cyclic forms of carbohydrates, the bonds or lines that are parts of the rings (hexagons or pentagons) form a flat surface. You should imagine that this surface comes out of the page, perpendicular to it. Bonds or lines that point downward from corners of these rings actually point below the plane of the ring. Those pointing up from a corner project above the plane. These relations hold even if we rotate the ring around an imaginary axis going through the center of the ring and perpendicular to the plane of the ring.

1. Just to make sure that the condensed structural symbols for the monosaccharides are understood, convert this symbol into its full structural formula with the atomic symbols given for all carbon, hydrogen, and oxygen atoms and with all bonds represented by lines.

2. Study the following Fischer projection structures and answer the questions. (Not all of these plane projections are in strict accord with all of the rules for making them, but the rearward projections of vertical lines and the forward projections of horizontal lines are intended to be as specified by these rules.) Remember that in imaginary operations with Fisher projections, they may be slid around on the plane, but they may not be taken out of the plane and tipped in any way.

    (a) Which, if any, are D-glyceraldehyde? _____
    (b) Which, if any, are L-glyceraldehyde? _____

3. Given the following Fischer projection structure for D-galactose, write the Fischer projection structures for L-galactose.

D-galactose   _____   L-galactose

4. Modify the structure of D-galactose given above to show the structure of D-2-deoxygalactose.

_____

SELF-TESTING QUESTIONS

COMPLETION

1. A carbohydrate whose molecules will react with water to produce two sugar units is a

_____.

2. The structural feature involved in the link between two glucose units in maltose is the

_____.

3. The hydrolysis of sucrose produces _____ and _____.

4. The partial hydrolysis of starch gives a mixture called _____.

5. The iodine test is used to detect the presence of _____.

6. Carbohydrates all have high melting points, which means that their molecules are strongly attracted to each other in the crystals. The force of attraction responsible for this property is the

_____.

7. The names of the two very broad families of optically active compounds, which are organized solely on the bases of "handedness," are the _____ family and the _____ family.

8. The hydrolysis of lactose produces _____ and _____.

9. The linear polymer of α-glucose is _____.

10. Starch is a mixture of _____ and _____.

11. The technical name for a potential aldehyde group is the _____.

12. The hydrolysis of maltose produces _____.

13. Because maltose and lactose have the _____ group, they are reducing disaccharides.

14. The storage form of glucose molecules in animals is _____, which structurally is similar to _____, one of the storage forms of glucose units in plants.

15. If a polysaccharide can be hydrolyzed into nothing but glucose, it might be any one of the following (make a complete list):_____

_____

MULTIPLE–CHOICE

1. Which is (are) the structure(s) of α-glucose?

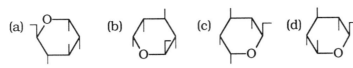

2. If the symbol Gl is used to represent a glucose unit, then the symbol:

    etc. —O—Gl—O—Gl—O—Gl—O—Gl—O—Gl—etc.

    could be a way of showing the basic structural feature of

    (a) amylopectin   (b) amylose

    (c) glycogen   (d) galactose

3. An example of a reducing carbohydrate is

    (a) sucrose   (b) maltose   (c) cellulose   (d) galactose

4. An aldohexose would have

    (a) a potential aldehyde group

    (b) five hydroxyl groups in its open form

    (c) six carbons

    (d) one CH_2OH group

5. The 50:50 mixture of glucose and fructose

    (a) is called dextromaltose

    (b) is obtainable by the hydrolysis of sucrose

    (c) is called invert sugar

    (d) is a disaccharide

6. If the molecules of a substance have the structure

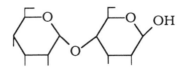

    (a) it is a disaccharide

    (b) it is in the D-family

    (c) it can mutarotate

    (d) it has a β-acetal oxygen bridge

7. If the molecules of a substance have the structure:

    (a) it can be hydrolyzed into two glucose units

    (b) it can be hydrolyzed into one glucose unit plus another monosaccharide

    (c) it will not give a Benedict's Test

    (d) it will not mutarotate

8.   If the molecules of a substance have the structure:

![structure with -O and OCH₂CH₃]

    (a) it is an α-glucoside
    (b) it is a β-glucoside
    (c) it will mutarotate
    (d) it will give a positive Benedict's Test

9.   If the molecules of a substance have the structure:

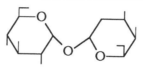

    (a) it is an α-glucoside
    (b) it is a β-glucoside
    (c) it will give a positive Benedict's Test
    (d) it is a disaccharide

10.   If the molecules of a substance have the structure:

![disaccharide structure]

    (a) it is a disaccharide
    (b) it will give a positive Benedict's Test
    (c) it will mutarotate
    (d) it will hydrolyze into two D-galactose units

ANSWERS

ANSWERS TO DRILL EXERCISES

Exercises in Carbohydrate Structures and Symbols

1.

2. (a) 1 and 2 (Note that 2 results from spinning 1 by 180° in the plane of the paper.)

   (b) 3 and 4 (3 is the mirror image of 1; 4 results from turning 3 by 180° in the plane of the paper.)

3.

4.

ANSWERS TO SELF–TESTING QUESTIONS

Completion

1. disaccharide
2. α-acetal bridge or an $\alpha(1\rightarrow4)$ glycosidic link)
3. fructose, glucose
4. dextrin
5. starch
6. hydrogen bond
7. D- and L- (families)

8. glucose, galactose
9. amylose
10 amylose, amylopectin
11. hemiacetal group
12. glucose
13. hemiacetal (or potential aldehyde)
14. glycogen, amylopectin
15. starch, amylose, amylopectin, glycogen, cellulose, dextrin

Multiple–Choice

1. a, b, and d
2. b
3. b and d
4. a, b, c, and d
5. b and c
6. a, b, and c
7. b
8. b (Only when the hemiacetal group—or the hemiketal group — is present will the substance mutarotate and give a positive Benedict's Test.)
9. c (To be a glucoside, the CH_3 group would have to be part of an acetal system at carbon number one.)
10. a (It has no hemiacetal system and therefore cannot mutarotate or give a positive Benedict's Test.)

9

LIPIDS

Just as with carbohydrates, so with lipids; we don't know these substances until we know their structures. With lipids, however, we learn the features of the structures held in common by all lipids in a given family, not the structures of specific lipids.

OBJECTIVES

Although all of these objectives are important, one through five constitute minimum goals in our preparation for later chapters. Objectives 11 and 12 are necessary for an understanding of how the membranes of animal cells are organized and how they can hold a cell together while letting substances in and out.

After you have studied this chapter and worked the exercises in it, you should be able to do the following.

1. Write the structure of a molecule that includes at least one carbon-carbon double bond and is typically found in a triacylglycerol.
2. Using that structure, write the products of its reaction with
   (a) water—catalyzed by an enzyme (as in digestion);
   (b) aqueous sodium hydroxide—saponification; and
   (c) hydrogen—hydrogenation with a catalyst and heat.
3. Write the names and structures for at least three saturated and three unsaturated fatty acids.
4. Describe the principal structural differences between animal fats and vegetable oils.
5. Explain what polyunsaturated means when used to describe vegetable oils.
6. Look at the structures of two triacylglycerols that have essentially identical formula masses and tell which has the higher degree of unsaturation.

7. Name two kinds of glycerophospholipids and tell where they may be found in the body.
8. Do the same for two kinds of sphingosine-based lipids.
9. Name several steroids and briefly describe their purposes.
10. Briefly explain why steroids are classified as lipids.
11. Describe the composition and the structure of a biological membrane.
12. Define each of the terms in the Glossary.

GLOSSARY

Amphipathic Molecule. A molecule with both hydrophilic and hydrophobic groups.

Fatty Acid. Any carboxylic acid that can be obtained by the hydrolysis of animal fats or vegetable oils.

Glycerophospholipid. A hydrolyzable lipid—a phosphatide or a plasmalogen—with an ester linkage between glycerol and one phosphoric acid unit that, in turn, is joined by a phosphate ester linkage to a small molecule

Glycolipid. A lipid whose molecules include a glucose unit, a galactose unit or some other carbohydrate unit.

Hydrolyzable Lipid. Any lipid with ester groups.

Hydrophilic Group. Any part of a molecular structure that attracts water molecules; a polar or ionic group such as OH, CO_2^-, NH_3^+, or NH_2.

Hydrophobic Group. Any part of a molecular structure that has no attraction for water molecules; a nonpolar group such as any alkyl group.

Lipid. A plant or animal product that tends to dissolve in such nonpolar solvents as ether, carbon tetrachloride, and benzene.

Lipid Bilayer. The sheetlike array of two layers of lipid molecules, interspersed with molecules of cholesterol and proteins, that make up the membranes of cells in animals.

Micelle. A globular arrangement of the molecules of an amphipathic compound in water in which hydrophobic parts intermingle inside the globule and hydrophilic parts are exposed to water.

Nonhydrolyzable Lipid. Any lipid, such as the steroids, that cannot be hydrolyzed or similarly broken down by aqueous alkali.

Phosphatide. A glycerophospholipid whose molecules are esters between glycerol, two fatty acids, phosphoric acid and a small alcohol.

Phosphoglyceride. (See *Glycerophospholipid*.)

Phospholipid. Lipids such as the glycerophospholipids and the sphingomyelins whose molecules include phosphate ester units.

Plasmalogens. Glycerophospholipids whose molecules also include an unsaturated fatty alcohol unit.

Saponifiable Lipid. (See *Hydrolyzable Lipid*.)

Sphingolipid. A lipid that, when hydrolyzed, gives sphingosine instead of glycerol, plus fatty acids, phosphoric acid, and a small alcohol or a monosaccharide; sphingomyelins and cerebrosides.

Steroids. Nonsaponifiable lipids such as cholesterol and several sex hormones whose molecules have the four fused rings of the steroid nucleus.

Triacylglycerol. A lipid that can be hydrolyzed to glycerol and fatty acids; a triglyceride; sometimes, simply called a glyceride.

Triglyceride. (See *Triacylglycerol.*)

Wax. A lipid whose molecules are esters of long-chain monohydric alcohols and long-chain fatty
acids.

SELF-TESTING QUESTIONS

COMPLETION

1. Write the structures of all of the products that would form from the complete hydrolysis of the
   following lipid.

$$
\begin{array}{l}
CH_2-O-\overset{\displaystyle O}{\overset{\|}{C}}(CH_2)_7CH=CHCH_2CH=CH(CH_2)_4CH_3 \\[2ex]
CH-O-\overset{\displaystyle O}{\overset{\|}{C}}(CH_2)_8CH_3 \\[2ex]
CH_2-O-\overset{\displaystyle O}{\overset{\|}{C}}(CH_2)_7CH=CH(CH_2)_7CH_3
\end{array}
$$

2. If the lipid of question 1 had been saponified by sodium hydroxide instead of hydrolyzed, the fatty
   acids would have been produced in the form of their _____.

3. Write the structure of the product of the hydrogenation of all the alkene groups in the lipid in
   question 1.

4. A molecule of vegetable oil will normally have more _____ than a
   molecule of animal fat.

5. The two functional groups in a simple lipid are _____ and _____.

6. An example of a nonsaponifiable lipid is any member of the family of _____.

7. Lipids with the general structure $R-O-\overset{\displaystyle O}{\overset{\|}{C}}-R'$ (where R and R' are both long-chain) are in the
   family of _____.

8. The acid, other than fatty acids, liberated when complex lipids are hydrolyzed is _____.

9. A vegetable oil is said to be more _____ than an animal fat (referring to double bonds).

10. The complete hydrolysis (digestion) of this phosphatidyl-choline would give what products?
    (Write their structures and names.)

$$CH_2-O-\overset{\overset{\displaystyle O}{\|}}{C}(CH_2)_7CH=CH(CH_2)_7CH_3$$

$$CH-O-\overset{\overset{\displaystyle O}{\|}}{C}(CH_2)_7CH=CHCH_2CH=CH(CH_2)_4CH_3$$

$$CH_2-O-\overset{\overset{\displaystyle O}{\|}}{\underset{\underset{\displaystyle O^-}{|}}{P}}-O-CH_2CH_2\overset{+}{N}(CH_3)_3$$

11. Another name for the structure in question 10 is _____.

12. Because the hydrocarbon chains in the structure in question 10 are water-avoiding, they are said to be _____. In the region of the phosphate group, however, the molecule is _____. The molecule in question 10, taken as a whole, is described as _____.

13. Because the hydrocarbon chains of the structure in question 10 have double bonds, the lipid would be described as poly-_____.

MULTIPLE–CHOICE

1. To reduce the degree of polyunsaturation of a triacylglycerol, a manufacturer might
   (a) hydrate it
   (b) hydrogenate it
   (c) hydrolyze it
   (d) dehydrogenate it

2. If a manufacturer hydrogenated a vegetable oil, the substance might
   (a) turn rancid
   (b) become a detergent
   (c) become a solid at room temperature
   (d) become a diglyceride

3. In glycerides, aqueous sodium hydroxide will attack
   (a) ester linkages
   (b) ether linkages
   (c) alkene groups
   (d) alkenelike portions

4. Fatty acids obtained from natural lipids are generally
   (a) of even carbon number
   (b) monocarboxylic
   (c) insoluble in water
   (d) long chain

5. The hydrolysis of a naturally occurring triacylglycerol will give
   (a) glycerol + RCO_2H + $R'CO_2H$ + $R''CO_2H$
   (b) glycerol + RCO_2^- + $R'CO_2^-$ + $R''CO_2^-$
   (c) glycerol + $3RCO_2H$
   (d) glycerol + $3RCO_2H$

6. The name of a C_{18} acid with three alkene groups is
   (a) stearic acid
   (b) oleic acid
   (c) linoleic acid
   (d) linolenic acid

7. Phospholipids are esters of
   (a) sphingosine only
   (b) either glycerol or sphingosine
   (c) cholesterol only
   (d) glycerol only

8. Glycerophospholipids include
   (a) esters of phosphatidic acids
   (b) cerebrosides
   (c) esters of sphingosine
   (d) glycolipids

9. Cholesterol is classified as a lipid because
   (a) it is an ester
   (b) it dissolves in fat solvents
   (c) it is present in gallstones
   (d) it can be saponified

10. The cell membranes of animals are made of
    (a) lipids
    (b) amylose
    (c) proteins
    (d) cholesterol

11. Molecules of hydrolyzable lipids have
    (a) hydrophobic groups
    (b) ester groups
    (c) amide groups
    (d) alkene groups

12. The cell membranes of animals are organized
    (a) with hydrophobic groups projecting outward away from the membrane
    (b) as lipid bilayers with imbedded proteins
    (c) with hydrophilic groups projecting inward
    (d) with cellulose molecules lending structural support

ANSWERS

ANSWERS TO SELF–TESTING QUESTIONS

Completion

1.

$$HOCH_2CHCH_2OH \; + \; CH_3(CH_2)_4CH{=}CHCH_2CH{=}CH(CH_2)_7CO_2H$$
$$\underset{OH}{|}$$
$$+ \; CH_3(CH_2)_8CO_2H$$
$$+ \; CH_3(CH_2)_7CH{=}CH(CH_2)_7CO_2H$$

2. sodium salts

3.

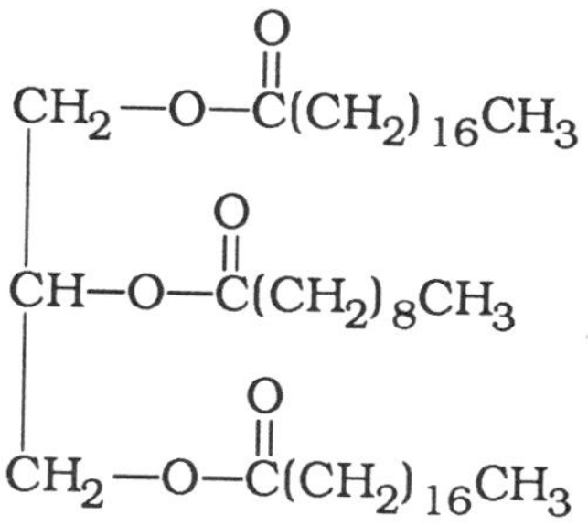

$$CH_2-O-\overset{\overset{\displaystyle O}{\|}}{C}(CH_2)_{16}CH_3$$

$$CH-O-\overset{\overset{\displaystyle O}{\|}}{C}(CH_2)_8CH_3$$

$$CH_2-O-\overset{\overset{\displaystyle O}{\|}}{C}(CH_2)_{16}CH_3$$

4. carbon-carbon double bonds (or alkene groups or unsaturation)
5. esters, alkenes
6. steroids
7. waxes
8. phosphoric acid
9. unsaturated

$$
\begin{array}{l}
CH_2OH \\
CH-OH \\
CH_2OH \\
\text{glycerol}
\end{array}
$$

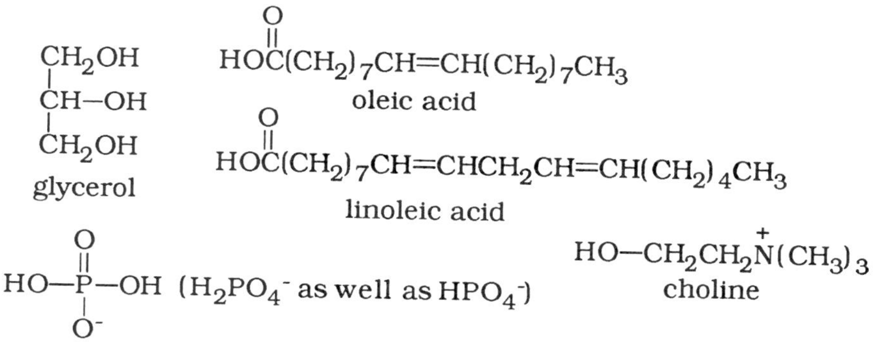

$$HO\overset{\overset{\displaystyle O}{\|}}{C}(CH_2)_7CH=CH(CH_2)_7CH_3$$
oleic acid

$$HO\overset{\overset{\displaystyle O}{\|}}{C}(CH_2)_7CH=CHCH_2CH=CH(CH_2)_4CH_3$$
linoleic acid

10.

$$HO-\overset{\overset{\displaystyle O}{\|}}{\underset{\underset{\displaystyle O^-}{|}}{P}}-OH \quad (H_2PO_4^- \text{ as well as } HPO_4^-)$$

dihydrogen phosphate ion

$$HO-CH_2CH_2\overset{+}{N}(CH_3)_3$$
choline

11. lecithin
12. hydrophobic; hydrophilic; amphipathic
13. unsaturated

Multiple–Choice

1. b	2. c	3. a
4. a, b, c, and d	5. a	6. d
7. b	8. a	9. b
10. a, c, and d	11. a, b, and d	12. b

10

PROTEINS

Proteins, like carbohydrates, are known only when their *structures* are known, so the main emphasis in this chapter is the structural features of proteins. Another emphasis is the importance of *molecular shape* and the factors that affect it. This carries over into the study of the protein (glycoprotein) components of animal cell membranes.

OBJECTIVES

These objectives are designed to promote general knowledge about proteins and are important for our understanding of later chapters. When you have completed your study of this chapter and have worked the Practice and Review Exercises in it, you should be able to do the following.

1. Write the names and structures of five representative amino acids, for example
   (a) glycine—because it is the simplest amino acid;
   (b) alanine—as representative of an amino acid with a hydrocarbon side chain (a hydrophobic group);
   (c) cysteine—because it has the important sulfhydryl group;
   (d) glutamic acid—to represent an amino acid with a side chain CO_2H (or CO_2^-) group;
   (e) lysine—to represent amino acids with a side chain NH_2 (or NH_3^+) group.
2. Based on the amino acids you have learned, write the structure of any di-, tri-, tetra-, or pentapeptide and identify the peptide bonds.
3. Translate a structure such as Gly-Ala-Glu into a condensed structural formula.
4. Examine the structure of a polypeptide and identify what parts are amino acid *residues* (peptide units) and what parts are *peptide groups*.

5. Write structures that illustrate how hydrogen bonds and salt bridges participate in the structure of proteins.

6. Name the four levels of structure in proteins and briefly describe each in terms of the kinds of forces that stabilize it and the kinds of geometric forms it takes.

7. Write the structures that show how disulfide bonds occur in proteins.

8. Describe and explain hydrophobic interactions and their origin.

9. Explain the relation between a polypeptide and a protein.

10. Explain how the solubility of a protein can be changed by changing the pH of its medium.

11. Given the structure of a small polypeptide, write the structures of the products of its digestion.

12. Describe what happens structurally when a protein is denatured.

13. Describe, in general terms, how denaturation affects the properties of a protein.

14. List at least four denaturing agents.

15. Explain in general terms how the proteins in a cell membrane participate in active transport to maintain concentration gradients.

16. Describe what gap junctions are and what purpose they serve.

17. In general terms, describe the services performed by the oligosaccharide components of membrane proteins.

18. In general terms, describe the composition and function of ground substance.

19. Name five fibrous proteins.

20. Name two globular proteins.

21. Define each of the terms in the Glossary.

GLOSSARY

Active Transport. The movement of a species across a cell membrane for which the needed energy is supplied by metabolism.

Albumin. One of a family of globular proteins that tend to dissolve in water, and that in blood contribute to the blood's colloidal osmotic pressure and aid in the transport of metal ions, fatty acids, cholesterol, triacylglycerols, and other water-insoluble substances.

Amino Acid. Any organic compound whose molecules have both an amino group and a carboxyl group.

Amino Acid Residue. A structural unit in a polypeptide,

$$NH-\underset{\underset{R}{|}}{CH}-CO-,$$

furnished by an amino acid, where R is a side chain group; a peptide unit.

Collagen. The fibrous protein of connective tissue that changes to gelatin in boiling water.

Denatured Protein. A protein whose molecules have suffered the loss of their native shape and form as well as their ability to function biologically.

Denaturation. The loss of the natural shape and form of a protein molecule together with its ability to function biologically, but not necessarily accompanied by the rupture of any of its peptide bonds.

Dipeptide. A compound whose molecules have two α-amino acid residues joined by a peptide (amide) bond.

Dipolar Ion. A molecule that carries one plus charge and one minus charge, such as an α-amino acid; a zwitterion.

Disulfide Link. The S—S unit in cystine.

Elastin. The fibrous protein of tendons and arteries.

Fibrin. The fibrous protein of a blood clot that forms from fibrinogen during clotting.

Fibrous Proteins. Water-insoluble proteins found in fibrous tissues.

Gap Junctions. Tubules made of membrane-bound proteins that interconnect one cell to neighboring cells and through which materials can pass directly.

Globular Proteins. Proteins that are soluble in water or in water that contains certain dissolved salts.

Globulins. Globular proteins in the blood that include γ-globulin, an agent in the body's defense against infectious diseases.

Glycolipids. A lipid whose molecules include a glucose unit, a galactose unit, or some other carbohydrate unit.

Glycoproteins. Proteins, often membrane-bound, whose molecules include a carbohydrate unit.

Gradient. The occurrence of a change in the value of some physical quantity with distance, as in a *concentration gradient* in which the concentration of a solute is different in different parts of the system.

Ground Substance. A gel-like material present in cartilage and other extracellular spaces that gives flexibility to collagen and other fibrous proteins.

α-Helix. One kind of secondary structure of a polypeptide in which its molecules are coiled.

Hemoglobin (HHb). The oxygen-carrying protein in red blood cells.

Hydrophobic Interaction. The water-avoidance by nonpolar groups or side chains that is partly responsible for the shape adopted by a polypeptide molecule in an aqueous environment.

Isoelectric Molecule. A molecule that has an equal number of positive and negative sites.

Isoelectric Point (pI). The pH of a solution in which a specified amino acid or a protein is in an isoelectric condition; the pH at which there is no net migration of the amino acid or protein in an electric field.

Keratin. The fibrous protein of hair, fur, fingernails, and hooves.

Myosins. Proteins in contractile muscle.

Native Protein. A protein whose molecules are in the configuration and shape they normally have within a living system.

Peptide Bond. The amide linkage in a protein; a carbonyl-to-nitrogen bond.

Peptide Unit. (See *Amino Acid Residue.*)

pI. (See *Isoelectric Point.*)

β-Pleated Sheet. A secondary structure for a polypeptide in which the molecules are aligned side by side in a sheetlike array with the sheet partially pleated.

Polypeptide. A polymer with repeating α-aminoacyl units joined by peptide (amide) bonds.

Primary Structure. The sequence of aminoacyl residues held together by bonds in a polypeptide.

Prosthetic Group. A nonprotein molecule joined to a polypeptide to make a biologically active protein.

Protein. A naturally occurring polymeric substance made up wholly or mostly of polypeptide molecules.

Quaternary Structure. An aggregation of two or more polypeptide strands each with its own primary, secondary, and tertiary structure.

Receptor Molecule. A molecule of a protein built into a cell membrane that can accept a molecule of a hormone or a neurotransmitter.

Salt Bridge. A force of attraction between (+) and (−) sites on polypeptide molecules.

Secondary Structure. A shape, such as the α-helix or a unit in a β-pleated sheet, that all or a large part of a polypeptide molecule adopts under the influence of hydrogen bonds or salt bridges after its peptide bonds have been made.

Side Chain. An organic group that can be appended to a main chain or to a ring.

Tertiary Structure. The shape of a polypeptide molecule that arises from further folding or coiling of secondary structures.

Triple Helix. The quaternary structure of tropocollagen in which three polypeptide chains are twisted together.

Zwitterion. (See *Dipolar Ion*)

SELF-TESTING QUESTIONS

COMPLETION

1. In their dipolar ionic forms, all amino acids have the same basic unit that (without side chains) has the structure:

_____

2. Objective 1 asks you to know the structures of five representative amino acids. The best way to learn them is to learn their side chains, because these are always affixed to the basic unit you wrote in question 1. On the lines provided, write the structures of these side chains.

_____   _____   _____

glycine side chain   alanine side chain   cysteine side chain

_____   _____

glutamic acid side chain   lysine side chain

3. Write the structure of the dipeptide that could form from two glycine units.

_____

4. In terms of the three-letter symbols for amino acids, the dipeptide in question 3 would be represented as _____.

5. Write the structure of a tripeptide that could be made from alanine, glutamic acid, and cysteine. Let alanine be the N-terminal residue and cysteine the C-terminal residue.

_____

6. In terms of the three-letter symbols, the tripeptide in question 5 has the formula _____ and it contains_____ peptide bonds.

7. Write the structure of the tetrapeptide: Ala-Gly-Lys-Cys

_____

8. Write the structure of cysteine in its dipolar ionic form.

_____

9. If the following polypeptide were subjected to mild reducing conditions, what would form? Write the structure(s) using the three-letter symbols.

Gly-Ala-Cys-Lys-Glu
          |
          S
          |
          S   _____
          |
Gly-Ala-Cys-Lys-Glu

10. In a discussion of how a polypeptide coils or otherwise assumes some geometric shape, we are talking about what structural level? _____

11. Three non-covalent forces that can determine the shape that will be adopted by a polypeptide are _____ .

12. When we speak of an α-helix itself undergoing folding or twisting, we are talking about what level of protein structure? _____

13. Which level of protein structure is not necessarily attacked by denaturing agents? _____

14. The side chains of which of the five amino acids in Objective 1 are the most susceptible to being altered by a change in the pH of the medium?(Name the amino acids.)

_____

15. When we speak of two or more polypeptides becoming associated in some way in a gigantic protein system, we are talking about which level of protein structure? _____

16. Because the concentration of dissolved calcium ion inside a cell is different from that outside the cell, we can say that a _____ for calcium ion exists across the cell membrane.

17. Tubules that connect cells to each other are called _____ and they are made of _____ molecules present in cell membranes.

18. The membrane-bound molecules of a _____ protein can uniquely recognize a particular hormone because something about the _____ of the protein molecule is complementary to that of the hormone.

19. Membrane-bound proteins are usually covalently linked to _____ molecules made from aminosugars.

20. Glycosaminoglycans are found in a gel-like material called _____, which helps to give flexibility to a kind of tissue known as _____. In this tissue, tensile strength is supplied by the fibrous protein called _____.

MULTIPLE-CHOICE

1. The partial hydrolysis of a protein produces a number of amino acids together with several dipeptides. Which of the following fragments would not be present?

    (a) $NH_2CH_2CH_2CO_2H$

    (c) $HO_2CCH_2NHCCH_3$ (with =O above the C)

    (b) $NH_2CHCNHCH_2CO_2H$ (with =O above the C, and CH_3 below)

    (d) NH_2CHCO_2H
        $(CH_2)_4NHCCH_2NH_2$ (with =O below the C)

2. If the dipolar ionic form of alanine neutralized H^+, it would be changed into

    (a) $\overset{+}{N}H_3CHCO_2^-$
        CH_3

    (c) $NH_2CHCO_2^-$
        CH_3

    (b) NH_2CHCO_2H
        CH_3

    (d) $\overset{+}{N}H_3CHCO_2H$
        CH_3

3. The peptide bonds in a protein can be broken by
    (a) hydrolysis
    (b) buffer action
    (c) denaturation
    (d) hydration

4. Two proteins might be joined together by the action of a mild oxidizing agent if among their amino acid units there is present
    (a) glutamic acid
    (b) cysteine
    (c) glycine
    (d) lysine

5. Ions of heavy metals (e.g., Hg^{2+} or Pb^{2+}) denature proteins by combining with
    (a) SH groups
    (b) CH_3 groups
    (c) peptide bonds
    (d) the protein backbones

6. In very strongly acidic solutions, the molecules of alanine would mostly be in what form?

(a) $NH_2\underset{\underset{\displaystyle CH_3}{|}}{C}HCO_2H$

(c) $NH_2\underset{\underset{\displaystyle CH_3}{|}}{C}HCO_2^-$

(b) $\overset{+}{N}H_3\underset{\underset{\displaystyle CH_3}{|}}{C}HCO_2H$

(d) $\overset{+}{N}H_3\underset{\underset{\displaystyle CH_3}{|}}{C}HCO_2^-$

7. In their dipolar ionic forms, amino acids are
    (a) electrically neutral
    (b) weak acids
    (c) weak bases
    (d) solids at room temperature

8. An amino acid with an isoelectric point of 6.8 would consist of molecules with side chains that have
    (a) an extra NH_2 group   (c) a hydrophobic group
    (b) an extra CO_2H group   (d) an extra SO_3H group

9. All the common, naturally occurring amino acids (except glycine) are members of the
    (a) D-family   (c) lipid family
    (b) L-family   (d) peptide family

10. The level of protein structure that has amide bonds is the
    (a) primary   (c) tertiary
    (b) secondary   (d) quaternary

11. If two molecules of alanine were joined as a dipeptide
    (a) two isomeric dipeptides are possible
    (b) one dipeptide is possible
    (c) the structural symbol would be Ala-Ala
    (d) one peptide bond would be present

12. When proteins are digested

    (a) $-\overset{\overset{\displaystyle O}{\|}}{C}-NH-$ units become $-CO_2H + NH_2-$ units

    (b) $-SH$ groups become $-S-S-$ groups

    (c) $-S-S-$ groups become $-SH$ groups

    (d) CO_2H and NH_2 units join to become $-\overset{\overset{\displaystyle O}{\|}}{C}-NH-$ units

13. If somehow the protein were changed to

    the protein would be

(a) less soluble in water

(b) more soluble in water

(c) at its isoelectric point

(d) digested

14. In the pentapeptide Gly-Ala-Lys-Cys-Glu, the N-terminal unit is

(a) glycine   (b) glutamic acid (c) lysine   (d) NH_2

15. If the conventions for writing the structures of proteins with the three-letter symbols of amino acids are properly obeyed, the symbol for this tripeptide would be

$$HOCCH_2NHCCHNHCCHNH_2$$

with O double bonds above the three C atoms, and CH_3 and CH_2SH side chains below.

(a) Gly-Ala-Cys   (c) Ala-Gly-Cys

(b) Cys-Ala-Gly   (d) Ala-Cys-Gly

16. The principal non-covalent force in protein structure is the

(a) salt bridge   (c) helix

(b) disulfide link   (d) hydrogen bond

17. An important secondary structural feature in proteins is the

(a) salt bridge   (c) disulfide link

(b) hydrogen bond   (d) α-helix

18. Nonprotein molecules that often are associated with proteins are called

(a) prosthetic groups   (c) side chains

(b) zwitterions   (d) 3° structures

19. A hormone molecule attaches to a cell at

(a) a gap junction   (c) a glycosaminoglycan

(b) a receptor molecule   (d) an N-link

20. Oligosaccharides are present in

(a) glycoproteins   (c) gradients

(b) hydrophobic groups   (d) collagen fibrils

21. Gap junctions are

(a) spaces between nerve cells

(b) spongy materials in cartilage

(c) tubules between adjacent cells

(d) regions between overlapping collagen fibrils

22. The principal protein in the walls of blood vessels is

(a) keratin   (b) myosin   (c) elastin   (d) fibrin

23. An important protein in contractile muscle is

(a) keratin   (b) myosin   (c) elastin   (d) fibrin

24. A protein in hair is

(a) keratin   (b) myosin   (c) elastin   (d) fibrin

25. One important transport protein is

(a) hemoglobin   (b) λ-globulin   (c) casein   (d) nucleoprotein

ANSWERS

ANSWERS TO SELF–TESTING QUESTIONS

Completion

1. 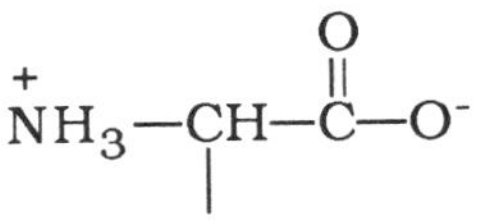 $\overset{+}{N}H_3-CH-\overset{\overset{\displaystyle O}{\displaystyle \|}}{C}-O^-$

2.   glycine   alanine   cysteine   glutamic acid   lysine
       H CH_3 CH_2SH $CH_2CH_2CO_2H$ $CH_2CH_2CH_2CH_2NH_2$

3. 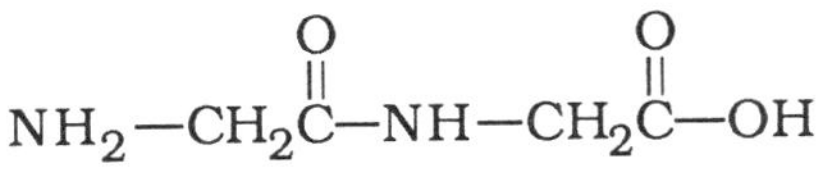 $NH_2-CH_2\overset{\overset{\displaystyle O}{\displaystyle \|}}{C}-NH-CH_2\overset{\overset{\displaystyle O}{\displaystyle \|}}{C}-OH$

   (Better yet, the dipolar ionic form:

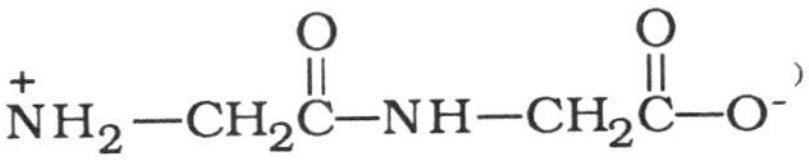

   $\overset{+}{N}H_2-CH_2\overset{\overset{\displaystyle O}{\displaystyle \|}}{C}-NH-CH_2\overset{\overset{\displaystyle O}{\displaystyle \|}}{C}-O^-$)

4.   Gly-Gly

5. 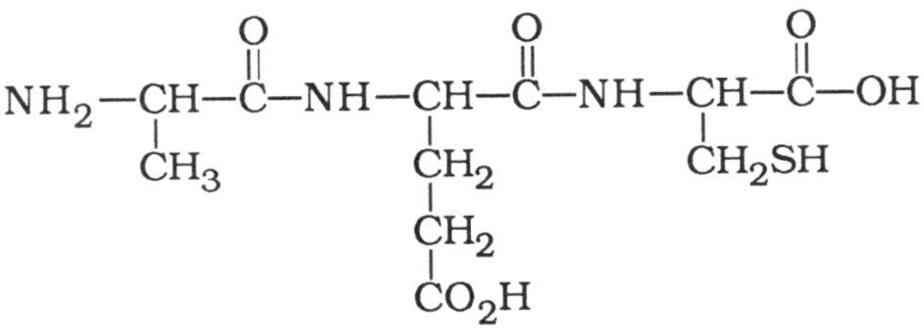 $NH_2-CH-\overset{\overset{\displaystyle O}{\displaystyle \|}}{C}-NH-CH-\overset{\overset{\displaystyle O}{\displaystyle \|}}{C}-NH-CH-\overset{\overset{\displaystyle O}{\displaystyle \|}}{C}-OH$
                $\overset{|}{CH_3}$ $\overset{|}{CH_2}$ $\overset{|}{CH_2SH}$
                                      $\overset{|}{CH_2}$
                                      $\overset{|}{CO_2H}$

6.   Ala-Glu-Cys, two

7. 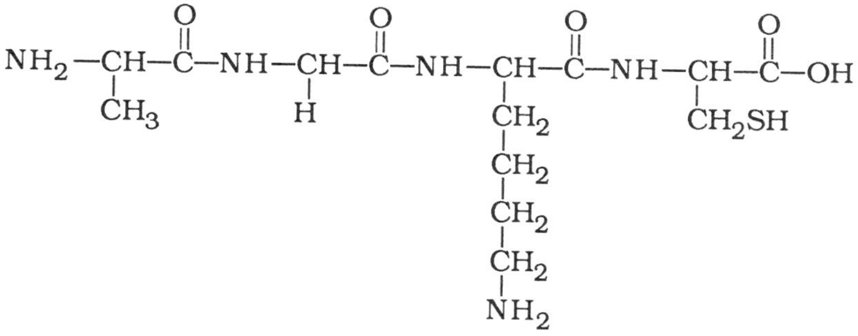 $NH_2-CH-\overset{\overset{O}{\|}}{C}-NH-CH-\overset{\overset{O}{\|}}{C}-NH-CH-\overset{\overset{O}{\|}}{C}-NH-CH-\overset{\overset{O}{\|}}{C}-OH$

with substituents:
- on first CH: CH_3
- on second CH: H
- on third CH: CH_2, CH_2, CH_2, CH_2, NH_2
- on fourth CH: CH_2SH

8. 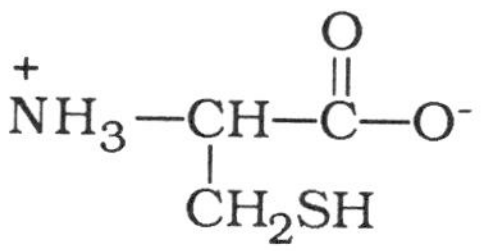 $\overset{+}{N}H_3-CH-\overset{\overset{O}{\|}}{C}-O^-$
      with CH_2SH on the CH

9.   Gly-Ala-Cys-Lys-Glu
      Gly-Ala-Cys-Lys-Glu (Two identical molecules would form.)

10.   the secondary structure

11.   salt bridges, hydrogen bonds, and the water-avoiding (or attracting) responses of hydrophobic (or hydrophilic) groups

12.   the tertiary structure

13.   the primary structure

14.   lysine and glutamic acid

15.   the quaternary structure

16.   concentration gradient

17.   gap junctions; protein

18.   receptor; shape

19.   oligosaccharide

20.   ground substance; cartilage; elastin

Multiple–Choice

1.   a, c, and d (Choice b is a dipeptide, Ala-Gly, and could be one of the several dipeptides formed. Choice a is not an α-amino acid. While choice c has an amide bond, the unit on the right is not an amino acid but rather an acetyl unit. Choice d involves lysine but not in a form found in proteins; the amide bond involves the side chain and not the α-amino group.)

2. d   3. a

4. b   5. a

6. b   7. a, b, c, and d

8. c   9. b

10. a   11. b, c, and d

12. a   13. b (It no longer is isoelectric.)

14. a

15. b (The N-terminal unit is Cys, and the conventions tell us to write the N-terminal unit on the left.)

16. d

17. d (All of the others stabilize secondary structures.)

18. a   19. b

20. a   21. c

22. c   23. b

24. a   25. a

11

ENZYMES, HORMONES, AND NEUROTRANSMITTERS

The heart of this chapter is the study of the ways in which the body controls which reactions go, which shut down, which accelerate, and which go slower. The different chemical and physiological functions of enzymes, hormones, and neurotransmitters are "must" topics of the chapter.

OBJECTIVES

Objectives 1 through 8 are mostly concerned with the basic vocabulary of enzymes; 9 and 10 are about how enzymes work in general. After you have studied this chapter and worked its Exercises, you should be able to do the following.

1. Describe the composition of enzymes and the kinds of cofactors in general terms.
2. Describe what enzymes do.
3. Explain how enzymes possess "specificity."
4. Explain why enzymes are sensitive to denaturing conditions and pH.
5. Describe in general terms how certain B vitamins are vital to some enzymes.
6. Recognize from the name of a substance whether it is an enzyme.
7. Tell what an enzyme's substrate (or type of reaction) is from its name.
8. Define and give an example of an isoenzyme and describe how an analysis of serum isoenzymes can be used in medical diagnosis.
9. Describe the "lock and key" mechanism of enzyme action and how "induced fit" is part of this mechanism.
10. Describe the relationship between the initial rate of an enzyme-catalyzed reaction, the initial enzyme concentration, and the substrate concentration, and explain why the rate levels off at sufficiently high substrate concentration.

11. Explain (using drawings to illustrate your explanation) how an allosteric activation of an enzyme with two active sites results in a sigmoid rate curve.
12. Explain how an effector influences enzyme activity.
13. Describe what happens in a zymogen-enzyme conversion and give an example.
14. Explain how feedback inhibition works and what homeostasis means.
15. Describe how allosteric inhibition works.
16. Explain how some poisons work, and name three kinds of poisons that act as irreversible enzyme inhibitors.
17. Explain in general terms what kind of substance is used to measure the concentration of an enzyme in some body fluid.
18. Name the enzymes whose serum levels are measured in (a) viral hepatitis and (b) myocardial infarction.
19. Name at least two blood-clot-dissolving enzymes and how they work.
20. Name the two kinds of primary chemical messengers.
21. Without necessarily reproducing molecular structures, describe what cyclic nucleotides are.
22. List the steps in the overall process that occurs when cyclic AMP is involved after a signal releases a hormone or a neurotransmitter.
23. Describe in general terms how the inositol phosphate system works.
24. Describe in general terms the four broad types of hormones.
25. Describe how a hormone finds its target cells and recognizes them.
26. In general terms, explain how neurotransmitters work and why they must eventually be deactivated.
27. Describe in general terms what the following do.
    (a) acetylcholine and cholinesterase
    (b) norepinephrine
    (c) monoamine oxidases
    (d) antidepressant drugs
    (e) dopamine
    (f) drugs for schizophrenia, like Thorazine and Haldol
    (g) GABA
    (h) mild tranquilizers, like Valium and Librium
    (i) enkaphalins and endorphins
    (j) substance P
28. Explain how calcium channel blockers work.

GLOSSARY

Agonist. A compound whose molecules can bind to a receptor on a cell membrane and cause a response by the cell.

Allosteric Activation. The activation of an enzyme's catalytic site by the binding of some molecule at a position elsewhere on the enzyme.

Allosteric Inhibition. The inhibition of the activity of an enzyme caused by the binding of an inhibitor molecule at some site other than the enzyme's catalytic site.

Antagonist. A compound that can bind to a membrane receptor but not cause any response by the cell.

Antibiotic. Antimetabolites made by bacteria and fungi.

Antimetabolite. A substance that inhibits the growth of bacteria.

Apoenzyme. The wholly polypeptide part of an enzyme.

Coenzyme. An organic compound needed to make a complete enzyme from an apoenzyme.

Cofactor. A nonprotein compound or ion that is an essential part of an enzyme.

Competitive Inhibition. The inhibition of an enzyme by the binding of a molecule that can compete with the substrate for the occupation of the catalytic site.

Effector. A chemical other than a substrate that can allosterically activate an enzyme.

Enzyme. A catalyst in a living system.

Enzyme-Substrate Complex. The temporary combination that an enzyme must form with its substrate before catalysis can occur.

Feedback Inhibition. The competitive inhibition of an enzyme by a product of its own action.

Homeostasis. The response of an organism to a stimulus such that the organism is restored to its pre-stimulated state.

Hormone. A primary chemical messenger made by an endocrine gland and carried by the bloodstream to a target organ where a particular chemical response is initiated.

Hydrolase. An enzyme that catalyzes a hydrolysis reaction.

Induced Fit Theory. Certain enzymes are induced by their substrate molecules to modify their shapes to accommodate the substrate.

Inhibitor. A substance that interacts with an enzyme to prevent its acting as a catalyst.

Isoenzymes. Enzymes that have identical catalytic functions but which are made of slightly different polypeptides.

Isomerase. An enzyme that catalyzes the conversion of a compound into an isomer.

Ligase. An enzyme that catalyzes the formation of covalent bonds at the expense of triphosphate energy.

Lock-and-Key Theory. The specificity of an enzyme for its substrate is caused by the need for the substrate molecule to fit to the enzyme's surface much as a key fits to and turns only one tumbler lock.

Lyase. An enzyme that catalyzes an elimination reaction to form a double bond.

Monoamine Oxidase. An enzyme that catalyzes the inactivation of neurotransmitters or other amino compounds of the nervous system.

Neurotransmitter. A substance released by one nerve cell to carry a signal to the next nerve cell.

Oxidoreductase. An enzyme that catalyzes the formation of an oxidation-reduction equilibrium.

Poison. A substance that reacts in some way in the body to cause changes in metabolism that threaten health or life.

Proenzyme. An inactive form of an enzyme; a zymogen.

Target Cell. A cell at which a hormone molecule finds a site where it can become attached and then cause some action that is associated with the hormone.

Target Tissue. The tissue where a particular hormone is taken up by target cells.

Transferase. An enzyme that catalyzes the transfer of some group.

Zymogen. A polypeptide that is changed into an enzyme by the loss of a few amino acid residues or by some other change in its structure; a proenzyme.

SELF-TESTING QUESTIONS

COMPLETION

1. An organic compound that is needed, in some cases, to complete an enzyme is called a _____.

2. The specific site on a large enzyme molecule where a substrate experiences the catalytic activity of the enzyme is called the _____.

3. An enzyme might lose its catalytic ability through the action of heat or a change in pH because an enzyme is made up mostly of a _____.

4. The temporary union of an enzyme with the compound on which it acts is called the _____.

5. The theory that accounts for the unusual specificity of an enzyme is the _____ theory.

6. Coenzymes can be made in the body, but in many cases it is essential that the body be supplied with _____via the diet because they are a necessary part of many coenzyme molecules.

7. Some metabolic sequences are shut down when molecules of their final product combine with and inactivate an _____ for one of the early steps in the sequence.

8. An enzyme that catalyzes the hydrolysis of an ester link would be called an _____; and an enzyme that helps in transferring a phosphate group is called a_____.

9. The enzyme sucrase catalyzes the hydrolysis of _____.

10. Some enzymes will remove a pair of electrons from a substrate; because of this particular kind of action, they are called _____.

11. What are the names of the vitamins needed to make each of these coenzymes?

    NAD^+ _____

    FAD _____

    thiamine pyrophosphate _____

12. For a fixed concentration of enzyme in an enzyme-catalyzed reaction, the initial rate doubles when the initial concentration of substrate, $[S]$, is doubled only when the value of $[S]$ is relatively _____(low or high). The initial rate becomes essentially independent of the initial value of $[S]$ when the latter is relatively _____(low or high). This is because at the higher initial values of $[S]$, all of the molecules of the _____are saturated with molecules of the _____.

13. The following statements describe ways in which enzymes might be activated or otherwise controlled. On the blank line write the name of the kind of control described.

    (a) An enzyme's catalytic site is activated by the binding of a nonsubstrate molecule elsewhere on the enzyme.

    _____

    (b) An enzyme with two (or more) catalytic sites has the second site activated by the binding of a substrate to the first site.

    _____

(c) The activity of an enzyme for its substrate is suppressed by the binding at its active site of a nonsubstrate molecule of similar shape to the substrate molecule.

_____

(d) The activity of an enzyme is suppressed by its binding at its active site a molecule of one of the products of the action of the enzyme.

_____

(e) The activity of an enzyme is suppressed by its binding another molecule at a place away from its active site

_____

(f) An enzyme forms when a small fragment is cut off from a polypeptide chain letting the active site emerge or unfold.

_____

14. Feedback inhibition is an example of a general kind of ability for self-regulation called _____.

15. Compounds that inhibit normal metabolism in disease-causing bacteria have the general name of _____.

16. By using the _____ for an enzyme as an analytical reagent, the concentration of the enzyme in some body fluid can be measured.

17. In diseases or injuries of the liver, one of the liver enzymes with the symbol _____ escapes into the _____ along with a lower level of another liver enzyme with the symbol _____. The ratio of these two enzymes is typically _____(higher or lower) in victims of viral hepatitis than in healthy individuals.

18. In a myocardial infarction, three enzymes with the short symbols of _____, _____, and _____ leak from heart tissue into general circulation. One of these can be found in the form of three isoenzymes which have the symbols _____ if present in skeletal muscle, _____ when found in heart muscle, and _____ when present in brain tissue.

19. In an infarct, the enzyme lactate dehydrogenase, symbolized as _____, appears in the blood, but there are _____isoenzymes for it. Normally, the first two of these isoenzymes appear in relative concentrations in which the level of the first is less than that of the second. This relative relationship is inverted in an infarct, and the phenomenon is called an_____.

20. Three enzymes that can be used to help dissolve a blood clot in a myocardial infarction are named _____, _____ and _____.

21. The protease that catalyzes the hydrolysis of _____, the protein of a blood clot, is named _____. In its inactive stage, as a zymogen, it is called _____.

22. The general names for the two kinds of primary chemical messengers are _____ and _____. Endocrine glands make the _____ and _____ make the other kind of messenger.

23. In the cyclic-AMP system for the delivery of chemical information, the hormone or neurotransmitter binds to its target cell by a _____ mechanism to form a complex. This complex alters a polypeptide called the_____. This then activates the enzyme _____, which catalyzes the conversion of _____ into _____. This product then activates an _____ inside the cell, and thus the message of the primary messenger is delivered. To bring this activation to its close, another cellular enzyme, _____, catalyzes the hydrolysis of _____ to _____.

24. Another receptor system that involves the alteration of the G-protein is called the _____ .

25. Hormones exert their influence in a variety of ways. Adrenaline, for example, is an _____ activator. Insulin and human growth hormone affect the _____ of the membranes of their target cells. Some sex hormones function by activating _____ which then direct the synthesis of _____. "Local" hormones or _____ work right where they are synthesized.

26. Chemical communication from one neuron to another occurs across a very narrow, fluid-filled space called a _____ , and the general name for substances that move across this space carrying a "message" is _____ .

27. The neuron from which this messenger moves is called the _____ neuron, and the neuron to which the messenger migrates is called the _____ neuron.

28. A complex forms between a receptor in the postsynaptic neuron and the _____ that works to activate_____. This enzyme then catalyzes the formation of _____ in the membrane of the postsynaptic neuron.

29. To switch a nerve signal off, enzyme-catalyzed reactions have to degrade or otherwise deactivate molecules of a_____ .

30. Cholinesterase catalyzes the hydrolysis of the neurotransmitter called _____. Nerve gases work by_____and the botulinus bacillus works by_____. Local anesthetics like nupercaine and procaine block pain signals by _____.

31. Several antidepressants work by interfering with the use of a neurotransmitter named _____. The deactivation of excess or unused amounts of this substance is catalyzed by enzymes called the_____, symbolized as _____. If this enzyme is inhibited by one kind of antidepressant, then the signal carried by the neurotransmitter is _____. (kept up or blocked)

32. The neurotransmitter called_____is believed to be involved with schizophrenia. Drugs that bind to receptors for this neurotransmitter inhibit its _____ to these receptors. Symptoms similar to those of schizophrenia can be induced by the abuse of drugs called _____ that work by triggering the _____ (release or decomposition) of this neurotransmitter.

33. The full name of GABA is _____ . If the work of this neurotransmitter is enhanced, the result is that nerve signals are _____ (accelerated or inhibited).

34. Two families of neurotransmitters that include powerful painkillers are named the _____ and the _____ .

35. The chemical symbol of the metal ion that is a major secondary chemical messenger is _____. Its concentration in the cytosol as the free ion is extremely low because at higher concentrations it could precipitate as a salt with the _____ ion. The metal ion is let into the cell through _____ every time it is needed to activate some cellular work (like muscle contraction). To tone down such activation, medications called _____ can be used.

36. The molecular basis of memory might involve a kind of chemical messenger called a _____messenger because it can move back into the cell from which a signal has been sent. Two compounds having the formulas _____and _____ are believed to constitute such messengers. Their mechanism of action apparently does not require a _____protein; their molecules can enter cells simply by _____.

MULTIPLE-CHOICE

1. The site on an enzyme that is able to accept the corresponding substrate is called the
   - (a) allosteric site
   - (c) catalytic site
   - (b) binding site
   - (d) effector site

2. A substance with the name protease is
   - (a) a prostaglandin
   - (c) an enzyme
   - (b) a coenzyme
   - (d) a hormone

3. An enzyme for the reaction: $A + B \rightleftharpoons C + D$
   - (a) increases the quantity of C that forms
   - (b) increases the quantity of A that remains unreacted
   - (c) shifts the equilibrium to favor the right side
   - (d) accelerates the establishment of the equilibrium

4. The catalytic activity of an enzyme will generally
   - (a) increase with increasing temperature
   - (b) increase with decreasing temperature
   - (c) be unaffected by temperature
   - (d) be greatest in a particular temperature range

5. Several coenzymes
   - (a) activate apoenzymes
   - (b) are made from various B-vitamins
   - (c) catalyze hydrolysis reactions
   - (d) are esters of carboxylic acids

6. The symbol NAD^+ stands for
   - (a) an enzyme
   - (c) a coenzyme
   - (b) a vitamin
   - (d) an isoenzyme

7. When an enzyme that contains FMN interacts with one having NADH, then which can form?
   - (a) $FMNH + NAD^+$
   - (c) $NADP^+ + FMNH_2$
   - (b) $NAD^+ + FMNH_2$
   - (d) $NAD^+ + H^+ + FMN$

8. Isoenzymes usually differ in
   - (a) the substrates they will accept
   - (b) their coenzymes
   - (c) their apoenzyme portions
   - (d) their cofactors

9. The symbol CK stands for
   - (a) creatine kinase
   - (c) cofactor kinase
   - (b) carbon-potassium
   - (d) an isoenzyme

10. A theory that is used to explain enzyme specificity is called the
    - (a) theory of allosteric activation
    - (b) the sigmoid rate curve theory
    - (c) induced fit theory
    - (d) lock-and-key theory

11. The Michaelis-Menton equation for the rate of a simple enzyme-catalyzed reaction
     (a) shows that a plot of rate versus substrate concentration is linear.
     (b) describes how the rate becomes constant when [S] is relatively high relative to $[E_0]$.
     (c) says that the rate is independent of initial enzyme concentration.
     (d) reduces to rate $\propto 1/2[S]$ at low values of [S].

Questions 12 through 14 refer to this sequence of reactions:

$$A \xrightarrow{E_a} B \xrightarrow{E_b} C \xrightarrow{E_c} D \xrightarrow{E_d} E$$

12. If the enzyme, E_a, for the conversion of A to B is activated by molecules of A, and if the graph or plot of the rate of the conversion versus the concentration of E_a has a sigmoid shape, then we probably have the operation of
     (a) feedback inhibition
     (b) gene activation
     (c) activation by an effector
     (d) allosteric activation
13. If molecules of final product, E, deactivate the enzyme E_a for the first step by binding to the active site of that enzyme, we are probably seeing the operation of
     (a) competitive inhibition by nonproduct
     (b) feedback inhibition
     (c) allosteric inhibition
     (d) inhibition by effector
14. If molecules of a substance, A', that resemble the molecules of A but that cannot be changed to B, deactivate the enzyme E_a by binding to the active site of that enzyme, then we are probably observing
     (a) competitive inhibition by nonproduct
     (b) allosteric inhibition by nonproduct
     (c) feedback inhibition by nonproduct
     (d) allosteric action by nonproduct
15. A polypeptide that changes into an enzyme when a small fragment is broken off by some activating reaction is
     (a) a prostaglandin   (c) a coenzyme
     (b) an apoenzyme   (d) a proenzyme
16. The general principle that makes possible the measurement of just one enzyme out of several that are present in blood serum is that of
     (a) enzyme specificity   (c) enzyme inhibition
     (b) electrophoresis   (d) allosteric activation
17. Symptoms of a myocardial infarction include
     (a) LD_1-LD_2 flip
     (b) a rise in the serum level of the CK(MB) isoenzyme
     (c) a drop in the serum level of GOT
     (d) a change in the serum level of glucose oxidase
18. Primary chemical messengers are
     (a) enzymes and hormones
     (b) hormones and neurotransmitters

(c)   neurotransmitters and cyclic nucleotides

(d)   cyclic nucleotides and axons

19.   The ability of a hormone to recognize its own target cells depends on a "lock-and-key" fit of the hormone molecule to

(a)   a receptor protein   (c)   cyclic AMP

(b)   a neurotransmitter   (d)   adenyl cyclase

20.   An example of a hormone that activates an enzyme is

(a)   adrenalin   (c)   testosterone

(b)   insulin   (d)   human growth hormone

21.   In a postsynaptic neuron, adenylate cyclase is activated by

(a)   a neurotransmitter

(b)   a receptor

(c)   a neurotransmitter-receptor complex

(d)   phosphodiesterase

22.   A substance that serves as both a neurotransmitter and a hormone is

(a)   acetylcholine   (c)   epinephrine

(b)   cholinesterase   (d)   norepinephrine

23.   Norepinephrine that is reabsorbed by a presynaptic neuron is degraded by

(a) GABA   (b) MAO   (c) LD   (d) CK(MB)

24.   A drug that binds to a postsynaptic neuron's receptor protein

(a)   inhibits the transmission of a nerve's signal

(b)   accelerates the activation of adenylate cyclase

(c)   prolongs the receipt of signals from the presynaptic neuron

(d)   inactivates MAO enzymes

25.   Amphetamines stimulate presynaptic neurons to release

(a) L-DOPA   (b) dopamine   (c) GABA   (d) MAO

26.   A pain-killer produced in the brain itself is

(a) substance P   (b) GABA   (c) enkephalin   (d) morphine

27.   To make heart muscle contractions less vigorous, some medications block

(a)   adrenaline   (c)   G-protein

(b)   plasminogen   (d)   calcium channels

ANSWERS

ANSWERS TO SELF–TESTING QUESTIONS

Completion

1.   coenzyme (Cofactor is not correct because it would include trace elements too. Vitamin as the answer is close, but coenzyme is better here because vitamins usually have to be changed into coenzymes first. Also, remember that just a few vitamins work this way.)

2.   catalytic site

3. protein (It is more specific than apoenzyme.)
4. enzyme-substrate complex
5. lock-and-key
6. vitamins
7. enzyme
8. esterase; kinase
9. sucrose
10. oxidases
11. nicotinamide (or nicotinic acid, but it is the amide that is used); riboflavin; thiamine
12. low; high; enzyme; substrate
13. (a) activation by an effector
    (b) allosteric activation
    (c) competitive inhibition by a nonsubstrate
    (d) feedback inhibition
    (e) allosteric inhibition
    (f) zymogen activation
14. homeostasis
15. antimetabolite
16. substrate
17. GPT; bloodstream; GOT; higher
18. CK, LD, and GOT; CK(MM); CK(MB); CK(BB)
19. LD; 5; LD_1–LD_2 flip
20. streptokinase, tissue plasminogen activator (TPA), and APSAC (acylated plasminogen-streptokinase-activator complex)
21. fibrin; plasmin; plasminogen
22. hormones and neurotransmitters; hormones; neurons (nerve cells)
23. lock-and-key; G-protein; adenylate cyclase; ATP into cyclic-AMP; enzyme; phosphodiesterase;cyclic-AMP into AMP
24. inositol phosphate system
25. enzyme; permeabilities; genes; enzymes; prostaglandins
26. synapse; neurotransmitter
27. presynaptic; postsynaptic
28. neurotransmitter; adenylate cyclase; cyclic AMP
29. neurotransmitter
30. acetylcholine; inhibiting cholinesterase; blocking the synthesis of acetylcholine; blocking the receptor protein for acetylcholine
31. norepinephrine; monoamine oxidases; MAO; kept up
32. dopamine; binding; amphetamines; release
33. gamma-aminobutyric acid; inhibited
34. endorphins and enkephalins
35. Ca^{2+}; phosphate; calcium channels; calcium channel blockers
36. retrograde; NO and CO; receptor; diffusion

Multiple–Choice

1.	b	10.	c, d	19.	a
2.	c	11.	b	20.	a
3.	d	12.	d	21.	c
4.	d	13.	b	22.	d
5.	b	14.	a	23.	b
6.	c	15.	d	24.	a
7.	b	16.	a	25.	b
8.	c	17.	a, b	26.	c
9.	a	18.	b	27.	d

12

EXTRACELLULAR FLUIDS OF THE BODY

In the professional health areas, the chemistry of respiration and the chemistry of the blood are areas of vital importance because *respiratory problems arise in a number of emergency situations*. To deal swiftly and correctly with the emergency, primary health care personnel both physicians and nurses must obtain and quickly evaluate several measurements, including the blood pH, its bicarbonate level, and the partial pressures of the blood gases. Early in your career, you will most likely have to learn about the chemistry of respiration and the chemistry of blood in order to increase your professional capabilities. If you use this occasion for a thorough study, your value as a professional will improve that much more quickly.

OBJECTIVES

Objectives 1 through 4 are concerned with the chemistry of digestion and the digestive juices. The first objective should be considered a minimum goal for this area. Objectives 5 through 17 deal with the chemistry of respiration and the chemistry of blood (although a study of the latter is not complete without a study of the function of the kidneys, the subject of objectives 19 and 20).

After you have studied this chapter and worked the exercises in it, you should be able to do the following.

1. Name the end products of the digestion of carbohydrates, proteins, and lipids.
2. List the major digestive reactions of the mouth, the stomach, and the duodenum.
3. Name the digestive juices and their principal enzymes and zymogens.
4. Describe the functions of bile salts.
5. Describe the functions of the principal components of the blood.
6. Give the names and formulas and general uses of the two Group IA cations that occur in the body.

7. Describe the levels of the Group IA cations in plasma and intracellular fluids.
8. Describe in general terms the problems associated with hypo- and hypernatremia as well as hypo- and hyperkalemia.
9. Give the names, formulas, and general uses of the two Group IIA cations present in the body, and describe their levels in blood and intracellular fluids.
10. Describe in general terms the problems associated with hypo- and hypercalcemia as well as hypo- and hypermagnesemia.
11. Describe the chemical composition of bone giving the name and the formula of its chief mineral and the name of its principal nonmineral substance.
12. State the range of values for the chloride ion level in blood and inside cells, and describe (in general terms) its function.
13. Describe what problems are associated with hypo- and hyperchloremia.
14. Explain how fluids and nutrients exchange at capillary loops.
15. Name two situations in which proteins leave the blood and describe the consequences.
16. Give three ways edema may arise.
17. Describe the main features of the composition of hemoglobin.
18. Explain what 2,3-bisphosphoglycerate (BPG) does.
19. Referring to the allosteric effect and the hemoglobin-oxygen dissociation curve, explain how oxygen binds cooperatively to hemoglobin.
20. Discuss how oxygen affinity varies with blood pH.
21. Discuss how oxygen affinity varies with the pCO_2 of blood.
22. Describe how partial pressure gradients aid in gas exchange.
23. Describe how localized changes in pH aid in gas exchange.
24. Describe the functions of the isohydric shift and the chloride shift.
25. Explain how waste CO_2 is transported in blood.
26. Explain how myoglobin in certain tissue aids in gas exchange.
27. Outline how values of blood pH, pCO_2, and $[HCO_3^-]$ change from the normal in clinical situations involving metabolic or respiratory acidosis and alkalosis.
28. Describe the role of the kidneys in preventing acidosis.
29. Describe what vasopressin, aldosterone, and renin do.
30. Define the terms in the Glossary.

GLOSSARY

Acidosis. A condition in which the pH of the blood is below normal. *Metabolic acidosis* is brought on by a defect in some metabolic pathway. *Respiratory acidosis* is caused by a defect in the respiratory centers or in the mechanisms of breathing.

Albumin. One of a family of globular proteins that tend to dissolve in water, and that in blood contribute to the blood's colloidal osmotic pressure and aid in the transport of metal ions, fatty acids, cholesterol, triacylglycerols, and other water-insoluble substances.

Aldosterone. A steroid hormone, made in the adrenal cortex, secreted into the bloodstream when the sodium ion level is low, and that signals the kidneys to leave sodium ions in the bloodstream.

Alkalosis. A condition in which the pH of the blood is above normal. *Metabolic alkalosis* is caused by a defect in metabolism. *Respiratory alkalosis* is caused by a defect in the respiratory centers of the brain or in the apparatus of breathing.

Bile. A secretion of the gall bladder that empties into the upper intestine and furnishes bile salts; a route of excretion for cholesterol and bile pigments.

Carbaminohemoglobin. Hemoglobin that carries chemically bound carbon dioxide.

Cardiovascular Compartment. The entire network of blood vessels and the heart.

Chloride Shift. An interchange of chloride ions and bicarbonate ions between a red blood cell and the surrounding blood serum.

Digestive Juice. A secretion into the digestive tract that consists of a dilute aqueous solution of digestive enzymes (or their zymogens) and inorganic ions.

2,3-Bisphosphoglycerate (BPG). An organic ion that nestles within the hemoglobin molecule in deoxygenated blood but is expelled from the hemoglobin molecule during oxygenation.

Edema. The swelling of tissue caused by the retention of water.

Electrolytes, Blood. The ionic substances dissolved in the blood.

Erythrocyte. A red blood cell.

Extracellular Fluids. Body fluids that are outside of cells.

Fibrin. The fibrous protein of a blood clot that forms from fibrinogen during clotting.

Fibrinogen. A protein in blood that is changed to fibrin during clotting.

Gastric Juice. The digestive juice secreted into the stomach and that contains pepsinogen, hydrochloric acid, and gastric lipase.

Globulins. Globular proteins in the blood that include gamma-globulin, an agent in the body's defense against infectious diseases.

Hemoglobin (HHb). The oxygen-carrying protein in red blood cells.

Hypercapnia. An elevated value of pCO_2 in blood.

Hyperkalemia. An elevated level of potassium ion in blood—above 5.0 meq/L.

Hypernatremia. An elevated level of sodium ion in blood—above 145 meq/L.

Hypocapnia. A lower than normal value of pCO_2 in blood.

Hypokalemia. A low level of potassium ion in blood—below 3.5 meq/L.

Hyponatremia. A low level of sodium ion in blood—below 135 meq/L.

Internal Environment. Everything enclosed within an organism.

Interstitial Fluids. Fluids in tissues but not inside cells.

Intestinal Juice. The digestive juice that empties into the duodenum from the intestinal mucosa and whose enzymes also work within the intestinal mucosa as molecules migrate through.

Isohydric Shift. In actively metabolizing tissue, the use of a hydrogen ion released from newly formed carbonic acid to react with and liberate oxygen from oxyhemoglobin; in the lungs, the use of hydrogen ion released when hemoglobin oxygenates to combine with bicarbonate ion and liberate carbon dioxide for exhaling.

Mucin. A viscous glycoprotein released in the mouth and the stomach that coats and lubricates food particles and protects the stomach from the acid and pepsin of gastric juice.

Oxygen Affinity. The percentage to which all of the hemoglobin molecules in the blood are saturated with oxygen molecules.

Oxyhemoglobin. Hemoglobin holding its maximum load of oxygen.

Pancreatic Juice. The digestive juice that empties into the duodenum from the pancreas.

Respiratory Gases. Oxygen and carbon dioxide.

Saliva. The digestive juice secreted in the mouth whose enzyme, amylase, catalyzes the partial digestion of starch.

Shock, Traumatic. A medical emergency in which relatively large volumes of blood fluid leave the vascular compartment and enter the interstitial spaces.

Vascular Compartment. The entire network of blood vessels and their contents.

Vasopressin. A hypophysis hormone that acts at the kidneys to help regulate the concentrations of solutes in the blood by instructing the kidneys to retain water (if the blood is too concentrated) or to excrete water (if the blood is too dilute).

SELF-TESTING QUESTIONS

COMPLETION

Questions 1 through 13 are concerned with the chemistry of digestion.

1. The first digestive juice to have much effect on proteins in the diet is _____.

2. Some digestive enzymes occur in their respective digestive juices initially in inactive forms called _____.

3. The proteolytic enzyme active in the stomach is called_____ and its zymogen is _____.

4. One kind of gastric cell contains a "pump" called the _____ ion pump. It moves _____ ion into the stomach and pumps _____ ion out of the stomach. However, the latter reenters the stomach together with _____ ion, so the net effect of all this action is the delivery of _____ acid to the stomach. One treatment for ulcers uses _____ to close down the _____ pump.

5. The partially digested material in the stomach that moves into the upper intestinal tract is called _____.

6. Starch in the diet is acted upon first in the _____ by the enzyme _____ found in _____.

7. All of the reactions of digestion are classified as _____ reactions.

8. For the most effective digestion of fats and oils in the diet, we depend on the emulsifying action of _____, compounds that are best described as _____ and that are released into the _____ of the digestive tract from an organ called the _____.

9. A special enzyme in intestinal juice called _____ helps convert trypsinogen to trypsin.

10. Besides helping to digest proteins in the diet, trypsin also catalyzes the conversion of _____ to chymotrypsin and of_____ to carboxypeptidase.

11. The arrival of acidic chyme in the duodenum causes the release of enzyme-rich fluids called _____ and _____.

12. Another fluid stimulated by the arrival of chyme in the upper intestinal tract is called _____.

13. The principal fat-digesting enzyme released in pancreatic juice is called _____.

14. Complete the following table by writing the names of the end products of the complete digestion of each family of foods given in the first column.

Food	End-Products of Digestion
Proteins	_____
Starch	_____
A mixture of lactose, maltose, and sucrose	_____
Triacylglycerols	_____

15. The chief cation in the blood has the formula _____, and its normal concentration is from _____ to _____ (include the unit). When its level exceeds the higher of these two values, the condition is called _____, and when it falls below the lower value, the condition is _____.

16. The chief cation inside cells has the formula _____, and its concentration is generally _____ (include the unit).

17. When the K^+ ion level of the blood drops below _____ (include the unit), the condition is called _____, and when its level rises above _____ (include the unit), the condition is _____.

18. Three health conditions that can cause severe hyperkalemia are _____, _____, and _____.

19. (Supply the prefixes, either hypo or hyper.) A general and unusual decrease in body fluids can cause

    _____-natremia as well as _____-kalemia.
    (hypo or hyper)   (hypo or hyper)

20. The formulas of the two Group IIA cations in the body are _____ and _____.

21. The second most abundant cation inside cells has the formula _____.

22. The chief cation in bones has the formula _____.

23. Besides being necessary in bones, calcium ions also participate in the _____ of muscles.

24. The overuse of milk of magnesia can cause _____-magnesemia.
    (hypo or hyper)

25. A deficiency of vitamin _____ can cause hypocalcemia.

26. The overuse of calcium-based antacids can cause _____.

27. Both Ca^{2+} and Mg^{2+} can activate certain _____.

28. From the standpoint of osmosis and dialysis, which is the more concentrated mixture, blood or interstitial fluid? _____

29. In which direction will water naturally have a net tendency to migrate: from a blood capillary into the interstitial compartment or from the interstitial compartment into the bloodstream?

    _____

30. Besides the forces generated by osmosis and dialysis, what force is present in the circulatory system that is especially important on the arterial side of a capillary loop?

    _____

31. Materials are taken away from tissues by both veins and the _____ ducts.

32. If fluids accumulate in some tissue (e.g., the lower limbs) a condition of _____ exists.

33. Within the red blood cells, called _____ are molecules of _____, which carry oxygen from the lungs to tissues needing oxygen.

34. Each molecule of _____ in red blood cells has _____ subunits and each subunit can carry _____ molecule(s) of oxygen.

35. The reaction whereby oxygen is picked up may be symbolized as follows: $HHb + O_2 \rightleftharpoons$ _____ + _____

36. To help force the reaction from left to right, the system acts to neutralize the _____ ion.

37. The ion largely responsible for doing this (question 36) has the formula _____, and the product of the neutralizing reaction subsequently breaks up into _____ and water.

38. If 95% of all the hemoglobin in a sample of blood is saturated with oxygen, then the _____ of the sample is 95%.

39. The first molecule of oxygen to bind to deoxygenated hemoglobin has an _____ effect that aids in bringing oxygen to the remaining oxygen-binding sites.

40. The symbol HbO_2^- is our symbol for _____, and that symbol is incorrect to the extent it does not show that each molecule of this substance carries _____ molecules of oxygen.

41. At active tissues needing oxygen, ions with the symbol _____ are generated which help HbO_2^- release oxygen. In other words, at a lower pH hemoglobin has a lower _____.

42. The equation for the formation of carbaminohemoglobin is:

$$\rule{3cm}{0.4pt} \rightleftharpoons \rule{3cm}{0.4pt}$$

43. The equation of question 42 is one way that newly formed molecules of carbon dioxide are taken up by the bloodstream. The other chemical change to carbon dioxide that occurs when it enters a red blood cell has the equation (an equilibrium):

$$\rule{3cm}{0.4pt} \rightleftharpoons \rule{3cm}{0.4pt}$$

44. The equations (or equilibria) of questions 42 and 43 occur at actively metabolizing tissue, and their operation shifts what equilibrium involving oxygen release?

$$\rule{3cm}{0.4pt} \rightleftharpoons \rule{3cm}{0.4pt}$$

As you have written this equilibrium, how does it shift at actively metabolizing cells?

_____
        (left or right)

45. To recapitulate, the equilibrium of question 44 shifts to the _____ at cells needing _____. They need it because they have done some chemical work that produces water and _____ which reacts somewhat with the water to give _____ according to the equilibrium of question _____. CO_2 also reacts somewhat with hemoglobin according to the equilibrium of question _____. These last two reactions generate _____ ions that aid in forcing the equilibrium of question 44 to the _____ and thereby help in releasing _____.

46. The use of the hydrogen ions generated in the equilibria of questions 42 and 43 to shift the equilibrium of 44 is called the _____.

47. Bicarbonate ions travel back to the lungs in the serum, not inside _____.

48. When bicarbonate ions leave the erythrocyte,_____ ions move in to replace them. This switch is called the _____.

49. The combined action of H^+ and CO_2 generated at actively metabolizing cells serves to _____ (raise or lower) the oxygen affinity of _____.

50. A hemoprotein at muscles that has a higher oxygen affinity than hemoglobin is called _____.

51. An acidosis brought on by an error in metabolism is called _____.

52. An acidosis caused by a breakdown in the respiratory centers or by some deterioration of the lungs is called _____.

53. Hyperventilation is a method used by a system to expel excess _____, the loss of which should help _____(raise or lower) the pH of the blood.

54. Prolonged vomiting may cause metabolic _____.

55. Shallow breathing, or _____, is the way the system tries to retain _____ which has the effect of retaining a neutralizer of _____ in the carbonate buffer. This helps _____ (raise or lower) the pH of the blood. Thus, shallow breathing may be a response to metabolic _____ (acidosis or alkalosis).

56. If shallow breathing occurs because the respiratory centers are not working, the individual cannot efficiently expel _____ and may experience respiratory _____.

57. The hyperventilation of a patient in hysterics causes an over-removal of_____, a loss of the _____ buffer, and a _____ in the pH of the blood. These responses result in respiratory _____. Rebreathing exhaled air helps suppress this because it supplies more _____ to the lungs and thence to the bloodstream.

58. If air too enriched in oxygen is given to a patient for too long a period, that individual will have trouble removing _____ from metabolizing cells.

59 The technical name for an increase in arterial pCO_2 is_____. The ventilation problem behind this condition has the technical name of _____ and the pH of the blood _____.
                    (increases or decreases)

60. What organ(s), in effect, removes acid from the blood?

_____

61. If the blood pressure drops greatly, the kidneys may not be able to _____ the blood because the blood flow through them is reduced.

62. The kidneys respond to a fall in blood pressure by secreting an enzyme called_____ into the blood which acts on a proenzyme called _____. This proenzyme is changed to _____ which helps to generate _____ the most powerful _____ known.

63. The endocrine gland called the_____ is sensitive to the _____ pressure of the blood. If this pressure goes up, it means that the concentration of dissolved and dispersed substances in blood is too _____(high or low). If this happens, the gland secretes the hormone _____, whose target organ is _____.

64. Secretion of this hormone (in question 63) eventually results in the formation of a _____(lesser or greater) volume of urine.

65. The hormone that helps the bloodstream to conserve its sodium ions called_____.

MULTIPLE-CHOICE

1. The end products of the digestion of milk sugar are
   (a) glucose and galactose   (c) glucose and fructose
   (b) only glucose   (d) glucose and maltose

2. If gastric juice were completely devoid of its hydrochloric acid, this would impair the digestion in the stomach of
   (a) lipids   (c) proteins
   (b) carbohydrates   (d) chyme

3. Removal of the gall bladder would reduce the efficiency of the digestion of
   (a) lipids   (c) proteins
   (b) carbohydrates   (d) fatty acids

4. The enzyme that catalyzes the conversion of trypsinogen to trypsin is
   (a) amylase   (b) pepsin   (c) mucin   (d) enteropeptidase

5. The enzyme in saliva is
   (a) mucin   (b) pepsin   (c) α-amylase   (d) amylose

6. Bile contains
   (a) a lipase   (c) a carbohydrase
   (b) a protease   (d) no enzymes

7. Without bile salts, which substances would not be as easily absorbed from the intestinal tract into the bloodstream?
   (a) fat-soluble vitamins   (c) glucose
   (b) water-soluble vitamins   (d) amino acids

8. The most abundant cation in the blood is
   (a) Na   (b) K   (c) K^+ (d) Na^+

9. Any injury that causes a large number of cells to break open can lead to
   (a) hypernatremia   (c) hypermagnesemia
   (b) hyperkalemia   (d) hyperchloremia

10. If the intake of K^+ ion is high, the body spontaneously works to lose
    (a) Cl^- (b) Ca^{2+} (c) Na^+ (d) Mg^{2+}

11. The blood can become hypocalcemic in
    (a) vitamin D deficiency
    (b) a misfunctioning thyroid gland
    (c) overdoses of antacids based on $Ca(OH)_2$
    (d) blood transfusions

12. The blood can become hypermagnesemic
    (a) in vitamin B_{12} deficiency
    (b) when it become hypercalcemic
    (c) if milk of magnesia is overused as a laxative
    (d) if the thyroid gland's activity becomes impaired

13. A normal value for the level of Cl^- in the blood is
    (a) 2 meq/L
    (c) 35.5 mg/L
    (b) 100 meq/L
    (d) 100 eq/L

14. If the level of Cl^- ion in blood drops, the body tends to retain
    (a) HCO_3^- (b) H^+ (c) Na^+ (d) H_2O

15. If the level of Cl^- in blood rises, one result can be
    (a) hypochloremia
    (c) acidosis
    (b) hyponatremia
    (d) alkalosis

16. Of all of the anions in blood, about two-thirds are
    (a) HPO_4^{2-} (b) $H_2PO_4^-$ (c) SO_4^{2-} (d) Cl^-

17. Plasma and interstitial fluids are most unlike in their concentrations of
    (a) electrolytes
    (c) lipids
    (b) proteins
    (d) carbohydrates

18. Oxygen is transported in the bloodstream chiefly as
    (a) molecules of O_2
    (c) oxyhemoglobin ions
    (b) hydronium ions
    (d) methemoglobin ions

19. Hemoglobin in blood has a relatively high oxygen affinity in the tissue capillaries of those tissues where
    (a) the pH is dropping
    (c) pCO_2 is rising
    (b) the pH is rising
    (d) pCO_2 is low

20. Hemoglobin more easily accepts its second, third, and fourth molecules of oxygen than it does its first because the first has
    (a) an allosteric effect
    (c) an isohydric effect
    (b) a Bohr effect
    (d) a releaser effect

21. The equilibrium $HHb + O_2 \rightleftharpoons HbO_2^- + H^+$ will respond to a drop in pH by
    (a) shifting to the right
    (c) remaining unchanged
    (b) shifting to the left
    (d) absorbing more O_2

22. If the equilibrium $CO_2 + H_2O \rightleftharpoons H^+ + HCO_3^-$ shift to the right, then the equilibrium of question 21 will tend to
    (a) shift to the right
    (c) remain unchanged
    (b) shift to the left
    (d) absorb oxygen

23. The equilibrium $HHb + CO_2 \rightleftharpoons Hb-CO_2^- + H^+$ shifts to the right in a region where
    (a) the oxygen affinity of HHb is high
    (b) the pO_2 is relatively high
    (c) the pH is relatively high
    (d) the pCO_2 is relatively high

24. When the carbonate buffer system in blood acts, a hydrogen ion is replaced by a water molecule at the expense of a
    (a) hydroxide ion
    (c) bicarbonate ion
    (b) hydronium ion
    (d) calcium ion

25. Something that has a higher oxygen affinity under identical conditions than adult hemoglobin is
    - (a) sickle cell hemoglobin
    - (b) myoglobin
    - (c) fetal hemoglobin
    - (d) none of these

26. Healthy kidneys respond to acidosis by
    - (a) putting H^+ into urine
    - (b) retaining HCO_3^-
    - (c) retaining Na^+
    - (d) removing ketone bodies

27. When respiratory centers are healthy, the body can respond to acidosis by
    - (a) hyperventilation
    - (b) hypoventilation
    - (c) retaining CO_2
    - (d) retaining H^+

28. Hyperventilation aids in controlling acidosis by
    - (a) removing CO_2 from blood
    - (b) bringing in more O_2
    - (c) promoting the chloride shift
    - (d) increasing the pCO_2 of blood

29. In severe emphysema, acidosis may develop because
    - (a) hyperventilation cannot be stopped
    - (b) shallow breathing cannot be used
    - (c) oxygen toxicity has become a problem
    - (d) the removal of CO_2 from the blood at the lungs is impaired

30. If for any reason hemoglobin molecules leave the lungs not fully saturated with oxygen, the condition is called
    - (a) oxygen affinity
    - (b) oxygen toxicity
    - (c) hypoxia
    - (d) anoxia

31. When a tissue cannot get oxygen, the condition is called
    - (a) oxygen affinity
    - (b) hypocapnia
    - (c) hypoxia
    - (d) anoxia

32. The nitrogen wastes present in urine is (are)
    - (a) urea
    - (b) creatine
    - (c) uric acid (and the urate ion)
    - (d) ammonia

33. If the osmotic pressure of the blood increases by even as little as 2%,
    - (a) the kidneys release renin
    - (b) the hypophysis releases vasopressin
    - (c) the adrenal cortex releases aldosterone
    - (d) the liver releases fibrinogen

34. The hormone that helps to regulate the level of sodium ions in the blood is
    - (a) angiotensin I
    - (b) vasopressin
    - (c) renin
    - (d) aldosterone

35.   On the arterial side of a capillary loop, the blood pressure is
   (a)   lower than on the venous side
   (b)   equal to that on the venous side
   (c)   lower than the osmotic pressure from the interstitial areas
   (d)   higher than the osmotic pressure from the interstitial areas

ANSWERS

ANSWERS TO SELF–TESTING QUESTIONS

Completion

1.   gastric juice
2.   zymogens
3.   pepsin, pepsinogen
4. $K^+ - H^+$; H^+; K^+; Cl^-; hydrochloric; cimetidine (Tagamet); $K^+ - H^+$
5.   chyme
6.   mouth, amylase, saliva
7.   hydrolysis (or hydrolytic)
8.   bile salts, detergents (or soaps or steroid-based detergents), upper intestine (or duodenum), gall bladder
9.   enteropeptidase
10.   chymotrypsinogen, procarboxypeptidase
11.   intestinal juice and pancreatic juice
12.   bile
13.   pancreatic lipase
14.   amino acids
   glucose
   glucose, fructose, galactose
   fatty acids and monoacylglycerols (some diacylglycerols)
15. Na^+; 135 to 145 meq/L; hypernatremia; hyponatremia
16. K^+; 125 meq/L
17.   3.5 meq/L; hypokalemia; 5.0 meq/L; hyperkalemia
18.   burns, crushing injuries, and heart attacks
19.   hyper-; hypo-
20. Mg^{2+}; Ca^{2+}
21. Mg^{2+}
22. Ca^{2+}
23.   contraction
24.   hyper-
25.   D

26. hypercalcemia
27. enzymes
28. blood
29. from the interstitial compartment into the bloodstream
30. simple blood pressure (from the pumping action of the heart)
31. lymph
32. edema
33. erythrocytes, hemoglobin
34. hemoglobin, four, one
35. $HbO_2^- + H^+$
36. hydrogen
37. HCO_3^-, carbon dioxide
38. oxygen affinity
39. allosteric
40. oxyhemoglobin, four
41. H^+ (H_3O^+ is better), oxygen affinity
42. $HHb + CO_2 \rightleftharpoons Hb-CO_2^- + H^+$
43. $CO_2 + H_2O \rightleftharpoons HCO_3^- + H^+$
44. $HHb + O_2 \rightleftharpoons HbO_2^- + H^+$
    to the left (or right to left)
45. left, oxygen; carbon dioxide, HCO_3^- and H^+, 43; 42 hydrogen (H^+), left, oxygen
46. isohydric shift
47. red blood cells (or erythrocytes)
48. chloride (Cl^-); chloride shift
49. lower, hemoglobin
50. myoglobin
51. metabolic acidosis
52. respiratory acidosis
53. carbon dioxide, raise
54. alkalosis
55. hypoventilation, carbon dioxide, base; lower; alkalosis
56. carbon dioxide, acidosis
57. carbon dioxide, carbonate, rise; alkalosis; carbon dioxide
58. carbon dioxide
59. hypercapnia, hypoventilation, decreases
60. the kidneys
61. filter or cleanse
62. renin, angiotensinogen; angiotensin I, angiotensin II, vasoconstrictor
63. hypophysis (or pituitary), osmotic; high; vasopressin, the kidneys
64. lesser than usual
65. aldosterone

Multiple–Choice

1. a	13. b	25. b and c
2. c	14. a	26. a, b, c, and d
3. a	15. c	27. a
4. d	16. d	28. a
5. c	17. b	29. d
6. d	18. c	30. c
7. a	19. b and d	31. d
8. d	20. a	32. a, b, c, and d
9. b	21. b	33. b
10. c	22. b	34. d
11. a, b	23. d	35. d
12. c	24. c	

13

NUCLEIC ACIDS

Heredity is not only involved with substances having certain functions; it is also concerned with molecules having particular structures that carry out these functions. Although the emphasis in this chapter is the relationship between structure and function, you are not expected to memorize the structures of specific compounds. You should, however, be able to name the hydrolysis products of the nucleic acids. You should also be prepared to give some of the condensed structural representations for these products using the symbols for the phosphate-pentose-phosphate-pentose chain and the letters— A, T, G, C, and U—that represent the side chains.

OBJECTIVES

After you have completed your study of this chapter, you should be able to do the following.

1. Give the name and the abbreviation for the chemical of an individual gene.
2. Give the general name for the monomer unit of a nucleic acid.
3. Give the names of the compounds produced when these monomeric units are hydrolyzed.
4. Describe in words the two main structural differences between DNA and RNA.
5. Using simple symbols, describe the features of a DNA strand.
6. Describe what in that structure is the genetic code.
7. Describe in words the contribution of Crick and Watson to the chemistry of heredity.
8. Describe in words the structure and shape of paired strands of DNA and the forces that stabilize them.
9. Explain why, regardless of species, A and T are always found in a ratio of 1:1, and why G and C are found in the same ratio.

10. Name and outline the process by which a gene copies itself.
11. Describe the functions of the four types of RNA and where they work.
12. Using words and simple drawings, explain how a gene specifies a unique amino acid sequence on a polypeptide.
13. Using letter symbols—A, T, C, G, and U—explain what a codon and its anticodon are and name the chemicals that bear them.
14. In general terms, explain how a virus works in a host cell.
15. In general terms, describe some of the things that can be done by recombinant DNA technology.
16. Give four examples of diseases related to genetic disorders.
17. Describe the chief goal of the Human Genome Project.
18. Define each of the terms in the Glossary.

GLOSSARY

Anticodon. A sequence of three adjacent side chain bases on a molecule of tRNA that is complementary to a codon and that fits to its codon on an mRNA chain during polypeptide synthesis.

Base, Heterocyclic. A heterocyclic amine obtained from the hydrolysis of nucleic acids: adenine, thymine, guanine, cytosine, or uracil.

Base Pairing. In nucleic acid chemistry, the association by means of hydrogen bonds of two heterocyclic, side-chain bases—adenine with thymine (or uracil) and guanine with cytosine.

Chromosome. Small threadlike bodies in a cell nucleus that carry genes in a linear array and that are microscopically visible during cell division.

Codon. A sequence of three adjacent side-chain bases in a molecule of mRNA that codes for a specific amino acid residue when the mRNA participates in polypeptide synthesis.

Deoxyribonucleic Acid (DNA). The chemical of a gene; one of a large number of polymers of deoxyribonucleotides and whose sequences of side-chain bases constitute the genetic messages of genes.

Double Helix DNA. A spiral arrangement of two intertwining DNA molecules held together by hydrogen bonds between side-chain bases; duplex DNA.

Duplex DNA. (See *Double Helix DNA*.)

Enzyme Induction. The process of switching on a gene to direct the synthesis of an enzyme.

Exon. A segment of a DNA strand that eventually becomes expressed as a corresponding sequence of aminoacyl residues in a polypeptide.

Gene. A unit of heredity carried on a cell's chromosomes and consisting of DNA.

Genetic Code. The set of correlations that specify which codons on mRNA chains are responsible for which aminoacyl residues when the latter are steered into place during the mRNA-directed synthesis of polypeptides.

Genetic Engineering. The use of recombinant DNA to make genes and the products of such genes.

Genome. The entire complement of the genetic information of a species; all of the genes of an individual.

Heterogeneous Nuclear RNA (hnRNA). RNA made directly at the guidance of DNA and from which messenger RNA (mRNA) is made.

Inducer. A substance whose molecules remove repressor molecules from operator genes and so open the way for structural genes to direct the overall syntheses of particular polypeptides.

Intron. A segment of a DNA strand that separates exons and that does not become expressed as a segment of a polypeptide.

Messenger RNA (mRNA). RNA that carries the genetic code as a specific series of codons for a specific polypeptide from the cell's nucleus to the cytoplasm.

Mitochondria. Cellular bodies in which a cell's ATP is made.

Nucleic Acid. A polymer of nucleotides in which the repeating units are pentose phosphate esters, each pentose unit bearing a side-chain base (one of four heterocyclic amines); polymeric compounds that are involved in the storage, transmission, and expression of genetic messages.

Plasmid. A circular molecule of supercoiled DNA in a bacterial cell.

Radiomimetic Substance. A substance whose chemical effect in a cell mimics the effect of ionizing radiation.

Recombinant DNA. DNA made by combining the natural DNA of plasmids in bacteria or the natural DNA in yeasts with DNA from external sources, such as the DNA for human insulin, and made as a step in a process that uses altered bacteria or yeasts to make specific proteins (e.g., interferons, human growth hormone, or insulin).

Replication. the reproductive duplication of a DNA double helix.

Repressor. A substance whose molecules can bind to a gene and prevent the gene from directing the synthesis of a polypeptide.

Ribonucleic Acids (RNA). Polymers of ribonucleotides that participate in the transcription and the translation of the genetic messages into polypeptides. (See also *Heterogeneous Nuclear RNA, Messenger RNA, Ribosomal RNA,* and *Transfer RNA*.)

Ribosomal RNA (rRNA). RNA that is incorporated into cytoplasmic bodies called ribosomes.

Ribosome. A granular complex of rRNA that becomes attached to a mRNA strand and that supplies some of the enzymes for mRNA-directed polypeptide synthesis.

Ribozyme. An enzyme whose molecules consist of ribonucleic acid rather than polypeptide.

Transcription. The synthesis of messenger RNA under the direction of DNA.

Transfer RNA (tRNA). RNA that serves to carry an aminoacyl group to a specific acceptor site of a mRNA molecule at a ribosome where the aminoacyl group is placed into a growing polypeptide chain.

Translation. The synthesis of a polypeptide under the direction of messenger RNA.

Virus. One of a large number of substances that consist of nucleic acid (usually RNA) surrounded (usually) by a protein overcoat and that can enter host cells, multiply, and destroy the host.

SELF-TESTING QUESTIONS

COMPLETION

1 The smallest unit of life in an organism is the _____. Everything it holds is called _____, which includes discrete bodies called _____, the principal sites of the synthesis of _____, a source of chemical energy.

2. The part of a cell outside its nucleus is the _____ and its liquid portion is the _____. Also present, however, are particles made of nucleoprotein called the _____.

3. Intertwined filaments of nucleoprotein in a cell nucleus make up a material called _____. Parts of this necklace-like material (the "pearls") consist of proteins called _____ which are wrapped with coils of one of the nucleic acids called _____ for short.

4. Prior to cell division, the cell's chromatin thickens and bodies called _____ become visible under a microscope. What is taking place to cause the thickening is the duplication of _____.

5. Nucleic acids are polymers whose monomers have the general name of _____. These monomers can be hydrolyzed to give one or the other of two sugars named _____ and _____, an inorganic _____ ion and a set of heterocyclic _____.

6. Nucleic acids that are made with ribose have the full name of _____ which is usually abbreviated _____. The four bases usually obtained from this kind of nucleic acid have the names and one-letter symbols of

_____ ____   _____ ____

_____ ____   _____ ____

7. Nucleic acids made with deoxyribose have the full name of _____ which is usually abbreviated _____. The four bases obtained from this kind of nucleic acid have the names and one-letter symbols of

_____ ____   _____ ____

_____ ____   _____ ____

8. The "backbones" of all nucleic acids contain a chain of diesters of _____ and the particular kind of sugar molecule. The bases are attached to the backbone, one base at each _____ unit.

9. The bases have functional groups and molecular geometries that permit them to form base-pairs with each other by means of _____ bonds. Base A always pairs with either _____ or with _____; G always pairs with _____.

10. Two strands of DNA form a twisting _____ according to evidence cited by _____ and _____. Base A of one strand pairs to base _____ opposite it on the other strand; and base C pairs to base _____.

11. An individual hereditary unit called a _____ consists of a particular series of triplets of nucleotides in one of the kinds of nucleic acids, _____.

12. In higher organisms, complete genes generally come in interrupted sequences of triplets. The interrupting segments in DNA molecules are called _____, and segments that make up a full gene are called _____.

13. The process whereby a gene becomes reproduced exactly, just prior to cell division, is called _____. The faithfulness of the copying depends on _____.

14. When DNA is used to direct the synthesis of a polypeptide, the DNA first directs the synthesis of a specific form of RNA called _____, abbreviated _____. This product is then processed to delete the triplets in its chain that correspond to the _____ segments of DNA. Then the remaining triplets are "knitted" together to give another form of RNA called _____, abbreviated _____.

15. Each triplet in mRNA is called a _____, and each is able to specify a particular _____ residue in a completed _____.

16. The overall process of using DNA to direct the formation of mRNA is called _____.

17. Following this process is another complicated process called _____, and its final product is a specific _____. When the latter is made, specific substances

called _____ help the product to adopt its final, correct native configuration.

18. To accomplish this last process, the cell needs two other kinds of RNA. One kind is used to make ribosomes and is called _____, abbreviated _____. The other kind is called _____, abbreviated _____, and its function is to carry _____to polypeptide assembly sites on _____.

19. One particular triplet of bases on a tRNA molecule can match a complementary triplet on mRNA. This tRNA triplet is called _____.

20. In some organisms at least, genes are in a switched-off status because a _____molecule has become bound to a segment of the gene. A molecule that can combine with this and remove it is called _____. Several drugs called _____ kill bacteria by interfering with bacterial gene-directed polypeptide synthesis.

21. Radiations such as X rays or gamma rays cause the most damage to a cell when they strike _____. Those that cause cancer are called _____. If a birth defect is the result, the agent is called a _____.

22. Substances that can invade particular "host" cells and take over the genetic machinery are called _____. To enter a cell, this substance uses an _____ to catalyze the breakdown of the _____ of the host cell.

23. Some RNA viruses carry an enzyme called _____ that can catalyze the synthesis of new RNA from directions encoded on the old viral _____.

24. Some RNA viruses carry a DNA polymerase, an enzyme called _____. This enzyme has the unusual ability to use _____ to direct the synthesis of _____, a flow of genetic information opposite the normal flow.

25. A family of viruses that uses its RNA and reverse transcriptase to make duplex DNA is called the _____. The duplex DNA than directs the synthesis of viral_____. This family of viruses includes oncogenic RNA viruses which can change normal genes to _____. Another virus in this family attacks the immune system, specifically the _____ lymphocytes, and so this virus has the abbreviated name of_____, the cause of _____.

26. In a technology called _____ DNA technology, particles called _____ in a bacterium are modified and given new genetic material. Then the bacteria will manufacture some _____ that corresponds to its additional genes.

27. To amplify (clone) small amounts of DNA, the _____ reaction is used.

28. Four diseases attributed to defective genes are

    (1) Associated with the overproduction of thick mucus in the lungs:_____

    (2) Associated with an impairment in the metabolism of the amino acid phenylalanine:

    _____

    (3) Associated with a blood disorder: _____

    (4) Associated with poor pigmentation of the eyes and the skin:_____

29. The defective gene in cystic fibrosis is one that directs the synthesis of a transmembrane _____ in cells of the lungs and the digestive tract that controls the movement of _____ ion through the cell membrane to the outside. When the movement of this ion is impaired, the movement of molecules of _____ is also impaired, leading to a thickening of the _____.

30. In a technology called _____, attempts are made to insert proper DNA into cells carrying defective _____.

MULTIPLE–CHOICE

1. What is transmitted from parents to offspring is a complete set of
   - (a) DNA molecules
   - (c) enzymes
   - (b) polypeptides
   - (d) hormones

2. The site of polypeptide synthesis in a cell is
   - (a) a chromosome
   - (c) a ribosome
   - (b) a gene
   - (d) a nuclear membrane

3. If DNA were fully hydrolyzed, the products would be
   - (a) ribose
   - (c) phosphoric acid
   - (b) deoxyribose
   - (d) a few heterocyclic amines

4. Molecules of tRNA differ from molecules of mRNA in being
   - (a) longer
   - (c) shorter
   - (b) triple helices
   - (d) inside ribosomes

5. The overall process of transcription proceeds in which order?
   - (a) exon to intron to polypeptide
   - (b) mRNA to polypeptide
   - (c) DNA to hnRNA to mRNA
   - (d) DNA to rRNA to tRNA to mRNA

6. DNA segments that appear to be uninvolved in polypeptide synthesis are called
   - (a) introns
   - (b) exons
   - (c) triplets
   - (d) anticodons

7. Human insulin can be made by bacteria or yeasts by a method called
   - (a) recombinant bacteria
   - (c) recombinant plasmids
   - (b) recombinant RNA
   - (d) recombinant DNA

8. Gene-directed polypeptide synthesis proceeds in which order of events?
   - (a) gene to mRNA to hnRNA to tRNA
   - (b) gene to hnRNA to mRNA to polypeptide
   - (c) gene to rRNA to tRNA to hnRNA
   - (d) gene to replicated gene to polypeptide

9. According to the Crick-Watson theory, the genetic message carried by a gene is related most particularly to
   - (a) the kinds of heterocyclic amines projecting from the phosphate-pentose chain in the gene
   - (b) the sequence in which the heterocyclic amines are lined up along the "backbone" of the gene
   - (c) the absence of one of the OH groups normally found in RNA
   - (d) the sequence of amino acids in gene molecule

10. The functional groups in adenine are geometrically arranged to enable adenine to pair by hydrogen bonding with
    - (a) adenine
    - (b) guanine
    - (c) thymine
    - (d) uracil

11. The molecular basis of a mutation is most closely linked to a
    - (a) defect in the transcription of a genetic message to mRNA
    - (b) change in the sequence or identity of heterocyclic amines on a DNA molecule
    - (c) defect in the arrival sequence of tRNA molecules at mRNA codon sites
    - (d) defect in the rRNA of ribosomes

12. If a codon triplet were U–C–G, the anticodon would be
(a) A–C–G   (b) C–G–A   (c) T–G–C   (d) A–G–C
13. The entire complement of genetic information of a species is called its
(a) exons   (c) chromatin
(b) chromosomes   (d) genome
14. To clone DNA in the lab, scientists can use
(a) the polymerase chain reaction
(b) restriction enzymes
(c) interferons
(d) chaperonines
15. A viral enzyme that uses RNA information to make DNA is
(a) RNA replicase   (c) restriction enzyme
(b) reverse transcriptase   (d) DNA replicase

ANSWERS

ANSWERS TO SELF–TESTING QUESTIONS

Completion

1. cell; protoplasm; mitochondria; ATP
2. cytoplasm; cytosol; ribosome
3. chromatin; histones; DNA
4. chromosomes; DNA
5. nucleotides; ribose and deoxyribose; phosphate; bases (or amines)
6. ribonucleic acids; RNA

| adenine | A | guanine | G |
| uracil | U | cytosine | C |

7. deoxyribonucleic acids; DNA

| adenine | A | guanine | G |
| thymine | T | cytosine | C |

8. phosphoric acid; sugar (or pentose)
9. hydrogen; T or U; C
10. double helix; Crick and Watson; T; G
11. gene; DNA
12. introns; exons
13. replication; base pairing of A to T and of G to C
14. heterogeneous nuclear RNA; hnRNA; intron; messenger RNA; mRNA
15. codon; amino acid; polypeptide
16. transcription
17. translation; polypeptide; chaperonines
18. ribosomal RNA; rRNA; transfer RNA; tRNA; aminoacyl units; a ribosome (or at an mRNA site at a ribosome)

19. an anticodon
20. repressor; an inducer; antibiotics
21. DNA in a cell nucleus; carcinogens; teratogen
22. viruses; enzyme; membrane
23. RNA replicase; RNA
24. reverse transcriptase; RNA; DNA
25. retroviruses; RNA; oncogenes; T_4; HIV, AIDS
26. recombinant; plasmids; polypeptide (or protein)
27. polymerase chain
28. (1) cystic fibrosis
    (2) PKU (phenylketonuria)
    (3) sickle cell anemia
    (4) albinism
29. protein; Cl⁻; water; mucous
30. gene therapy; genes

Multiple–Choice

1. a
2. c
3. b, c, d
4. c
5. c
6. a
7. d
8. b
9. b
10. c, d
11. b
12. b
13. d
14. a
15. b

14

BIOCHEMICAL ENERGETICS

This chapter has two broad objectives—to provide an overview of the major metabolic pathways that the body can tap for chemical energy, and to go into detail about how the cell makes ATP by the citric acid cycle coupled to the respiratory chain. As minimum goals, be sure to be able to name the major pathways introduced in this chapter, name their starting and ending materials, and describe their overall place in the scheme of biochemical energetics.

Pay careful attention to what your instructor tells you concerning what details must be learned and what can be skimmed over. Don't worry about any details, however, until you have the broad picture well in hand.

OBJECTIVES

After you have studied this chapter and worked the Exercises in it, you should be able to do the following.

1. Name the two products of digestion that are most frequently used for chemical energy in living systems.
2. Name the end products of the complete catabolism of carbohydrates and simple lipids.
3. Write the structures of the triphosphate and diphosphate networks.
4. Explain the basis for classifying organophosphates as high or low energy.
5. Name the principal triphosphate used as an immediate source of chemical energy in cells.
6. Outline (by a flow sheet) the principal pathways in biochemical energetics between products of digestion and the synthesis of ATP.
7. Describe the place acetyl coenzyme A has in biochemical energetics.
8. Give the general purpose of the citric acid cycle (Kreb's Cycle; Tricarboxylic Acid Cycle).

9. Give the symbols for the chief hydride-accepting coenzymes and write equations (in two ways) illustrating their activity.

10. In general terms, explain what kinds of gradients are forced into existence by the operation of the respiratory chain, and where they occur.

11. In general terms, describe the connections between these gradients and the synthesis of ATP (using the chemiosmotic theory).

13. Define the terms in the Glossary.

GLOSSARY

Acetyl Coenzyme A. The molecule from which acetyl groups are transferred into the citric acid cycle or into the fatty acid synthesis.

$$CH_3\overset{\overset{\displaystyle O}{\|}}{C}-S-CoA$$

Adenosine Diphosphate (ADP). A high-energy diphosphate ester obtained from adenosine triphosphate (ATP) when part of the chemical energy in ATP is tapped for some purpose in a cell.

Adenosine Monophosphate (AMP). A low-energy phosphate ester that can be obtained by the hydrolysis of ATP or ADP; a monomer for the biosynthesis of nucleic acids.

Adenosine Triphosphate (ATP). A high-energy triphosphate ester used in living systems to provide chemical energy for metabolic needs.

Aerobic Sequence. An oxygen-consuming sequence of catabolism that starts with glucose or with glucose units in glycogen, and proceeds through glycolysis, the citric acid cycle, and the respiratory chain.

Anaerobic Sequence. The oxygen-independent catabolism of glucose or of glucose units in glycogen to lactate ion.

Catabolism. Metabolic reactions that break up larger molecules.

Chemiosmotic Theory. An explanation of how oxidative phosphorylation is related to a flow of protons in a proton gradient that is established by the respiratory chain, and that extends across the inner membrane of a mitochondrion.

Citric Acid Cycle. A series of reactions that dismantle acetyl units and send electrons (and protons) into the respiratory chain; a major source of metabolites for the respiratory chain.

Fatty Acid Cycle. (See *β-Oxidation Pathway*.)

Glycolysis. A series of chemical reactions that break down glucose or glucose units in glycogen until pyruvate remains (when the series is operated aerobically) or lactate forms (when the conditions are anaerobic).

High-Energy Phosphate. An organophosphate with a phosphate group transfer potential equal to or higher than that of ADP or ATP.

β-Oxidation Pathway. The catabolism of a fatty acid by a series of steps that produce acetyl units (in acetyl CoA).

Oxidative Phosphorylation. The synthesis of high energy phosphates such as ATP from lower energy phosphates and inorganic phosphate by the reactions that involve the respiratory chain.

Phosphate Group Transfer Potential. The relative ability of an organophosphate to transfer a phosphate group to some acceptor.

Proton-Pumping ATPase. The enzyme in the inner mitochondrial membrane that catalyzes the formation of ATP from ADP and P_i under the influence of a flow of protons across the membrane through a tubular part of this enzyme.

Respiratory Chain. The reactions that transfer electrons from the intermediates made by other pathways to oxygen; the mechanism that creates a proton gradient across the inner membrane of a mitochondrion and that leads to ATP-synthesis; the enzymes that handle these reactions.

Respiratory Enzymes. The enzymes of the respiratory chain.

Substrate Phosphorylation. The direct transfer of a phosphate unit from an organophosphate to a receptor molecule.

SELF-TESTING QUESTIONS

COMPLETION

1. When glucose is burned in air, the elements present in the glucose emerge in molecules of _____ and _____.

2. When glucose is carried through both the anaerobic and aerobic sequences of metabolism in the body, its elements emerge in molecules of _____ and _____.

3. The energy produced by the combustion of glucose appears largely as _____.

4. Some of the energy produced by the breakdown of glucose in the body appears as _____, but of greater importance to the body, it also appears in the form of molecules of _____.

5. The structure of ATP in its electrically neutral, un-ionized form is (complete the following):

$$\boxed{\text{adenosine}}\text{—}$$

6. When adenosine triphosphate reacts, provides energy for some chemical change, and loses one phosphate, the remainder of its molecule is called _____ and has the symbol _____.

7. The symbol, P_i, stands for a mixture having (give the formulas) _____ and _____ as its two principal ions. The exact proportion of these ions in the mixture is largely a function of the _____ of the medium.

8. If structure II in the following reaction has a higher negative phosphate group transfer potential than structure I, where should the arrowhead be placed on the incomplete arrow?

$$\text{R—O—PO}_3^{2-} + \text{R'OH} \xrightarrow{\hspace{3cm}} \text{ROH} + \text{R'—O—PO}_3^{2-}$$
$$\text{I} \qquad\qquad \text{(incomplete arrow)} \qquad\qquad \text{II}$$

9. When proteins of a relaxed muscle interact with ATP, the muscle contracts as _____ an _____ are produced.

10. In a cell receiving an insufficient supply of oxygen, fresh ATP can be made from glucose by a series of reactions called _____, also called the _____ sequence, for which lactate is an end product.

11. In a cell receiving a sufficient supply of oxygen, the catabolism of glucose can occur by the _____ sequence, which changes glucose to _____, which is then converted to _____ coenzyme A. The latter is catabolized by the _____ cycle, and intermediates of this cycle interact with the _____ chain.

12. The respiratory chain is a series of reactions whereby intermediates of the _____ cycle send electrons through complexes of enzymes until _____ is reduced to water.

13. The organic group that is joined to coenzyme A has the structure: _____, and the symbol we use for the combination of this with coenzyme A is

_____.

14. The catabolism of fatty acids by a series of reactions called _____ generates a succession of molecules of _____, which enter the _____ and the _____ pathways.

15. The name of the high energy phosphate in muscles that can quickly remake ATP from ADP is _____.

16. The chief location in the cell for the synthesis of ATP is the _____ of a mitochondrion. When the respiratory chain operates, the net effect is to change MH_2 and half a molecule of oxygen to M and _____ and to generate a gradient of _____ ions as well as a gradient of _____ across the _____ of the mitochondrion.

17. As _____ ions move across the inner membrane of the mitochondrion at those places where such movement can occur, other changes are initiated that lead to the synthesis of _____ from _____ and P_i.

18. Our symbol for a donor of H:⁻ units to the respiratory chain is _____, and the metabolic pathway that is the richest supplier of H:⁻ is called the _____.

19. The first receptor of H:⁻ in the respiratory chain is an enzyme whose coenzyme has the short symbol of _____; the symbol for its reduced form is _____. This reduced form can pass H:⁻ to another enzyme with a coenzyme having the short symbol of _____ and whose reduced form is _____.

20. When $FMNH_2$ passes H:⁻ to the next enzyme in the chain having the short symbol _____, only the electrons in H:⁻ go to this next enzyme and the hydrogen nuclei, H^+, are put into _____. In this way, the first "installment" of the two gradients, the _____ gradient and the _____ gradient that extends from relatively high concentrations of these species in the _____ space to lower concentrations in the mitochondrial _____. The enzyme package that accomplishes this is called _____.

21. Following the iron-sulfur enzyme and the enzyme containing coenzyme Q is a series of enzymes collectively called the _____ that are present in two enzyme packages called _____ and _____. The final acceptor of the electrons being transferred from one of these enzymes to the next is _____, and water forms.

22. The respiratory chain has a "branch," and it involves one acceptor of H:⁻ from certain kinds of metabolites. The coenzyme for this acceptor has the short symbol of _____, and its reduced form is symbolized as _____. The latter can be reoxidized by interacting with an enzyme of the main respiratory chain with the coenzyme having the symbol _____. The package of enzymes of this branch is called _____.

MULTIPLE–CHOICE

1. Of the following substances, which has the most potential for generating molecules of ATP?
   (a) glucose
   (b) $CH_3(CH_2)_{16}CO_2H$
   (c) glycine
   (d) oxygen

2. Among the substances essential to the aerobic sequence of glucose catabolism is (are)
   (a) NAD^+
   (b) O_2
   (c) ATP
   (d) ADP

3. The product of the operation of the respiratory chain that is most needed by the body is
    (a) ATP
    (b) H_2O
    (c) $[H:^- + H^+]$
    (d) FAD

4. Among the accomplishments of the citric acid cycle is (are)
    (a) the synthesis of active acetyl
    (b) supplying $H:^-$ and H^+ to the respiratory chain
    (c) glycogenesis
    (d) glycolysis

5. To help get glucose inside cells of certain tissues, the blood should carry
    (a) epinephrine   (b) insulin   (c) cyclic AMP   (d) NAD

6. If a reaction is symbolized by

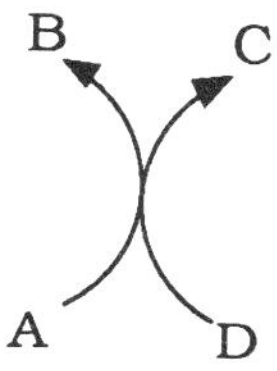

$$\begin{array}{ccc} B & & C \\ & \diagdown \diagup & \\ & \diagup \diagdown & \\ A & & D \end{array}$$

    it could be rewritten as
    (a) $A + B \rightarrow C + D$
    (b) $B + C \rightarrow A + D$
    (c) $A + C \rightarrow B + D$
    (d) $A + D \rightarrow B + C$

7. To complete the following reaction, we should include as a product
$$ATP + H_2O \rightarrow \underline{\hspace{2cm}} + PP_i$$
    (a) ADP
    (b) AMP
    (c) H_2O
    (d) $CO_2 + H_2O$

8. The chief use of acetyl CoA is to provide "fuel" for
    (a) the citric acid cycle
    (b) glycolysis
    (c) the anaerobic sequence
    (d) homeostasis

9. The chief purpose of the respiratory chain is to
    (a) accept $H:^-$
    (b) use up acetyl CoA
    (c) resupply ATP
    (d) use O_2

10. The complete catabolism of pyruvate by the citric acid cycle gives a maximum ATP yield of
    (a) 15   (b) 18   (c) 36   (d) 38

11. Between the respiratory chain and glycolysis in the catabolism of glucose occurs the
    (a) anaerobic sequence
    (b) oxidative phosphorylation
    (c) citric acid cycle
    (d) homeostasis

12. The operation of the respiratory chain establishes two gradients within a mitochondrion. These involve specifically
    (a) OH^- and Cl^-
    (b) H^+ and (+) charge
    (c) NAD^+ and $NADH_2$
    (d) electron pairs and ATP

13. One of the (incomplete) sequences of enzymes or coenzymes in the respiratory chain is (in the correct order)
    (a) FMN → FAD → Cyt c → Cyt a
    (b) FAD → NAD$^+$ → FeS—P → Q
    (c) Q → Cyt c$_1$ → Cyt b → FeS—P
    (d) NAD$^+$ → FMN → Q → Cyt a

14. The symbol FMN stands for
    (a) an enzyme
    (b) the reduced form of a flavin enzyme
    (c) the oxidized form of a flavoprotein
    (d) a coenzyme

15. The chemiosmotic theory was proposed by
    (a) Hans Krebs   (c) Charles MacMunn
    (b) Peter Mitchell   (d) David Keilin

16. The proton-pumping ATPase
    (a) is in the outer membrane of the mitochondrion.
    (b) includes a "gate" for protons, which is made of polypeptides.
    (c) is an integral component of complex III of the respiratory chain of enzymes.
    (d) is an enzyme of the citric acid cycle.

17. One important source of acetyl CoA is
    (a) pyruvate + CoA—SH + NAD$^+$
    (b) citrate
    (c) the respiratory chain
    (d) oxidative phosphorylation

18. When glucose is catabolized to CO_2 and H_2O, oxygen atoms from O_2 in the air end up as parts of molecules of
    (a) CO_2 (b) H_2O (c) ATP   (d) P$_i$

19. Which of the following types of reactions occur in the citric acid cycle?
    (a) dehydration
    (b) hydrogenation
    (c) dehydrogenation
    (d) addition of water to a double bond

20. When acetyl CoA adds to the keto group of oxaloacetate, the product is
    (a) pyruvate   (b) ATP   (c) lactate   (d) citrate

ANSWERS

ANSWERS TO SELF–TESTING QUESTIONS

Completion

1. carbon dioxide, water
2. carbon dioxide, water
3. heat

4. heat, ATP

5. $\text{adenosine}-\overset{\overset{\displaystyle O}{\|}}{P}-O-\overset{\overset{\displaystyle O}{\|}}{P}-O-\overset{\overset{\displaystyle O}{\|}}{P}-OH$
    with OH groups below each P

6. adenosine diphosphate, ADP
7. $H_2PO_4^-$, HPO_4^{2-}; pH
8. Place the arrow at the left end of the line
9. ADP; P_i
10. glycolysis; anaerobic
11. aerobic; pyruvate; acetyl; citric acid; respiratory
12. citric acid; oxygen

13. $CH_3-\overset{\overset{\displaystyle O}{\|}}{C}-$; $CH_3-\overset{\overset{\displaystyle O}{\|}}{C}-S-CoA$

14. the β-oxidation pathway; acetyl coenzyme A; citric acid cycle; respiratory chain.
15. creatine phosphate
16. matrix; H_2O; H^+; (+) charge; inner membrane
17. H^+; ATP; ADP
18. MH_2; citric acid cycle
19. NAD^+; NADH; FMN; $FMNH_2$
20. FeS—P; the intermembrane space of the mitochondrion; H^+ ion; (+) charge; intermembrane; matrix; complex I
21. cytochromes; complexes III and IV; oxygen
22. FAD; $FADH_2$; Q (or coenzyme Q); complex II

Multiple–Choice

1.	b	11.	c
2.	a, b, and d	12.	b
3.	a	13.	d
4.	b	14.	d
5.	b	15.	b
6.	d	16.	b
7.	b	17.	a
8.	a	18.	b
9.	c	19.	a, c, and d
10.	a	20.	d

15

METABOLISM OF CARBOHYDRATES

This chapter continues our study of the molecular basis of energy for living and introduces the molecular basis of a widespread disease—diabetes. We cannot go into the molecular basis of very many diseases, but we need to study at least some examples that illustrate the intimate involvement of fundamental chemical principles with life in both wellness and illness. Acid-base theory, buffers, properties of gases, chemical equilibria, organic reactions, enzymes, and chemical messengers all have major or supporting roles in the tragic drama of diabetes. This chapter and the next provide the final acts.

OBJECTIVES

After you have studied this chapter and worked the Review Exercises in it, you should be able to do the following.

1. Discuss the factors that influence the blood sugar level, using the technical terms for high and low values.
2. List some of the effects of hypoglycemia.
3. Describe the glucose tolerance test and its purpose.
4. Describe the influences that the hormones epinephrine, glucagon, insulin, and somatostatin have on glucose metabolism.
5. Describe in general terms what diabetes mellitus is.
6. Outline the Cori Cycle.
7. Write an overall equation for glycolysis—glucose to lactate.
8. Explain the importance of glycolysis.
9. Explain why anaerobic glycolysis ends at lactate, not pyruvate.
10. Give the general purpose of the pentose phosphate pathway of glucose catabolism.

11. Describe in general terms what the liver can do if the brain's chief nutrient is not available in the diet.
12. Define all of the terms in the Glossary.

GLOSSARY

Anaerobic Sequence. The oxygen-independent catabolism of glucose or of glucose units in glycogen to lactate ion.

Blood Sugar Level. The concentration of carbohydrate (mostly glucose) in the blood; usually stated in units of mg/dL.

Cori Cycle. The sequence of chemical events and transfers of substances in the body that describes the distribution, storage, and mobilization of blood sugar, including the reconversion of lactate to glycogen.

Diabetes Mellitus. A disease in which there is an insufficiency of effective insulin and an impairment of glucose tolerance.

Epinephrine. A hormone of the adrenal medulla that activates the enzymes needed to release glucose from glycogen.

Glucagon. A hormone secreted by the α-cells of the pancreas in response to a decrease in the blood sugar level that stimulates the liver to release glucose from its glycogen stores.

Gluconeogenesis. The synthesis of glucose from compounds with smaller molecules or ions.

Glucose Tolerance. The ability of the body to manage the intake of dietary glucose while keeping the blood sugar level from fluctuating widely.

Glucose Tolerance Test. A series of measurements of the blood sugar level after the ingestion of a considerable amount of glucose; used to obtain information about an individual's glucose tolerance.

Glucosuria. The presence of glucose in urine.

Glycogenesis. The synthesis of glycogen.

Glycogenolysis. The breakdown of glycogen to glucose.

Glycolysis. A series of chemical reactions that break down glucose or glucose units in glycogen until pyruvate remains (when the series is operated aerobically) or lactate forms (when the conditions are anaerobic).

Human Growth Hormone. One of the hormones that affects the blood sugar level; a stimulator of the release of the hormone glucagon.

Hyperglycemia. An elevated level of sugar in the blood—above 110 mg/dL in whole blood.

Hypoglycemia. A low level of glucose in blood—below 70 mg/dL of whole blood.

Insulin. A protein hormone made by the pancreas, released in response to a rise in the blood sugar level, and used by certain tissues to help them take up glucose from circulation.

Normal Fasting Level. The normal concentration of something in the blood, such as blood sugar, after about four hours without food.

Oxygen Debt. The condition in a tissue when anaerobic glycolysis has operated and lactate has been excessively produced.

Pentose Phosphate Pathway. The synthesis of NADPH that uses chemical energy in glucose-6-phosphate and that involves pentoses as intermediates.

Renal Threshold. That concentration of a substance in blood above which it appears in the urine.

Somatostatin. A hormone of the hypothalamus that inhibits or slows the release of glucagon and insulin from the pancreas.

SELF-TESTING QUESTIONS

COMPLETION

1. The synthesis of glucose from starting materials that do not include any mono-, di-, or polysaccharides is called_____.
2. The synthesis of glycogen may be called _____.
3. The release of glucose from glycogen is called _____.
4. A hormone that can help release glucose from glycogen in a hurry is _____.
5. Events involved in the distribution, storage, mobilization, and usage of glucose may be summarized by the _____ Cycle.
6. If the _____ rises above normal, a condition of hyperglycemia is said to exist.
7. The appearance of glucose in urine is called _____.
8. Hyperinsulinism or starvation may produce a condition of_____-emia.
9. Clinically, an insufficiency of effective insulin is associated with _____.
10. If the kidneys are releasing glucose into the urine, the _____ for glucose has probably been exceeded.
11. In a situation of sudden stress, the adrenal medulla secretes a trace of the hormone _____. Its target cells are in _____ and to some extent in_____. At these cells it launches a series of enzyme activations that lead finally to the enzyme _____that catalyzes the _____.
12. Because liver cells have an enzyme that catalyzes the hydrolysis of_____ to glucose, the glucose stored as _____ in liver can be made available for release into the _____ and thereby help to raise the _____.
13. A hormone that is an even better than epinephrine as an activator of glycogenolysis at the liver is called _____.
14. The hormone secreted by the pancreas in response to an increase in the blood sugar level is _____, and it stimulates the removal of_____ from the bloodstream.
15. The "signal" for the release of insulin is an increase in _____.
16. The ability of the pancreas to respond to glucose is measured by the _____ test.
17. The effect of reduced insulin activity while glucagon activity remains normal is to reduce the supply of _____ inside cells. In this situation, the cells increase their rate of _____.
18. A sequence in glucose catabolism that can be run anaerobically is also called _____. Its end product is _____ which (when there is enough oxygen available) can lose the pieces of the element _____and be thereby oxidized to _____. This, in turn, can be broken down oxidatively to acetyl _____.
19. The anaerobic sequence enables a tissue to make a fresh supply of _____ for energy-demanding processes without oxygen being immediately available.
20. When a tissue is operating anaerobically, we say it is running an oxygen _____, and an accumulation of _____an end-product of _____, occurs.

21. When a tissue is operating aerobically, the change of an NADH-enzyme in glycolysis to its NAD+-enzyme form is accomplished by passing the H:⁻ from the NADH to _____; but when the tissue is operating anaerobically, the H:⁻ (plus H⁺) is transferred to _____ to make _____.

22. The H:⁻ unit in NADH is used eventually to reduce _____, whereas the H:⁻ in _____ is used to reduce organic substances that later become fatty acids.

23. The body's chief source of NADPH is the _____.

24. A large fraction of the lactate produced in the anaerobic sequence is converted to _____. The energy for accomplishing this is provided by sending the smaller fraction of the lactate (or some other metabolite) through the _____ sequence.

MULTIPLE-CHOICE

1. The end product of glycolysis under anaerobic conditions is
   (a) phosphoenolpyruvate   (c) lactate
   (b) pyruvate   (d) acetyl CoA

2. The end product of glycolysis under aerobic conditions is
   (a) phosphoenolpyruvate   (c) lactate
   (b) pyruvate   (d) acetyl CoA

3. All dietary monosaccharides undergo anaerobic catabolism in which the steps are for the most part the same as the steps in the catabolism of
   (a) glucose   (c) insulin
   (b) acetyl CoA   (d) pyruvate

4. The principal reducing agent needed to make fatty acids from carbohydrates is
   (a) NADH   (b) NADPH   (c) NAD⁺   (d) NADP⁺

5. If lactate is generated by hard work,
   (a) glycolysis has occurred
   (b) an oxygen debt exists
   (c) a fraction is eventually converted back to glycogen
   (d) the blood will become hyperglycemic

6. One important purpose of the pentose phosphate pathway of glucose catabolism is to make
   (a) ATP   (b) NADPH   (c) NADH   (d) glucose-1-phosphate

7. A hormone made in the α-cells of the pancreas that is a powerful activator of adenylate cyclase at liver cells is
   (a) insulin   (c) somatostatin
   (b) epinephrine   (d) glucagon

8. A hormone whose target tissue is the pancreas and that acts to slow down the release of insulin from the pancreas is
   (a) glucagon   (c) epinephrine
   (b) somatostatin   (d) human growth hormone

9. The carbon atoms in glucose most directly made by gluconeogenesis come from
   (a) HCO_3^- (c) acetyl coenzyme A
   (b) lactate   (d) CO_2

10. The breakdown of glucose to lactate is called
(a) gluconeogenesis   (c) glycolysis
(b) lactolysis   (d) glycogenolysis

11. The principal site of gluconeogenesis is the
(a) muscle   (c) pancreas
(b) liver   (d) adrenal medulla

12. The series of changes that move from glucose to lactate and back to glucose is called the
(a) Cori Cycle   (c) Krebs' Cycle
(b) citric acid cycle   (d) tricarboxylic acid cycle

13. The concentration of monosaccharides in blood is called
(a) diabetes mellitus   (c) the blood sugar level
(b) the renal threshold   (d) the normal fasting level

14. Hormones that tend to raise the blood sugar level include
(a) insulin   (c) glucagon
(b) epinephrine   (d) somatostatin

15. If too much insulin somehow appears in the bloodstream,
(a) glucosuria will result
(b) the blood will become hypoglycemic
(c) the blood will become hyperglycemic
(d) the renal threshold for glucose will be exceeded

ANSWERS

ANSWERS TO SELF–TESTING QUESTIONS

Completion

1. gluconeogenesis
2. glycogenesis
3. glycogenolysis
4. epinephrine
5. Cori
6. blood sugar level
7. glucosuria
8. hypoglycemia
9. diabetes mellitus
10. renal threshold
11. epinephrine; muscles; liver; phosphorylase kinase; conversion of glycogen into glucose-1-phosphate
12. glucose-6-phosphate; glycogen; bloodstream; blood sugar level
13. glucagon
14. insulin; glucose

15. the blood sugar level
16. glucose tolerance
17. glucose; gluconeogenesis
18. glycolysis; lactate, hydrogen, pyruvate; coenzyme A
19. ATP
20. debt, lactate, glycolysis
21. the respiratory chain (or, to FMN in the respiratory chain); pyruvate, lactate
22. oxygen, NADPH
23. pentose phosphate pathway of glucose catabolism
24. glucose; aerobic

Multiple–Choice

1.	c	9.	b
2.	b	10.	c
3.	a	11.	b
4.	b	12.	a
5.	a, b, and c	13.	c
6.	b	14.	b and c
7.	d	15.	b
8.	b		

16

METABOLISM OF LIPIDS

Our study of the molecular basis of the energy required for living is essentially completed in this chapter, as is our study of the molecular basis of diabetes.

OBJECTIVES

When you have finished studying the chapter and have answered the Review Exercises, you should be able to do the following.

1. Describe how lipids are transported in the bloodstream.
2. State the function of chylomicrons.
3. Describe what occurs to chylomicrons in adipose tissue and muscle.
4. Name the ways in which cholesterol is used outside of the liver.
5. Give the names and symbols of the lipoprotein complexes that handle endogenous lipids.
6. Describe the function of VLDL complexes.
7. Explain how VLDL changes to IDL.
8. State how IDL changes to LDL.
9. Explain the uses of LDL and how it becomes HDL.
10. Describe the function of HDL.
11. Explain the function of the liver receptors for IDL and LDL and how they help the body regulate the blood cholesterol level.
12. Contrast human lipid reserves and carbohydrate reserves in terms of roughly how long they would last during a period of starvation.
13. Outline the steps in the mobilization of the energy reserves in adipose tissue.

14. Using palmitic acid as an example, describe the overall result of the degradation of such an acid by the β-oxidation pathway.

15. Describe by equations how acetyl CoA is converted to butyryl ACP.

16. Describe by equations how acetyl CoA is used to lengthen the chain of butyryl ACP.

17. Explain how cholesterol participates in the feedback control of its own synthesis.

18. Explain how lovastatin (or compactin) work to inhibit cholesterol synthesis.

19. Name the ketone bodies and explain how they are produced in greater than normal amounts when effective insulin is missing.

20. Explain the relations between (a) ketonemia and ketonuria, (b) ketonemia and ketone breath, and (c) ketonemia and ketoacidosis.

21. Discuss the step-by-step progression of events from a condition in which effective insulin is lacking to the coma that results if the condition remains untreated.

GLOSSARY

β-Oxidation Pathway. The catabolism of a fatty acid by a series of repeating steps that produce acetyl units (in acetyl CoA).

Fatty Acid Cycle. (See *β-Oxidation Pathway.*)

Ketoacidosis. The acidosis caused by untreated ketonemia.

Ketone Bodies. Acetoacetate, β-hydroxybutyrate, or their parent acids, and acetone.

Ketonemia. An elevated concentration of ketone bodies in the blood.

Ketonuria. An elevated concentration of ketone bodies in the urine.

Ketosis. The combination of ketonemia, ketonuria, and acetone breath.

Lipoprotein Complex. A combination of a lipid molecule with a protein molecule that serves as the vehicle for carrying the lipid in the bloodstream.

SELF-TESTING QUESTIONS

COMPLETION

1. After triacylglycerols have been digested, the products are largely a mixture of _____ and _____. As these migrate out of the digestive tract they are changed into _____ that then become incorporated into particles called _____, which are carried in the blood.

2. The principal storage site for triacylglycerols is _____, but when they are in the bloodstream they are carried by complexes called _____.

3. In muscle and adipose tissue, chylomicrons unload some of their molecules of _____, and this leaves the chylomicron remnants richer in molecules of the more dense, nonhydrolyzable lipid, _____.

4. The organ that has receptors that can recognize chylomicron remnants is the _____.

5. When these receptors are defective or absent, the individual is likely to have _____.

6. The lipoprotein complex that the liver makes to export endogenous cholesterol into circulation is symbolized as _____, which stands for _____.

7. Tissues that can remove triacylglycerols from VLDL are _____ and _____.

8. This removal transforms the VLDL into _____, which stands for _____.

9. The IDL can be reabsorbed by the _____, but some experiences further loss of _____ so that the IDL changes into _____, which stands for _____.

10. The chief purpose of LDL is to carry cholesterol to _____ tissue and to glands that make _____.

11. Left-over cholesterol is carried back to the liver as _____, which stands for _____.

12. Storing chemical energy as lipid rather than as wet glycogen or dissolved glucose is advantageous because lipids have a particularly low _____, which means a low quantity of _____ for each calorie stored. In a 70-kg adult male, there is roughly _____ of triacylglycerol, enough to last about _____ if it has to serve as the sole source of caloric needs.

13. When insulin is in circulation, the activity of the cellular _____ that catalyzes the release of fatty acids from adipose tissue is _____(suppressed or activated). Otherwise the hormone _____can act to activate this enzyme.

14. When the acyl group of a fatty acid is catabolized inside a mitochondrion, it is attached to _____, a coenzyme. The first step in the oxidation is the loss of _____ from the α- and β-carbon atoms to give a functional group adjacent to the carbonyl group having the name _____. Then a molecule of _____adds to this functional group to give a β-_____ acyl system. This is then oxidized in the third step to give the _____-group. Finally a molecule of coenzyme A interacts with this compound to split out _____ and leaves an acyl-CoA system having _____fewer carbon atoms than it originally had. Then this shorter version is subjected to the same series of steps, and the process is called the _____.

15. The hydrogen removed from the acyl CoA in the β-oxidation pathway to give the α, β-unsaturated acyl CoA derivative is accepted by an enzyme having as its coenzyme _____whose reduced form is _____. This enzyme is part of the _____.

16. The hydrogen removed in the oxidation of the 2° alcohol group in the third step of the β-oxidation pathway is accepted by an enzyme having the coenzyme _____, whose reduced form is symbolized as _____.

17. The acetyl CoA units produced by the β-oxidation pathway enter the metabolic pathway having the name _____.

18. If glycolysis produces acetyl CoA that the cell doesn't need for energy, this excess acetyl CoA can be made into _____by a series of reactions catalyzed by the enzyme of a complex called _____ and fatty acid synthesis takes place in the _____ of the cell.

19. In fatty acid synthesis, one acetyl CoA molecule first combines with HCO_3^- to give _____, which has the structure:

_____

Another molecule of acetyl CoA is hooked to a unit called _____ of the synthase complex. The malonyl unit of malonyl CoA is joined to another part of the synthase and become malonyl ACP, where ACP stands for _____. Malonyl ACP now reacts with acetyl-E to give _____which has the structure:

_____

20. The keto group of this compound is next _____ by NADPH + H⁺, which changes the keto group into _____.

21. In the next step this group is removed to leave the ACP derivative of an unsaturated acid with the structure:

_____

22. This unsaturated acyl derivative of coenzyme A is now hydrogenated by the action of an enzyme with _____ as its coenzyme. We now have the ACP derivative of a simple fatty acid with the name _____.

23. The raw material for making the steroid nucleus is _____. One principal end product in steroid synthesis is an alcohol called _____ from which such other steroids as the_____hormones are made.

24. One way that the body controls how much cholesterol it makes is by regulating the activity of an enzyme called _____. One of the inhibitors of this enzyme is _____, so if the diet is rich in this substance, the body itself makes little if any more. A drug that successfully lowers the cholesterol level of the blood is _____ and it acts by _____.

25. An enhanced rate of fatty acid oxidation in the liver may produce excessive amounts of_____ in the blood, a condition known as _____. This will slowly lower the pH of the blood, a condition known as _____.

26. Both starvation and untreated _____ can lead to acidosis.

27. Efforts by the body to eliminate ketone bodies and associated positive ions, nitrogen wastes, and excess glucose from the bloodstream mean that more _____ than usual is also removed.

28. If insufficient water is drunk per day, the blood may _____, its circulation in the brain in sufficient quantities may become more difficult, and some blood flow may be diverted from the _____ in order to supply the brain. This diversion only makes it more difficult for the body to eliminate wastes.

29. As the anions from ketonemia become part of the urine being made, the chief cation to leave with them has the formula _____.The loss of this cation is sometimes called "the loss of _____," but in reality this ion is not a base. However, each of these cations that is lost represents the neutralization of one _____ ion, so the net effect is the loss of base within the system.

30. A simplified statement of one sequence of events in undetected (and therefore untreated) diabetes mellitus would be as follows. The excessive release of _____ from adipose tissue and the
                                                    (a)
increased rate of catabolism of fatty acids in the _____ cause the production
                                                    (b)
of_____ at a rate faster than can normally be handled. As a result there is a slow
      (c)
increase in the concentration of _____ in the blood, a condition called
                                        (d)
_____. Because two of them are acids, the _____ of the blood
      (e)   (f)
slowly drops.

MULTIPLE–CHOICE

1. The β-oxidation pathway
    (a) produces lactic acid
    (b) requires the β-form of oxygen
    (c) involves extensive glycogenolysis
    (d) produces units of acetyl CoA

2. If the amount of energy taken into a healthy body in the form of carbohydrates is greater than the body needs for energy, the body will experience
    (a) glucosuria
    (b) a greater rate of fatty acid synthesis
    (c) a greater rate of running the β-oxidation pathway
    (d) an enhanced rate of glycogenolysis

3. The citric acid cycle is fed two carbon units from
    (a) the catabolism of glucose
    (b) the β-oxidation pathway
    (c) the cholesterol degradation
    (d) the Cori Cycle

4. The average, adequately nourished adult male has enough chemical energy in storage as lipids to sustain life for how long:
    (a) 1 day   (b) 1 week   (c) 1 month   (d) 2 months

5. Fatty acids are transported in the bloodstream bound as complexes to molecules of
    (a) protein   (b) triacylglycerol   (c) FFA   (d) cholesterol

6. The densities of the various lipoprotein complexes increase in the order:
    (a) chylomicron < LDL < VLDL < IDL < HDL
    (b) chylomicron < VLDL < IDL < LDL < HDL
    (c) VLDL < chylomicron
    (d) LDL < chylomicron < VLDL < IDL < HDL

7. The chief carriers of exogenous triacylglycerols to adipose tissue are
    (a) chylomicrons (b) HDL   (c) LDL   (d) IDL

8. The chief carrier of cholesterol from peripheral tissue to the liver is
    (a) LDL   (b) IDL   (c) chylomicron (d) HDL

9. During each "turn," the β-oxidation pathway produces
    (a) $FADH_2$, NADH, acetyl CoA
    (b) $FADH_2$, NADH, acetoacetyl CoA
    (c) $FADH_2$, NADP, ADP
    (d) $FADH_2$, NADH, β-hydroxybutyrate

10. Once the substance $R{-}\overset{\overset{\textstyle O}{\|}}{C}{-}S{-}CoA$ has been formed, the next step in the β-oxidation pathway is
    (a) the addition of water   (c) dehydration
    (b) an attack by CoASH   (d) dehydrogenation

11. Once the substance R—CH=CH—$\overset{\displaystyle O}{\overset{\displaystyle \|}{C}}$—S—CoA has formed in the β-oxidation pathway, the next step is
    (a) the addition of water   (c) dehydration
    (b) an attack by CoASH   (d) dehydrogenation

12. The β-oxidation pathway degrades fatty acids by how many carbons per "turn"?
    (a) 1   (b) 2   (c) 3   (d) 4

13. To make the butyryl-ACP needed for the synthesis of fatty acids, acetoacetyl-ACP is first made from
    (a) acetoacetic acid
    (b) two molecules of acetyl-ACP
    (c) butyric acid
    (d) malonyl-ACP and acetyl-E

14. A fat-free, high carbohydrate diet promotes
    (a) the synthesis of fatty acids.
    (b) an elevation of the serum VLDL level.
    (c) the β-oxidation pathway.
    (d) lipolysis.

15. Which one of these structures is not a ketone body?

    (a) $CH_3\overset{\displaystyle O}{\overset{\displaystyle \|}{C}}CH_2\overset{\displaystyle O}{\overset{\displaystyle \|}{C}}O^-$

    (c) $CH_3\overset{\displaystyle O}{\overset{\displaystyle \|}{C}}CH_3$

    (b) $CH_3\overset{\displaystyle O}{\overset{\displaystyle \|}{C}}O^-$

    (d) $CH_3\overset{\displaystyle HO}{\overset{\displaystyle |}{C}}HCH_2\overset{\displaystyle O}{\overset{\displaystyle \|}{C}}O^-$

16. Which one of the conditions given is most closely linked to acidosis?
    (a) acetone breath   (c) ketonuria
    (b) ketonemia   (d) glycosuria

17. The uses of acetyl CoA in the body include
    (a) the synthesis of fatty acids
    (b) the synthesis of cholesterol
    (c) the synthesis of certain amino acids
    (d) fuel for the citric acid cycle

ANSWERS

ANSWERS TO SELF–TESTING QUESTIONS

Completion

1. fatty acids; monoacylglycerols; triacylglycerols; lipoprotein complexes
2. adipose tissue; lipoprotein complexes
3. triacylglycerols; cholesterol
4. liver
5. hypercholesterolemia
6. VLDL; very low density lipoprotein complex
7. adipose tissue; muscle tissue
8. IDL; intermediate density lipoprotein complex
9. liver; triacylglycerol; LDL; low density lipoprotein complex
10. peripheral (extrahepatic); steroid hormones
11. HDL; high density lipoprotein complexes
12. energy density; material mass (in grams); 12 kg; 43 days
13. lipase; suppressed; epinephrine
14. coenzyme A; hydrogen; alkene group; water; hydroxy; keto; acetyl CoA; two; fatty acid; β-oxidation pathway
15. FAD; FADH$_2$; respiratory chain
16. NAD$^+$, NADH
17. citric acid cycle
18. fatty acids; fatty acid synthase; cytosol
19. malonyl CoA; $^-OCCH_2C$—S—CoA (with two C=O groups); E; acyl carrier protein

    acetoacetyl ACP; CH_3CCH_2C—S—ACP (with two C=O groups)
20. reduced (or hydrogenated); a 2° alcohol group
21. $CH_3CH{=}CHC$—S—ACP (with C=O group)
22. NADPH; butyryl ACP
23. acetyl CoA; cholesterol; sex
24. HMG-CoA reductase; cholesterol; lovastatin (or compactin); inhibiting HMG-CoA synthase
25. ketone bodies; ketonemia; acidosis (or ketoacidosis)
26. diabetes
27. water
28. thicken (or become more viscous); kidneys

29. Na$^+$; base; H$^+$
30. (a) fatty acids
    (b) liver
    (c) ketone bodies
    (d) hydrogen ions (or, hydronium ions)
    (e) acidosis (or, in this case, ketoacidosis)
    (f) pH

Multiple–Choice

1. d
2. b
3. a and b
4. c
5. a
6. b
7. a
8. d

9. a
10. d
11. a
12. b
13. d
14. a
15. b
16. b
17. a, b, c, and d

17

METABOLISM OF NITROGEN COMPOUNDS

Although the catabolism of amino acids inevitably helps generate ATP and therefore contributes to the molecular basis of energy for living, this chapter brings us back to the molecular basis of materials for living.

OBJECTIVES

After you have studied the material in this chapter and have answered the Review Exercises, you should be able to do the following.

1. List the four main fates of amino acids in the body.
2. Describe the nitrogen pool.
3. By means of illustrative equations, show how reductive amination and transamination contribute to the synthesis of some amino acids.
4. Name the principal end products of the catabolism of amino acids.
5. Write an equation illustrating a specific example for each of the catabolic reactions of oxidative deamination, direct deamination, and decarboxylation.
6. Describe the overall result of the urea cycle.
7. Describe an origin of hyperammonemia and state the principal problem associated with it.
8. Describe the overall result of the catabolism of the purine bases (A or G).
9. In general terms, explain how gout and kidney stones are related to purine metabolism.
10. Give the relations between: (a) heme and biliverdin, (b) biliverdin and bilirubin, (c) bilirubin and bilinogen, and (d) bilirubin and jaundice.
11. Define each of the terms in the Glossary.

GLOSSARY

Bile Pigments. Colored products of the partial catabolism of heme that are transferred from the liver to the gall bladder for secretion via the bile.

Bilin. The brownish pigment that is the end product of the catabolism of heme and that contributes to the characteristic colors of feces and urine.

Bilinogen. A product of the catabolism of heme that contributes to the characteristic colors of feces and urine and some of which is oxidized to bilin.

Bilirubin. A reddish-orange substance that forms from biliverdin during the catabolism of heme and which enters the intestinal tract via the bile and is eventually changed into bilinogen and bilin.

Biliverdin. A greenish pigment that forms when partly catabolized hemoglobin (as verdohemoglobin) is further broken down, and which is changed in the liver to bilirubin.

Deamination, Direct. The removal of an amino group from an amino acid.

Decarboxylation. The removal of a carboxyl group.

Hyperammonemia. An elevated level of ammonia in blood.

Nitrogen Pool. The sum total of all nitrogen compounds in the body.

Oxidative Deamination. The change of an amino group to a keto group with loss of nitrogen.

Reductive Amination. The conversion of a keto group to an amino group by the action of ammonia and a reducing agent.

Transamination. The transfer of an amino group from an amino acid to a receptor with a keto group such that the keto group changes to an amino group.

Urea Cycle. The reactions by which urea is made from amino acids.

SELF-TESTING QUESTIONS

COMPLETION

1. Amino acids and other nitrogenous substances in the body, wherever they are, make up the _____.

2. Nitrogen enters the system largely as nitrogen compounds that are products of the digestion of _____.

3. Nitrogen leaves the system largely in the form of compounds called _____, _____, and a trace of _____.

4. The reducing agent in reductive amination has the short formula of _____, and the source of nitrogen is _____. If the compound: $^-O_2CCH_2\overset{\overset{\displaystyle O}{\|}}{C}CO_2^-$ underwent reductive amination, the following amino acid would form: _____

5. The same amino acid could be made by _____ using the same keto acid but using glutamate instead of NH_4^+ as the source of nitrogen. It leaves glutamate in the form of the following: _____

6. The reverse of reductive amination is called _____. The nitrogen of the amino acids undergoing this process emerges in the form of the _____, which can be changed to the chief nitrogen waste, _____.

7. When serine undergoes the following change:

$$\underset{\overset{|}{CH_2OH}}{\overset{+}{N}H_3CHCO_2^-} \longrightarrow \longrightarrow CH_3\overset{\overset{O}{\|}}{C}CO_2^- + NH_3$$

        serine   pyruvate

the overall reaction is called _____.

8. When tyrosine undergoes the following change:

$$NH_3CHCO_2^- \longrightarrow NH_2CH_2CH_2 - \underset{}{\bigcirc} - OH + CO_2$$

(with CH_2 branch and OH on ring for tyrosine)

      tyrosine   tyramine

the change is called _____.

9. When transamination occurs to alanine, the product is called _____, and it can be used to make acetyl CoA, the "fuel" for the _____ or as a building block for making _____.

10. When the amino group of an amino acid is destined to become part of urea, it is first removed from the amino acid by the process of _____. The oxaloacetate ion is the amino group acceptor, and it changes to _____. Then, a second transamination puts the amino group into the _____cycle. If enzymes needed for this cycle have reduced activity, a condition of _____ results.

11. The production of urea is accomplished by a series of reactions known as the _____ cycle.

12. The nitrogen in the purine bases_____ and _____ is excreted in the form of_____.

13. The formation of sodium urate at a rate faster than it is excreted may lead to a condition known as _____.

14. When heme is catabolized, the products are colored compounds generally known as the _____.The one that is reddish-orange is called_____.

15. Bilinogen that is excreted in feces is called _____.

16. Bilin that is excreted in urine is called _____.

MULTIPLE–CHOICE

1. The following reaction is an example of

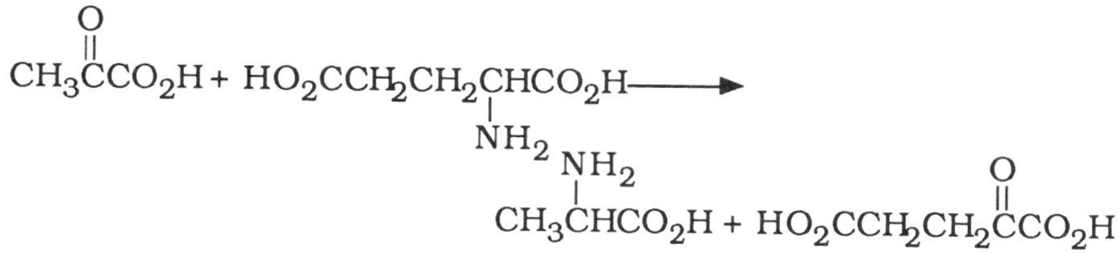

$$CH_3\overset{\overset{\displaystyle O}{||}}{C}CO_2H + HO_2CCH_2CH_2\underset{\underset{\displaystyle NH_2}{|}}{C}HCO_2H \longrightarrow$$

$$CH_3\underset{\underset{\displaystyle NH_2}{|}}{C}HCO_2H + HO_2CCH_2CH_2\overset{\overset{\displaystyle O}{||}}{C}CO_2H$$

   (a) oxidative deamination   (c) decarboxylation
   (b) transamination   (d) gluconeogenesis

2. The following reaction is an example of

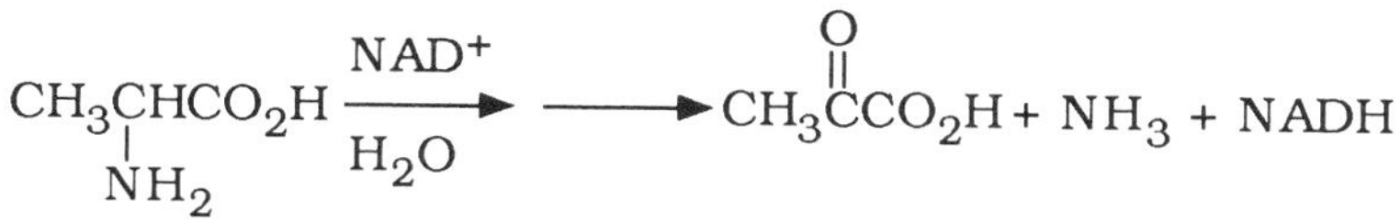

$$CH_3\underset{\underset{\displaystyle NH_2}{|}}{C}HCO_2H \xrightarrow[H_2O]{NAD^+} \longrightarrow CH_3\overset{\overset{\displaystyle O}{||}}{C}CO_2H + NH_3 + NADH$$

   (a) oxidative deamination   (c) decarboxylation
   (b) transamination   (d) gluconeogenesis

3. The following reaction illustrates

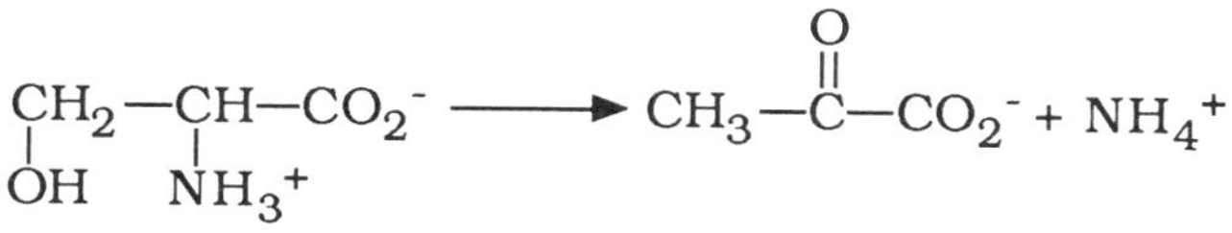

$$\underset{\underset{\displaystyle OH}{|}}{C}H_2-\underset{\underset{\displaystyle NH_3^+}{|}}{C}H-CO_2^- \longrightarrow CH_3-\overset{\overset{\displaystyle O}{||}}{C}-CO_2^- + NH_4^+$$

   (a) oxidative deamination   (c) decarboxylation
   (b) direct deamination   (d) transamination

4. If transamination occurred to $CH_3CH_2\underset{\underset{\displaystyle NH_3^+}{|}}{\overset{\overset{\displaystyle CH_3}{|}}{C}}HCHCO_2^-$, it would become

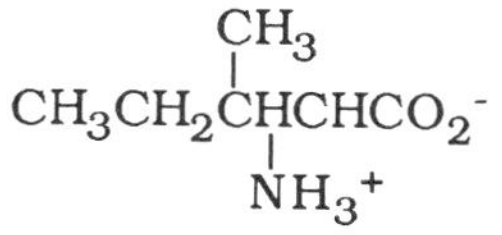

   (a) 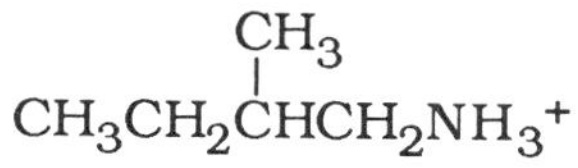 $CH_3CH_2\overset{\overset{\displaystyle CH_3}{|}}{C}HCH_2NH_3^+$

   (c) $CH_3CH_2\underset{\underset{\displaystyle H_3C\ \ NH_3^+}{|\ \ \ \ |}}{C}HCHCO_2^-$

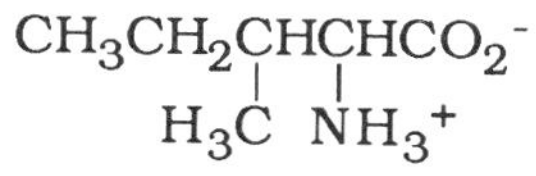

   (b) 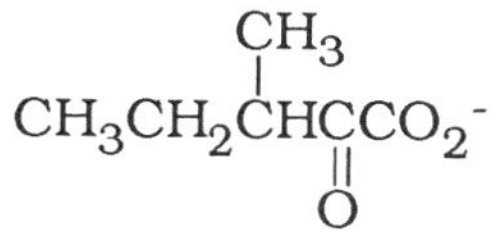 $CH_3CH_2\underset{\underset{\displaystyle O}{||}}{\overset{\overset{\displaystyle CH_3}{|}}{C}}HCCO_2^-$

   (d) 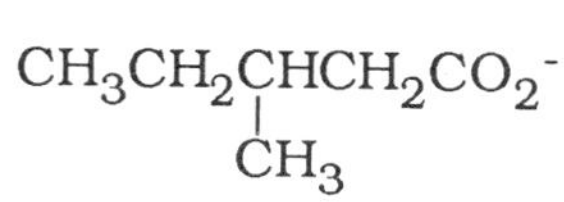 $CH_3CH_2\underset{\underset{\displaystyle CH_3}{|}}{C}HCH_2CO_2^-$

5. The nitrogen waste made from the purine bases of nucleic acids is
   (a) ammonia   (b) urea   (c) uric acid   (d) bilinogen

6. The end products in the catabolism of proteins are
   (a) amino acids
   (b) nitrogen, water, and carbon dioxide
   (c) ammonia, water, and carbon dioxide
   (d) urea, water, and carbon dioxide

7. In a dietary sense, alanine is classified as a nonessential amino acid. This is because
   (a) the body can make its own alanine
   (b) the body has no need for alanine
   (c) the body excretes alanine as fast as it can be ingested
   (d) alanine cannot be catabolized

8. A non-protein nitrogen compound made from molecules of amino acids is
   (a) triacylglycerol   (c) nucleic acid
   (b) creatine   (d) heme

9. Intermediates in the catabolism of amino acids may be used to make
   (a) glucose   (c) ketone bodies
   (b) fatty acids   (d) other amino acids

10. The major site of the catabolism of amino acids is the
    (a) liver   (b) kidneys   (c) adipose tissue   (d) gall bladder

11. In a condition of hyperammonemia, the level of concentration of what substance rises in the blood?
    (a) ammonia   (b) amino acids   (c) uric acid   (d) DOPA

12. The greenish pigment produced from heme is called
    (a) urobilinogen   (c) chlorophyll
    (b) urobilin   (d) biliverdin

ANSWERS

ANSWERS TO SELF–TESTING QUESTIONS

Completion

1. nitrogen pool
2. protein
3. urea, uric acid, ammonia
4. NADPH; ammonium ion; $^-O_2CCH_2CHCO_2^-$ with NH_3^+
5. transamination: $^-O_2CCH_2CH_2\overset{O}{\overset{\|}{C}}CO_2^-$
6. oxidative deamination; ammonium ion; urea
7. direct deamination

8. decarboxylation
9. pyruvate; citric acid cycle; fatty acids
10. transamination; aspartate; urea; hyperammonemia
11. urea (or Krebs' ornithine)
12. adenine, guanine, uric acid (or urate ions)
13. gout (also, kidney stones or the aggravation of arthritis)
14. tetrapyrrole pigments (or bile pigments); bilirubin
15. stercobilinogen
16. urobilin

Multiple–Choice

1. b
2. a
3. b
4. b
5. c
6. d
7. a
8. b, c, and d
9. a, b, c, and d
10. a
11. a
12. d

18

NUTRITION

Health, happiness, and life are all greatly influenced by what we eat and drink. When most people learn *why* various things must be in the diet in their proper proportions, they usually make an effort to ensure that they have a balanced diet.

OBJECTIVES

The objectives that follow, which you should be able to do once you have studied the chapter and have worked its Exercises and Review Questions, are designed to emphasize the knowledge that will help you have the best physical well-being possible.

1. Explain why the recommended daily allowances of the National Academy of Sciences are higher than the minimum daily requirements.
2. Explain why the best nutrition is obtained from a variety of foods.
3. Explain why food energy should come from a mix of lipids and carbohydrates.
4. Explain why several amino acids, but not all, are called "essential."
5. Compare meat, cereal, and fruit proteins in their digestibility.
6. Give both the positive and negative consequences of milling grains.
7. Compare the proteins of meats, dairy products, cereals, and nuts in their biological values.
8. Describe the major factor affecting the biological values of proteins.
9. Explain why one could not eat enough maize (corn) or cassava per day to satisfy one's needs for amino acids.
10. Describe the problems vegetarians must solve in order to have good nutrition.
11. Name the essential fatty acids and tell why they are important.

12. In general ways, explain how vitamins are different from other nutrients.
13. Name the vitamins and, where known, identify the human deficiency diseases or syndromes associated with each.
14. Give one good source for each vitamin.
15. Name and give the chemical formulas of the six minerals needed at levels above 100 mg/day.
16. Name and give the principal functions of the ten chief trace elements.
17. Define the terms in the Glossary.

GLOSSARY

Adequate Protein. A protein that, when digested, makes available all of the essential amino acids in suitable proportions to satisfy both the amino acid and total nitrogen requirements of good nutrition without providing excessive calories.

Biological Value. In nutrition, the percentage of the nitrogen of ingested protein that is absorbed from the digestive tract and retained by the body when the total protein intake is less than normally required.

Biotin. A water-soluble vitamin needed to make enzymes used in fatty acid synthesis.

Choline. A compound needed to make complex lipids and acetylcholine; classified as a vitamin.

Coefficient of Digestibility. The proportion of an ingested protein's nitrogen that enters circulation rather than elimination (in feces); the difference between the nitrogen ingested and the nitrogen in the feces divided by the nitrogen ingested.

Dietetics. The application of the findings of the science of nutrition to the feeding of individual humans, whether well or ill.

Essential Amino Acid. An α-amino acid that the body cannot make from other amino acids and that must be supplied by the diet.

Essential Fatty Acid. A fatty acid that must be supplied by the diet.

Folate. A vitamin supplied by folic acid or pteroylglutamic acid and that is needed to prevent megaloblastic anemia.

Food. A material that supplies one or more nutrients without contributing materials that, either in kind or quantity, would be harmful to most healthy people.

Limiting Amino Acid. The essential amino acid most poorly provided by a dietary protein.

Minerals. Ions that must be provided in the diet at levels of 100 mg/day or more; Ca^{2+}, Mg^{2+}, Na^+, K^+, Cl^-, and phosphate.

Niacin. A water-soluble vitamin needed to prevent pellagra and essential to the coenzymes in NAD^+ and $NADP^+$; nicotinic acid or nicotinamide.

Nutrients. Chemical substances that take part in normal, healthy metabolism.

Nitrogen Balance. A condition of the body in which it excretes as much nitrogen as it receives in the diet.

Nutrition. The science of the substances of the diet that are necessary for growth, operation, energy, and repair of bodily tissues.

Pantothenic Acid. A water-soluble vitamin needed to make coenzyme A.

Recommended Dietary Allowance (RDA). The level of intake of a particular nutrient as determined by the Food and Nutrition Board of the National Research Council of the National Academy of Sciences to meet the know nutritional needs of most healthy individuals.

Riboflavin. A B vitamin needed to give protection against the breakdown of tissue around the mouth, the nose, and the tongue, as well as to aid in wound healing.

Thiamin. A B vitamin needed to prevent beri beri.

Trace Element. Any element that the body needs each day in an amount of no more than 20 mg.

Vitamin. An organic substance that must be in the diet; whose absence causes a deficiency disease; which is present in foods in trace concentrations; and that isn't a carbohydrate, lipid, protein, or amino acid.

Vitamin A. Retinol; a fat-soluble vitamin in yellow-colored foods and needed to prevent night blindness and certain conditions of the mucous membranes.

Vitamin B_6. Pyridoxine, pyridoxal, or pyridoxamine; a vitamin needed to prevent hypochromic microcytic anemia and used in enzymes of amino acid catabolism.

Vitamin B_{12}. Cobalamin; a vitamin needed to prevent pernicious anemia.

Vitamin C. Ascorbic acid; a vitamin needed to prevent scurvy.

Vitamin D. Cholecalciferol (D_3) or ergocalciferol (D_2); a fat-soluble vitamin needed to prevent rickets and to ensure the formation of healthy bones and teeth.

Vitamin Deficiency Diseases. Diseases caused not by bacteria or viruses but by the absence of specific vitamins, such as pernicious anemia (B_{12}), hypochromic microcytic anemia (B_6), pellagra (niacin), the breakdown of certain tissues (riboflavin), megaloblastic anemia (folate), beri beri (thiamin), scurvy (C), hemorrhagic disease (K), rickets (D), and night blindness (A).

Vitamin E. A mixture of tocopherols; a fat-soluble vitamin apparently needed for protection against edema and anemia (in infants) and possibly against dystrophy, paralysis, and atherosclerosis.

Vitamin K. The antihemorrhagic vitamin that serves as a cofactor in the formation of a blood clot.

SELF-TESTING QUESTIONS

COMPLETION

1. With respect to the terms "nutrient" and "food," we may say that bread is a _____ and the wheat protein in bread is a _____.

2. The symbol RDA stands for _____.

3. A quantitative measure of the digestibility of a given protein is its _____, and the equation that defines it is _____.

4. The extent to which a given protein has a high biological value is determined by the extent to which it supplies _____.

5. A protein with little if any lysine would have a low _____.

6. The essential amino acid most poorly supplied by a given protein is called the _____ of that protein.

7. A protein with all of the essential amino acids in approximately the right proportions for humans is called an _____.

8. An individual whose intake of nitrogen in the diet equals the quantity of nitrogen excreted is said to have a _____.

9. In order to grow and develop an infant should have a _____ nitrogen balance.

10. The fatty acids about which all are agreed is an essential fatty acid is _____.

11. List four criteria that must be satisfied by a substance if it is to be considered a vitamin.

(a) _____

(b) _____

(c) _____

(d) _____

12. The two broad classes of vitamins are the _____ vitamins and the _____ vitamins.

13. The four vitamins that are most hydrocarbon-like are vitamins _____, _____, _____, and _____.

14. Because the molecules of some of the hydrocarbon-like vitamins have carbon-carbon _____ bonds, these vitamins can be slowly destroyed by contact with air.

15. Lack of vitamin D leads to _____, a disorder of _____.

16. Lack of vitamin _____ may lead to impaired vision in dim light.

17. Lack of vitamin _____ may lead to problems in controlling hemorrhaging.

18. In the absence of ascorbic acid, vitamin _____, the disease known as _____ would develop.

19. Because thiamin is a part of an enzyme essential to the catabolism of _____, the daily requirement for it is related to the daily intake of _____.

20. Beri-beri is a disease that occurs when _____ is lacking in the diet.

21. Pantothenic acid is needed to make coenzyme _____.

22. The principal minerals in the body are (6) _____, _____, _____, _____, _____, and _____. (Give their correct formulas.)

23. To qualify as a trace element, the mass that the element contributes to the total mass of the body does not exceed _____.

24. The names of eleven trace elements known or believed to be needed by the body are (in any order)

_____   _____   _____

_____   _____   _____

_____   _____   _____

_____   _____

25. All trace metal elements occur as their _____, not as their atoms.

26. The central ion in hemoglobin has the name _____ and the formula _____.

MULTIPLE-CHOICE

1. A zero-carbohydrate diet will cause the body to make its own
   (a) linoleic acid   (c) glucose
   (b) sucrose   (d) nonessential amino acids

2. The proteins with the highest digestibility coefficients are generally those of
   (a) vegetables   (b) meat   (c) fruit   (d) cereals

3. One of the essential amino acids that often is a limiting amino acid is
   (a) lysine   (b) alanine   (c) glycine   (d) hydroxyproline

4. If 102 g of wheat protein has to be eaten to result in 80.4 g being actually digested and absorbed, the digestibility coefficient of wheat protein is
   (a) 80.4   (b) 82   (c) 1.27   (d) 0.79

5. The vegetarian diet supplies sufficient and adequate protein each day without also furnishing large amount of calories provided that the diet emphasizes
    (a) brown rice
    (b) beans
    (c) two different vegetables about equally
    (d) cassava

6. Even though phosphate is essential in the diet, it is not classified as a vitamin because it is not
    (a) organic
    (b) required for normal growth
    (c) present in foods
    (d) a carbohydrate

7. The vitamin supplied by yellow-colored vegetables is vitamin
    (a) A   (b) B   (c) C   (d) D

8. A vitamin that can be stored in adipose tissue is
    (a) vitamin K   (c) vitamin D
    (b) vitamin B_6 (d) thiamine

9. A vitamin that detoxifies peroxides and so inhibits the oxidation of the lipids of lipoprotein complexes is
    (a) vitamin K   (c) vitamin D
    (b) vitamin B_6 (d) vitamin E

10. The trace element that serves as a cofactor for the action of insulin is
    (a) Cr^{3+} (b) K^+ (c) F^- (d) Mn^{2+}

11. The trace element needed to make heme is
    (a) Cu^{2+} (b) Na^+ (c) Fe^{2+} (d) K^+

ANSWERS

ANSWERS TO SELF–TESTING QUESTIONS

Completion

1. food, nutrient
2. recommended dietary allowances
3. coefficient of digestibility,
    coefficient of digestibility = ([N in food eaten - N in feces]/(N in food eaten)
4. essential amino acids
5. biological value
6. limiting amino acid
7. adequate protein
8. nitrogen balance
9. positive

10. linoleic acid
11. (a) an organic compound that cannot be made in the body
    (b) its absence causes a deficiency disease
    (c) its presence is required for growth and health
    (d) found only in trace concentrations in foods and is not a carbohydrate, a lipid, or a protein
12. fat-soluble, water-soluble
13. A, D, E, and K
14. double
15. rickets, bone metabolism
16. A
17. K
18. C, scurvy
19. carbohydrate, calories
20. thiamin
21. A
22. Ca^{2+}, Mg^{2+}, Na^+, K^+, Cl^-, and inorganic phosphate ion (P)
23. 20 mg
24. iron, cobalt, zinc, chromium, molybdenum, copper, manganese, nickel, tin, vanadium, and silicon
25. ions
26. iron(II) ion (or the ferrous ion); Fe^{2+}

Multiple–Choice

1. c
2. b
3. a
4. d
5. c
6. a
7. a
8. c
9. d
10. a
10. c

ANSWER BOOK

INTRODUCTION

This *Answer Book* is a supplement to *Fundamentals of General, Organic, and Biological Chemistry,* 5th edition by John R. Holum. Although the text itself supplies answers to many Exercises, this supplement provides answers for all simply for the convenience of having them all in one place.

If not used wisely, an answer book will give a false sense of security that will be revealed all too sadly at the time of an examination. *The mistake many students make is to go to the answer book too soon.* Consult this supplement only after you have tried your best to determine the answer *in writing.* Then the answer book can supply positive reinforcement by showing you that you are right, or it can get you over those "impossible" spots.

The answers have been checked by me and by Dr. Melinda Lee of St. Cloud State University. From many years of experience, however, I am sure that some errors have still slipped by. Feel free to write me about them. Don't be concerned by small differences between your answers and mine to problems having numerical, calculated results. Unless there are parts to a question, separately labeled as (a), (b), etc., I always use a chain calculation. This might causes differences attributable to rounding off answers at times other than mine. Be sure to follow the guidelines of the text concerning the rules for rounding. Also remember that in calculations involving atomic masses, the rule in the text is to round all atomic masses to their first decimal place *before* using them (except round the atomic mass of H to 1.01).

John R. Holum
3352 47th Ave. S
Minneapolis, MN 55406

Chapter 1

Practice Exercises, Chapter 1

1. (a) $CH_3-CH_2-CH_3$ (b) $CH_3-\underset{\underset{\displaystyle CH_3}{|}}{CH}-CH_3$

   (c) $CH_3-\underset{\underset{\displaystyle CH_3}{|}}{\overset{\overset{\displaystyle CH_3}{|}}{C}}-\overset{\overset{\displaystyle CH_3}{|}}{CH}-\overset{\overset{\displaystyle CH_3}{|}}{CH}-CH_3$

2. (a) $CH_3CH_2CH_3$ (b) $CH_3\underset{\underset{\displaystyle CH_3}{|}}{CH}CH_3$

   (c) $CH_3\underset{\underset{\displaystyle CH_3}{|}}{\overset{\overset{\displaystyle CH_3}{|}}{C}}-\overset{\overset{\displaystyle CH_3}{|}}{CH}-\overset{\overset{\displaystyle CH_3}{|}}{CH}CH_3$

3. (a)
$$H-\underset{\underset{\displaystyle H}{|}}{\overset{\overset{\displaystyle H}{|}}{C}}-\underset{\underset{\displaystyle H}{|}}{\overset{\overset{\displaystyle H}{|}}{C}}-H$$

   (b)
$$H-\underset{\underset{\displaystyle H}{|}}{\overset{\overset{\displaystyle H\;H-\overset{\overset{\displaystyle H}{|}}{C}-H\;H}{|}}{C}}-\underset{\underset{\displaystyle H\;H-\overset{\overset{\displaystyle H}{|}}{C}-H\;H}{|}}{\overset{\overset{\displaystyle H}{|}}{C}}-\overset{\overset{\displaystyle H}{|}}{C}-\overset{\overset{\displaystyle H}{|}}{C}-H$$

(c)

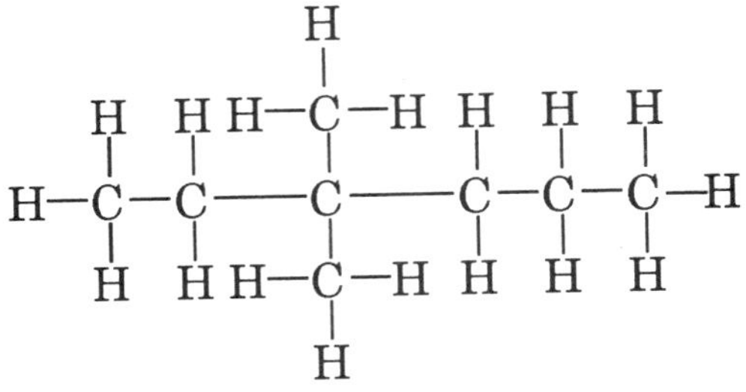

4. Structure (b) and (c) violate the tetravalences of carbon at one point.

5. (a)

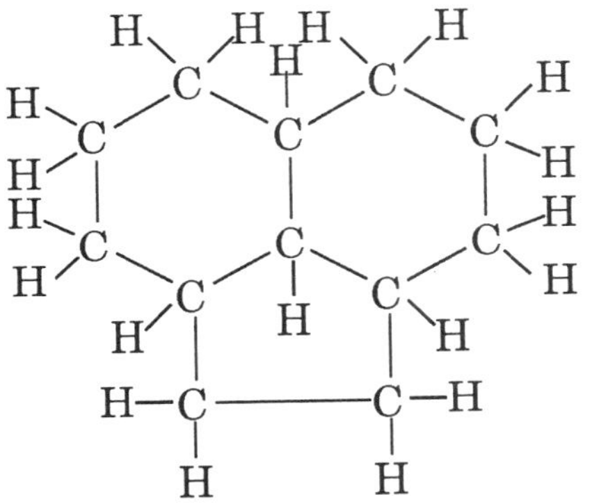

(b)

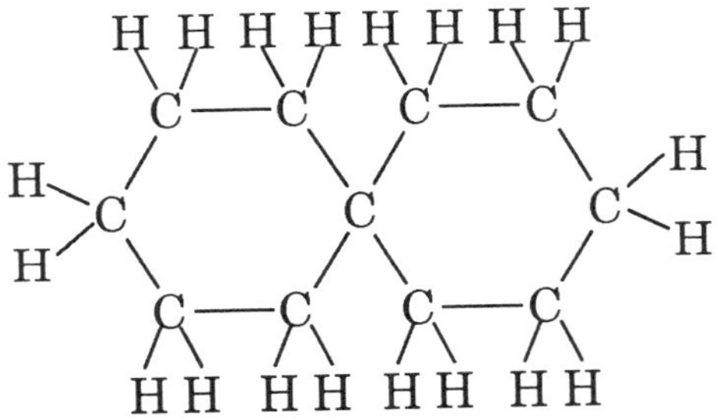

6. CH₃CH₂

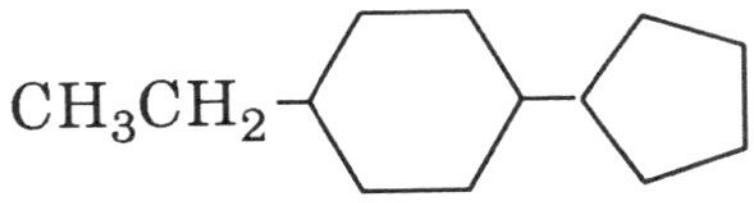

7. (a) identical
   (b) isomers
   (c) identical
   (d) isomers
   (e) different in another way
8. 2-methyl-1-butanol (less polar)
9. (a) 3-methylhexane
   (b) 4-ethyl-2,3-dimethylheptane
   (c) 5-ethyl-2,4,6-trimethyloctane

10.   (a) $BrCH_2CHCH_2CH_2CH_3$
$$\qquad\qquad\quad\underset{\displaystyle NO_2}{|}$$

(b)

$$\underset{\displaystyle \overset{|}{CH_3}\ \overset{|}{CH_3}\ \overset{|}{CH_3}}{CH_3\overset{\overset{\displaystyle CH_3}{|}}{C}-\overset{\overset{\displaystyle CH_3}{|}}{C}-\overset{\overset{\displaystyle CH_3}{|}}{C}-\overset{\overset{\displaystyle CH(CH_3)_2}{|}}{CH}CH_2CH_2CH_3}$$

(c)

$$CH_3\overset{\overset{\displaystyle I}{|}}{\underset{\underset{\displaystyle I}{|}}{C}}\overset{\overset{\displaystyle CH_3}{|}}{CH}\underset{\underset{\displaystyle CH(CH_3)_2}{|}}{CH}-\overset{\overset{\displaystyle CH_3CHCH_2CH_3}{|}}{CH}\underset{\underset{\displaystyle C(CH_3)_3}{|}}{CH}CH_2CH_2CH_3$$

(d) $BrCHCHCH_3$
$$\qquad\quad\underset{\displaystyle Cl}{|}\ \underset{\displaystyle CH_3}{|}$$

(e)

$$CH_3CH_2CH_2CH_2\underset{\underset{\displaystyle CH_3CHCH_2CH_3}{|}}{\overset{\overset{\displaystyle CH_3CHCH_2CH_3}{|}}{C}}CH_2CH_2CH_2CH_2CH_3$$

11.

$$\underline{CH_3}\overset{\displaystyle \downarrow}{C}\overset{\displaystyle \downarrow}{H}\dot{C}H_2\underline{C}H_3$$
$$\underline{CH_3}\overset{\displaystyle \downarrow}{\dot{C}}H_2\overset{\displaystyle \downarrow}{\dot{C}}H_2\overset{\displaystyle \downarrow}{\dot{C}}H_2\overset{\displaystyle \downarrow}{C}CH_2\overset{\displaystyle \downarrow}{\dot{C}}H_2\overset{\displaystyle \downarrow}{\dot{C}}H_2\overset{\displaystyle \downarrow}{\dot{C}}H_2\underline{C}H_3$$
$$\underline{CH_3}C H\overset{\displaystyle \uparrow}{C}H_2\underline{C}H_3$$

12.   (a)   ethyl chloride
      (b)   butyl bromide
      (c)   isobutyl chloride
      (d)   *t*-butyl bromide
13.   (a)   butyl chloride, 1-chlorobutane
            sec-butyl chloride, 2-chlorobutane
      (b)   isobutyl chloride, 2-methyl-1-chloropropane
            t-butyl chloride, 2-methyl-2-chloropropane

Review Exercises, Chapter 1

1.1   Carbon atoms are able to form strong bonds to each other while also forming strong bonds to other nonmetal atoms. Carbon atoms can join to each other in an infinite number of ways—by single, double, or triple bonds; by chains (straight and branched) and rings of three or more carbon atoms in size.

1.2   Chemists had been unable to synthesize organic compounds from minerals. The vital force theory held that such a synthesis was inherently impossible without the presence of a "vital force."

1.3   To prepare crystalline ammonium cyanate. He obtained an organic compound, urea, from an inorganic compound and this contradicted the vital force theory.

1.4   Covalent bonds

1.5   (a)   4   (b) 2   (c) 3   (d) 1   (e) 1

1.6   Compounds b, d, and e are considered to be inorganic.

1.7   Ionic and inorganic

1.8   Organic compounds generally consist of *molecules*, not ions, and their molecules generally do not react with water to give ions.

1.9   (a)   Molecular; low melting and flammable
      (b)   Ionic; water-soluble (and likely a carbonate or a bicarbonate)
      (c)   Molecular; no ionic compound is a gas (or a liquid) at room temperature.
      (d)   Ionic; high melting and nonflammable
      (e)   Molecular; most liquid organic compounds are insoluble in water but will burn.
      (f)   Molecular; no ionic compound is a liquid at room temperature.

1.10 Straight chain. It has no 3° carbons as do all branched-chain, open-chain compounds.

1.11 (a)   Compounds a, d, and e are possible

1.12   (a)

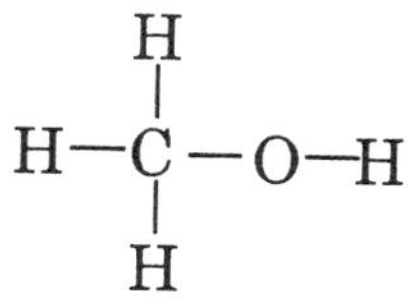

$$H-\overset{\displaystyle H}{\underset{\displaystyle H}{C}}-O-H$$

(b)

$$H-\overset{\displaystyle Cl}{\underset{\displaystyle Cl}{C}}-H$$

(c)

$$H-\overset{\displaystyle H}{N}-\overset{\displaystyle H}{N}-H$$

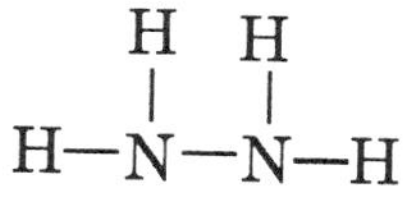

(d)

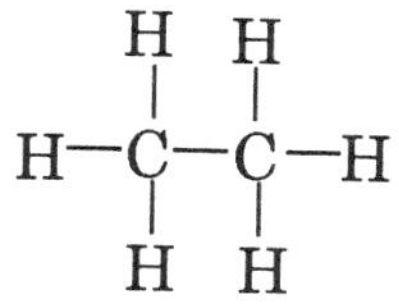

$$H-\overset{\displaystyle H}{\underset{\displaystyle H}{C}}-\overset{\displaystyle H}{\underset{\displaystyle H}{C}}-H$$

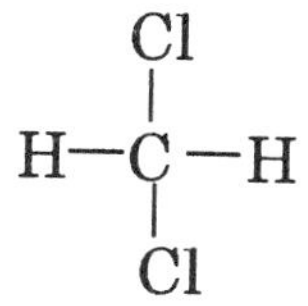

(e)

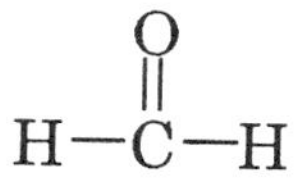

$$H-\overset{\displaystyle O}{\overset{\|}{C}}-H$$

(f)

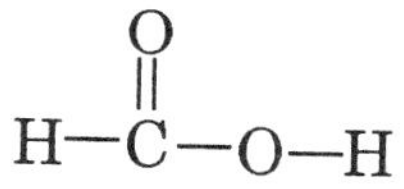

$$H-\overset{\displaystyle O}{\overset{\|}{C}}-O-H$$

(g)

$$H-\overset{\displaystyle H}{N}-O-H$$

(h)

$$H-\overset{\displaystyle H}{C}=\overset{\displaystyle H}{C}-H$$

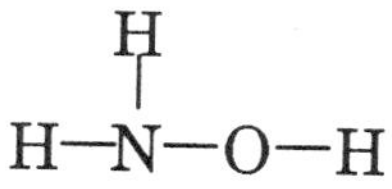

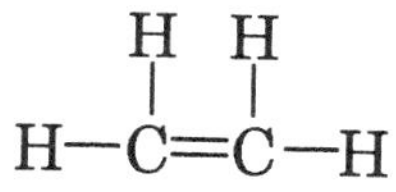

(i)

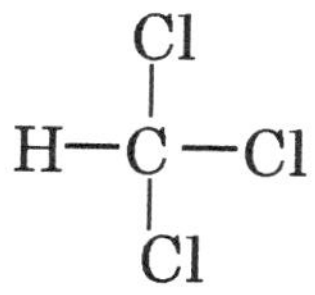

$$H-\overset{\displaystyle Cl}{\underset{\displaystyle Cl}{C}}-Cl$$

(j) $H-C\equiv N$

(k)

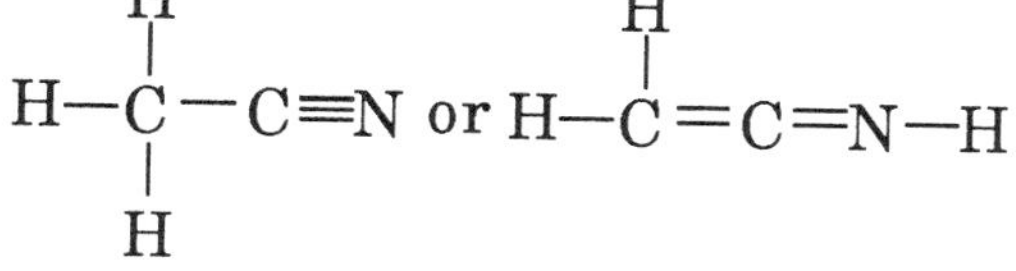

$$H-\overset{\displaystyle H}{\underset{\displaystyle H}{C}}-C\equiv N \text{ or } H-\overset{\displaystyle H}{C}=C=N-H$$

(l)

$$H-\overset{\displaystyle H}{\underset{\displaystyle H}{C}}-\overset{\displaystyle H}{N}-H$$

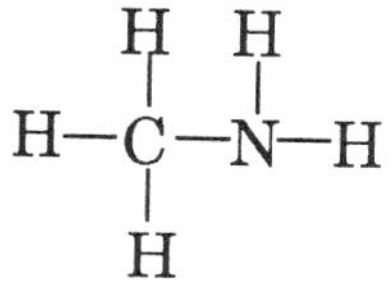

1.13

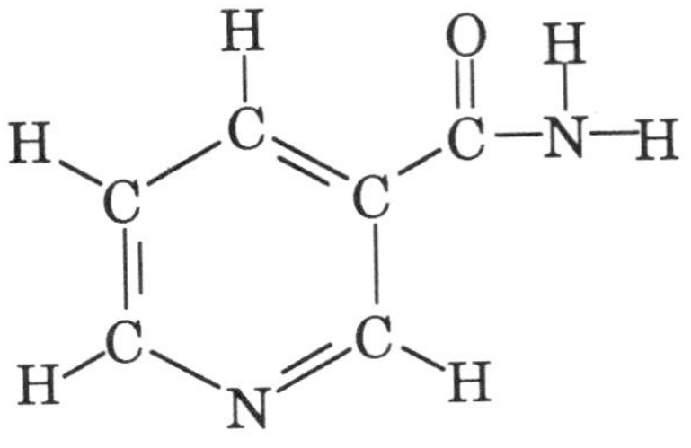

1.14

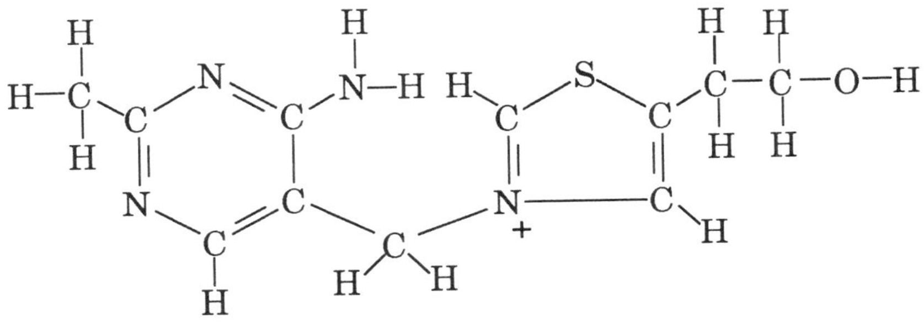

1.15   (a) $CH_3CH_2CH_2CH_2CH_3$

(b) CH_3CH_2 ... CH_3 H_3C

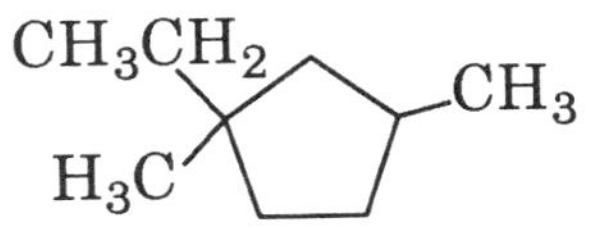

(c) CH_3

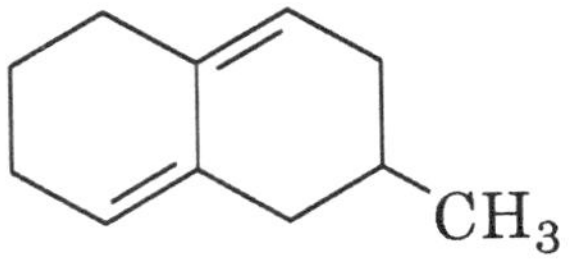

1.16   The geometry of a molecule is as important to the ability of an enzyme to work as other aspects of structure.

1.17 sp^3 hybrid orbitals

1.18   The rotation does not reduce the degree of the overlap of the hybrid atomic orbitals and so does not weaken the molecule.

1.19   Unsaturated compounds are (b), (c), and (d).

1.20 C_4H_{10} (c) and cyclohexanol (d) are saturated. (C_4H_{10} can only be butane or isobutane.)

1.21   Identical compounds are a, b, e, f, and m
        Isomers are d, g, h, i, j, k, and l.
        Structures that are unrelated occur at c and n.

1.22   (a)   alkane
        (b)   alcohol
        (c)   thioalcohol
        (d)   alkyne
        (e)   aldehyde
        (f)   ester
        (g)   ester
        (h)   ketone
        (i)   amine
        (j)   ether

1.23   (a)   alcohol
        (b)   alcohol
        (c)   both thioalcohols
        (d)   first, alkene; second, cycloalkane
        (e)   ketone
        (f)   alkane
        (g)   both amines
        (h)   first, carboxylic acid; second, alcohol + ketone
        (i)   first, ester; second, carboxylic acid. (If you called the first an aldehyde, instead of an ester, don't worry about it now. Why it is an ester sill be explained later.)
        (j)   first, ester + alcohol; second, alcohol + carboxylic acid
        (k)   first, ether + ketone; second, ester
        (l)   both alkanes
        (m)   ketone

1.24   B. It consists of larger molecules and so there would be larger London forces of attraction between molecules.

1.25   B. Its molecules have water-like OH groups.

1.26   Add a drop of the liquid to water. If it does not dissolve, it is pentane.

1.27   Add a drop of the liquid to water. If it dissolves, it is methyl alcohol.

1.28

$$CH_3CH_2CH_2CH_2CH_2CH_3 \text{ hexane}$$

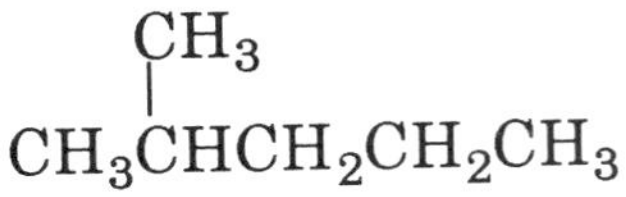

$$\begin{array}{c} CH_3 \\ | \\ CH_3CHCH_2CH_2CH_3 \end{array} \qquad \text{2-methylpentane}$$

$$\begin{array}{c} CH_3 \\ | \\ CH_3CH_2CHCH_2CH_3 \end{array} \qquad \text{3-methylpentane}$$

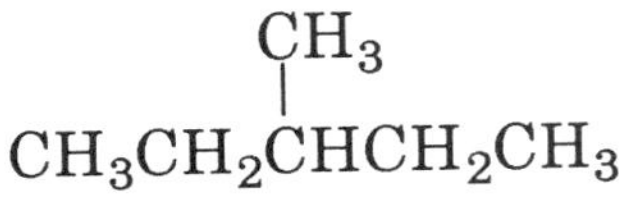

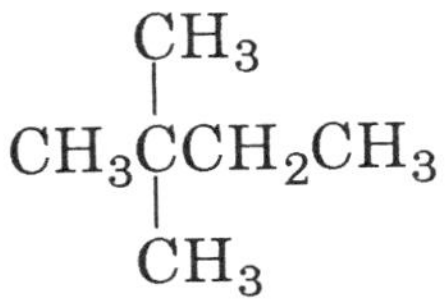

$$\begin{array}{c} CH_3 \\ | \\ CH_3CCH_2CH_3 \\ | \\ CH_3 \end{array} \qquad \text{2,2-dimethylbutane}$$

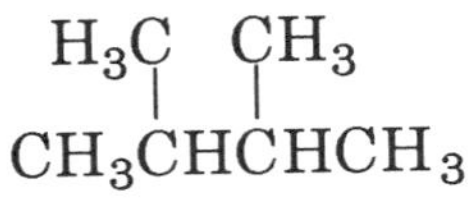

$$\begin{array}{c} H_3C \quad CH_3 \\ | \quad\; | \\ CH_3CHCHCH_3 \end{array} \qquad \text{2,3-dimethylbutane}$$

1.29

$$CH_3CH_2CH_2CH_2CH_2CH_3 \text{ hexane}$$

1.30

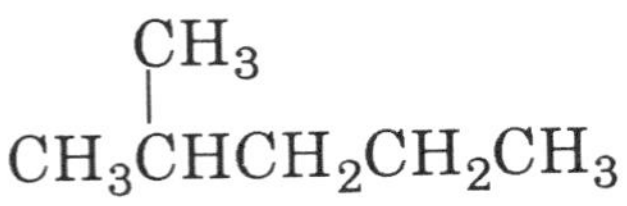

$$\begin{array}{c} CH_3 \\ | \\ CH_3CHCH_2CH_2CH_3 \end{array} \quad \text{2-methylpentane}$$

1.31

$$CH_3CH_2CH_2CH_2CH_2CH_2CH_3 \quad \text{heptane}$$

$$\underset{\underset{CH_3CHCH_2CH_2CH_2CH_3}{|}}{CH_3}$$ 2-methylhexane

$$\underset{\underset{CH_3CH_2CHCH_2CH_2CH_3}{|}}{CH_3}$$ 3-methylhexane

$$\underset{\underset{CH_3CH_2CHCH_2CH_3}{|}}{CH_2CH_3}$$ 3-ethylpentane

$$\overset{\overset{CH_3}{|}}{\underset{\underset{CH_3}{|}}{CH_3CCH_2CH_2CH_3}}$$ 2,2-dimethylpentane

$$\overset{\overset{CH_3}{|}}{\underset{\underset{CH_3}{|}}{CH_3CH_2CCH_2CH_3}}$$ 3,3-dimethylpentane

$$\underset{\underset{CH_3CHCHCH_2CH_3}{|\quad|}}{H_3C\ \ CH_3}$$ 2,3-dimethylpentane

$$\underset{\underset{CH_3CHCH_2CHCH_3}{|\qquad|}}{CH_3\quad\ CH_3}$$ 2,4-dimethylpentane

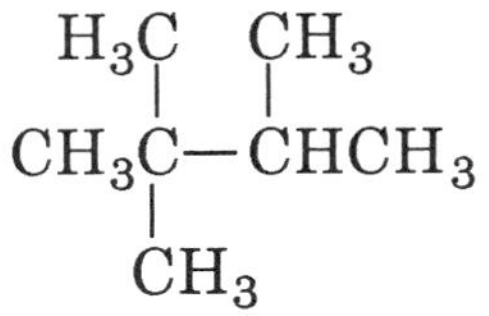

$$\begin{array}{c} \overset{\displaystyle H_3C}{|} \quad \overset{\displaystyle CH_3}{|} \\ CH_3C-CHCH_3 \\ \overset{\displaystyle |}{CH_3} \end{array}$$
2,2,3-trimethylbutane

1.32   (a) $CH_3CH_2CH_2CH_2CH_2CH_2CH_3$

(b) 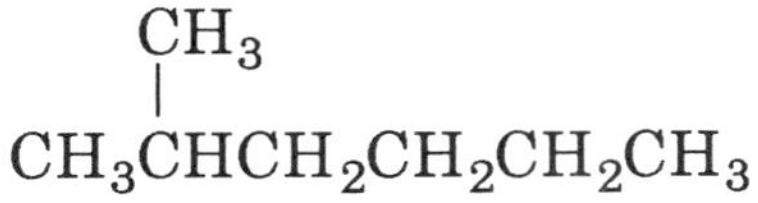
$$\begin{array}{c} \overset{\displaystyle CH_3}{|} \\ CH_3CHCH_2CH_2CH_2CH_3 \end{array}$$

1.33   (a)   5-sec-butyl-5-ethyl-2,3,3,9-tetramethyldecane
       (b)   7-t-butyl-5-isobutyl-2-methyl-6-propyldecane

1.34   (a) $CH_3CH_2CH_2CH_2Br$

(b) 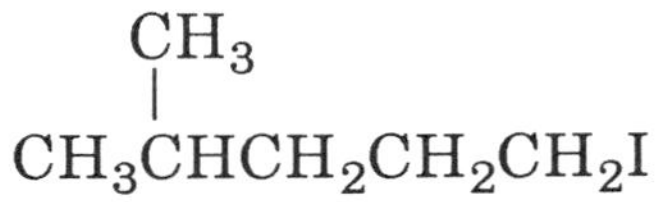
$$\begin{array}{c} \overset{\displaystyle CH_3}{|} \\ CH_3CHCH_2CH_2CH_2I \end{array}$$

(c)
$$\begin{array}{c} CH_3CHCH_2CH_3 \\ \overset{\displaystyle |}{Cl} \end{array}$$

(d) 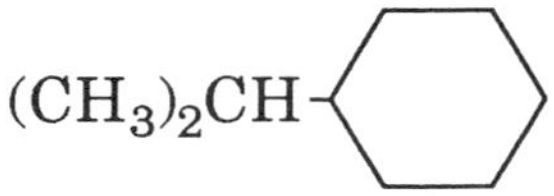 $(CH_3)_2CH-$⬡

1.35   (a) $CH_3CH_2CH_2Cl$

(b) 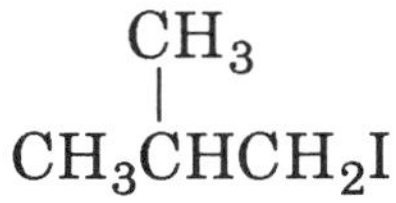
$$\begin{array}{c} \overset{\displaystyle CH_3}{|} \\ CH_3CHCH_2I \end{array}$$

(c)
$$\begin{array}{c} \overset{\displaystyle CH_3}{|} \\ CH_3CBr \\ \overset{\displaystyle |}{CH_3} \end{array}$$

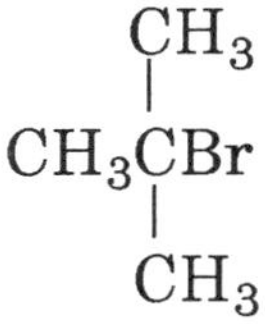

(d) CH_3CH_2Br

1.36   (a)

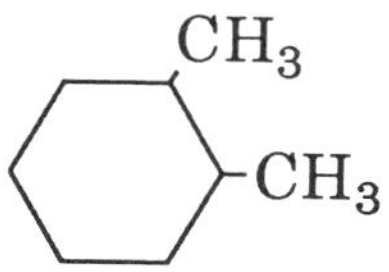

CH$_3$

-CH$_3$

1,2-dimethylcyclohexane

(b)

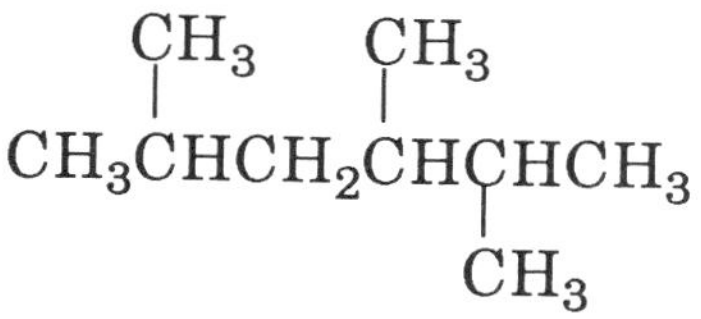

$$CH_3 \quad CH_3$$
$$CH_3CHCH_2CHCHCH_3$$
$$CH_3$$

2,3,5-trimethylhexane

(c) $CH_3CH_2CH_2CH_2Cl$

1-chlorobutane

(d)

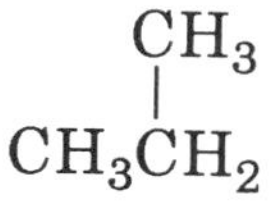

$$CH_3$$
$$CH_3CH_2$$

propane

1.37   (a)

$$CH_3$$
$$CH_3CHCH_2Cl$$

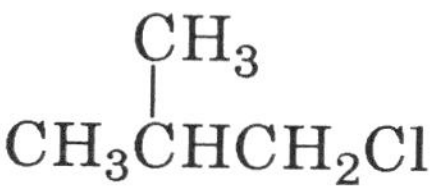

1-chloro-2-methylpropane

(b)

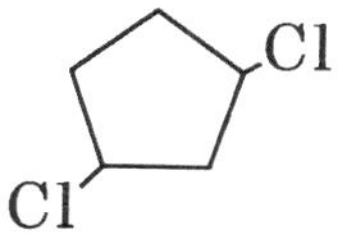

Cl

Cl

1,3-dichlorocyclopentane

(c)

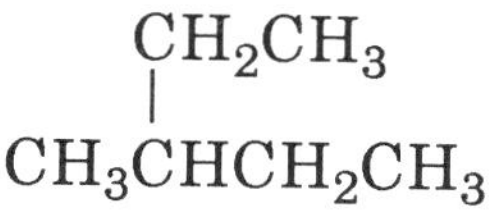

$$CH_2CH_3$$
$$CH_3CHCH_2CH_3$$

3-methylpentane

(d)

CH$_3$

-CH$_3$

CH$_3$

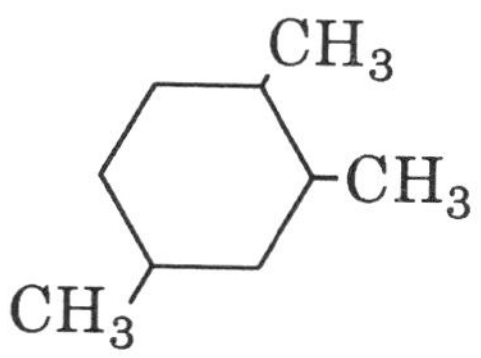

1,2,4-trimethylcyclohexane

1.38 $C_7H_{16} + 11O_2 \rightarrow 7CO_2 + 8H_2O$

1.39 $CH_3CH_2OH + 3O_2 \rightarrow 2CO_2 + 3H_2O$

1.40 CH_3Cl, methyl chloride
CH_2Cl_2, methylene chloride
$CHCl_3$, chloroform
CCl_4, carbon tetrachloride

1.41 CH_3CHCl_2, 1,1-dichloroethane
$ClCH_2CH_2Cl$, 1,2-dichloroethane

1.42

$$CH_3CH_2CHCl_2$$ $$ClCH_2CH_2CH_2Cl$$

1,1-dichloropropane   1,3-dichloropropane

$$CH_3\overset{\displaystyle Cl}{\underset{\displaystyle |}{C}}HCH_2Cl$$ $$CH_3\overset{\displaystyle Cl}{\underset{\displaystyle \underset{\displaystyle Cl}{|}}{\overset{\displaystyle |}{C}}}CH_3$$

1,2-dichloropropane   2,2-dichloropropane

1.43   Coal, oil, and natural gas
1.44   Petroleum is a mixture of liquid (crude oil) and gas.
1.45   The action of increasing pressure as sedimentary rock formations grew in size over the deposits of peat.
1.46   A portion of the liquid material that boils between two temperatures—the range of temperature defining the fraction.
1.47   Alkanes and cycloalkanes having up to 7 carbons per molecule.
1.48   They are able to "crack" large alkanes into those of the C_5 to C_8 range or they take smaller molecules and form them into larger alkanes.
1.49   The atom has one unpaired electron; the molecule has no unpaired electrons.
1.50   (a) Br_2 + UV or heat $\rightarrow$ 2Br· (bromine atoms, showing only the unpaired electron)
    (b)   Br· + $CH_4 \rightarrow$ HBr + CH_3·
           CH_3· + $Br_2 \rightarrow CH_3Br$ + Br·
    (c)   Br· + Br· $\rightarrow Br_2$
           Br· + CH_3· $\rightarrow CH_3Br$
           CH_3· + CH_3· $\rightarrow CH_3CH_3$
1.51   (a) $H_2 + Cl_2 \rightarrow$ 2HCl
    (b) Cl_2 + UV or heat $\rightarrow$ 2Cl·
           Cl· + $H_2 \rightarrow$ HCl + H·
           H· + $Cl_2 \rightarrow$ HCl + Cl·
1.52   A mixture of products (CH_3Cl, CH_2Cl_2, $CHCl_3$, and CCl_4) forms but the *mole* ratios do not bear any constant relationship to the mole ratio of the reactants.
1.53

One, cyclohexyl chloride:

1.54   No reaction with any of the given reactants.
1.55   *t*-Butyl
1.56   11.1 g chlorocyclopentane

Chapter 2

Practice Exercises, Chapter 2

1.
   (a)   2-methylpropene (or 2-methyl-1-propene)
   (b)   4-isobutyl-3,6-dimethyl-3-heptene
   (c)   1-chloropropene (or 1-chloro-1-propene)
   (d)   3-bromopropene (or 3-bromo-1-propene)
   (e)   4-methyl-1-hexene
   (f)   4-methylcyclohexene

2.   (a)

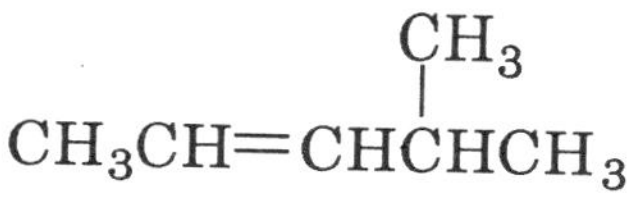

$$CH_3CH{=}CHCHCH_3$$
with CH_3 branch on the fourth carbon

   (b)

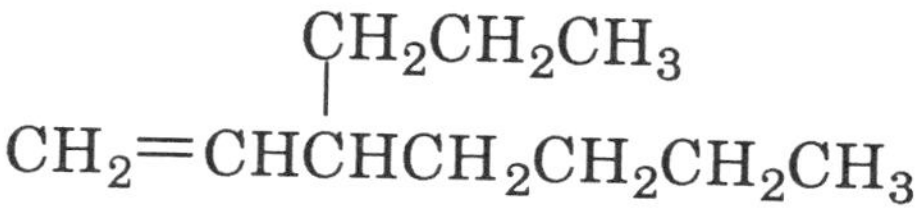

$$CH_2{=}CHCHCH_2CH_2CH_2CH_3$$
with $CH_2CH_2CH_3$ branch

   (c)

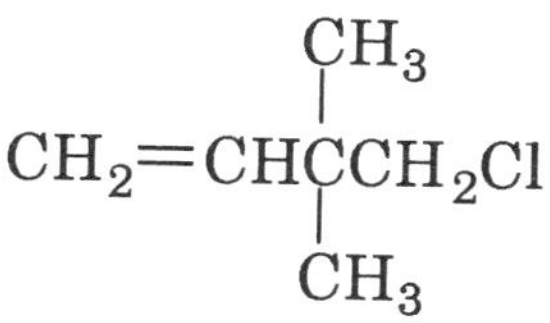

$$CH_2{=}CHCCH_2Cl$$
with CH_3 and CH_3 branches

   (d)

$$CH_3C{=}CCH_3$$
with H_3C and CH_3 branches

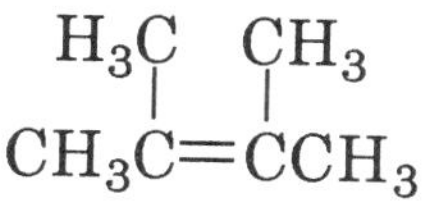

3.   Cis-trans isomerism is possible for a and b. For part (a), cis versus trans is based on the way the main chain passes through the double bond.

(a)

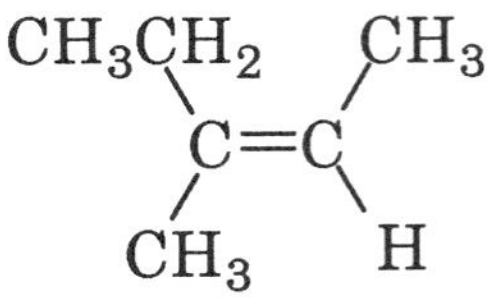

CH_3CH_2 CH_3
$\diagdown$ $\diagup$
$C=C$
$\diagup$ $\diagdown$
CH_3 H

cis isomer

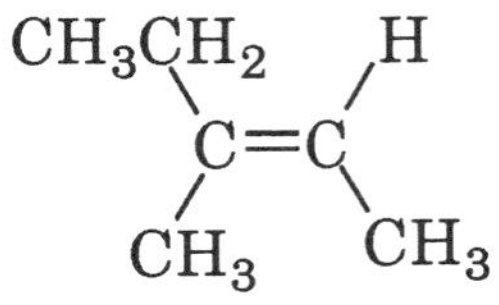

CH_3CH_2 H
$\diagdown$ $\diagup$
$C=C$
$\diagup$ $\diagdown$
CH_3 CH_3

trans isomer

(b)

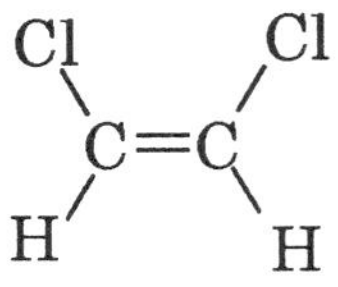

Cl Cl
$\diagdown$ $\diagup$
$C=C$
$\diagup$ $\diagdown$
H H

cis isomer

Cl H
$\diagdown$ $\diagup$
$C=C$
$\diagup$ $\diagdown$
H Cl

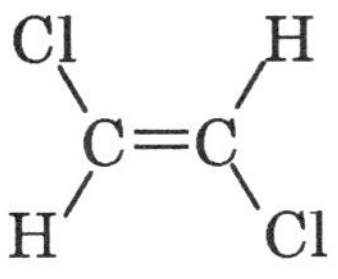

trans isomer

4.   (a) $CH_3CH_2CH_3$
       (b)   No reaction

(c)

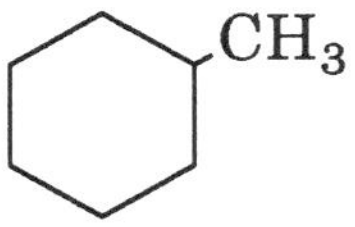

(d) $CH_3(CH_2)_{16}CO_2H$

5.   (a)

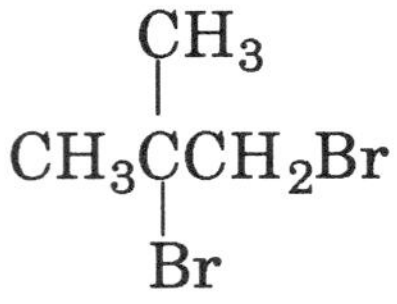

$\qquad\quad CH_3$
$\qquad\quad |$
$\qquad CH_3CCH_2Br$
$\qquad\quad |$
$\qquad\quad Br$

(b)   No reaction (in the absence of heat or UV radiation)

(c) 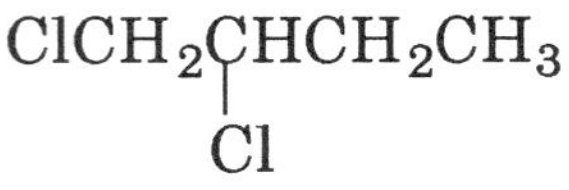 $ClCH_2CHCH_2CH_3$
$\qquad\qquad\quad |$
$\qquad\qquad\quad Cl$

(d) $CH_3CH_2CH_2CH_3$

6. (a)   CH$_3$CHCH$_2$CH$_3$
                |
                Cl

    (b)   CH$_3$
                |
          CH$_3$CCH$_3$
                |
                Br

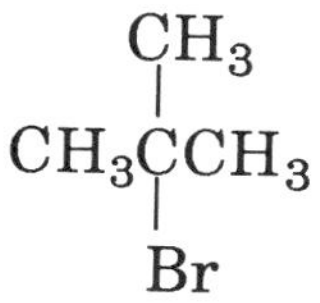

    (c)   CH$_3$
                  |
          CH$_3$CH$_2$C—⬡
                  |
                 OH

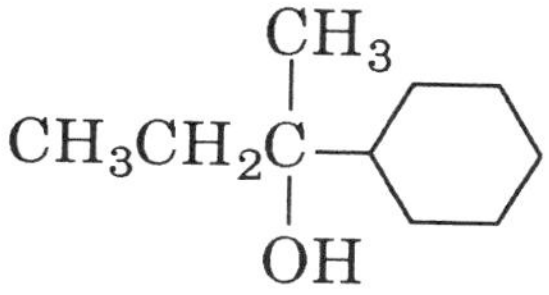

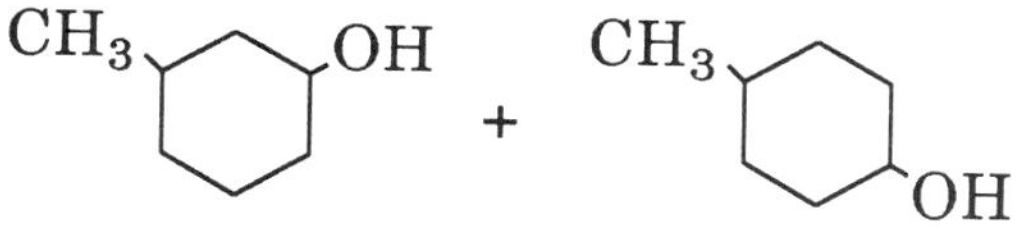

    (d)   CH$_3$⬡OH   +   CH$_3$⬡OH

    (e)   No reaction

7. (a)   CH$_3$CH$_2$CH$_2$$\overset{+}{\text{C}}H_2$ and (CH$_3$CH$_2$$\overset{+}{\text{C}}HCH_3$)   CH$_3CH_2$CHCH$_3$
                                                                                                  |
                                                                                                  Cl

    (b)   CH$_3$ CH$_3$ CH$_3$
                | | |
          CH$_3$CHCH$_2^+$ and   (CH$_3$$\overset{+}{\text{C}}CH_3$)   CH$_3CCH_3$
                                                                              |
                                                                              Cl

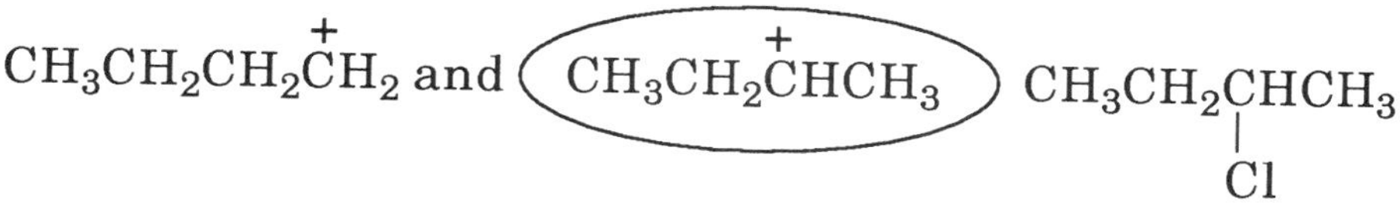

    (c)   ⬡—$\overset{+}{\text{C}}$H$_2$ and   (⬡$^+$—CH$_3$)   ⬡(Cl)(CH$_3$)

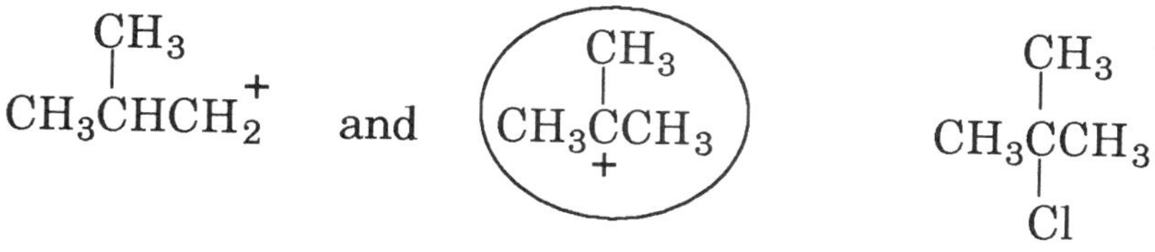

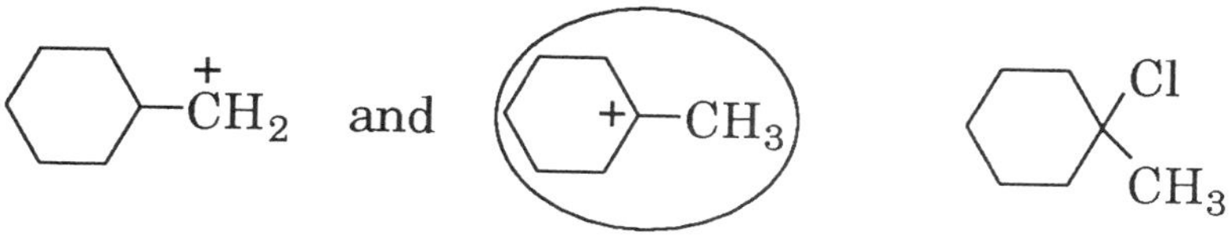

(d)   Only one carbocation is possible:

(e)

CH$_3$CHCH$_2$CH$_2$CH$_3$ CH$_3$CHCH$_2$CH$_2$CH$_3$
             |
             Cl

and

$\overset{+}{\text{CH}_3}$CH$_2$CHCH$_2$CH$_3$ CH$_3$CH$_2$CHCH$_2$CH$_3$
             |
             Cl

(Both carbocations are 2°, so both are equally possible an both form. Thus two isomeric chloropentanes form.)

(f)   CH$_3$CHCH$_2$CH$_2$CH$_3$ and CH$_3$CH$_2$CHCH$_2$CH$_3$
   | |
   OH   OH

Two 2 ° carbocations are can form, so two isomeric pentanols form. Only one carbocation can form from propene, namely the more stable isopropyl carbocatio

Review Exercises, Chapter 2

2.1   (a)   A, B, E
  (b)   C
  (c)   E (Structure C has "-pent-" in "cyclopentane.")
  (d)   All are insoluble in water.
  (e)   B, and C. No, C is not an alkene.
  (f)   A, D, and E. Only D is an alkyne.
  (g)   Not well; the presence of rings or multiple bonds can reduce the ratio of carbon atoms to hydrogen atoms of a hydrocarbon.

2.2   The alkene group is widely present in edible fats and oils and in related compounds that make up every animal cell membrane.

2.3   (a)

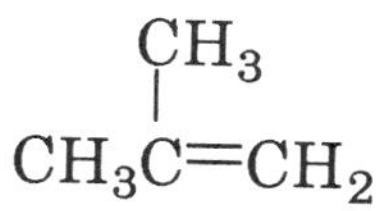

$$\underset{\underset{\text{CH}_3\text{C}=\text{CH}_2}{|}}{\text{CH}_3}$$

(b) $\text{CH}_3\text{CH}{=}\text{CH}_2$

(c)

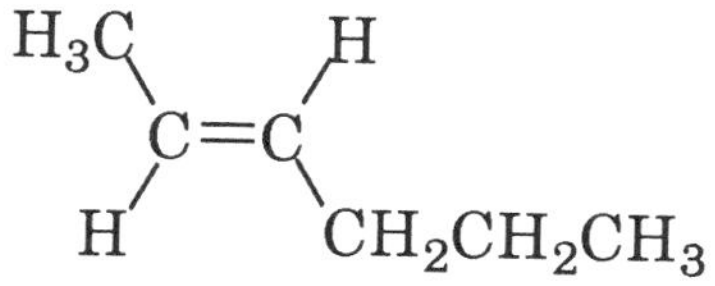

$$\underset{\text{H}}{\overset{\text{H}_3\text{C}}{}}\text{C}{=}\text{C}\underset{\text{CH}_2\text{CH}_2\text{CH}_3}{\overset{\text{H}}{}}$$

(d) 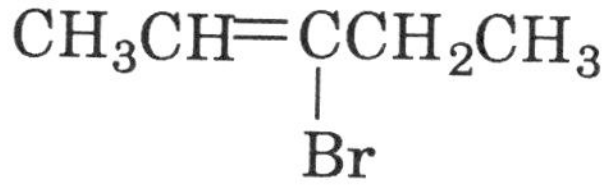 $\underset{\underset{\text{Br}}{|}}{\text{CH}_3\text{CH}{=}\text{CCH}_2\text{CH}_3}$

(e)

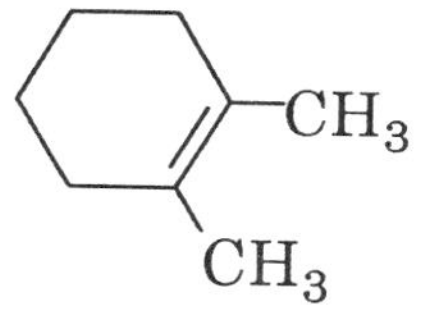

—CH$_3$
CH$_3$

(f)

CH$_3$
CH$_3$

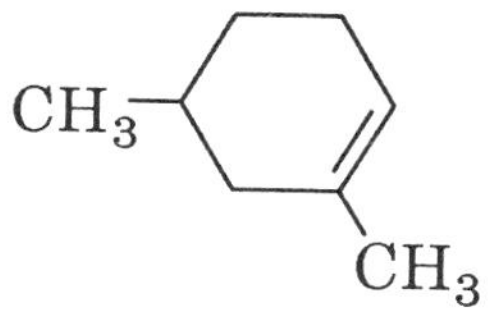

2.4   (a)   1-octene
(b)   1-bromo-3-methyl-1-butene
(c)   4-methyl-2-propyl-1-hexene
(d)   4,4-dimethyl-2-hexene

2.5

$\text{CH}_2{=}\text{CHCH}_2\text{CH}_2\text{CH}_3$ 1-pentene

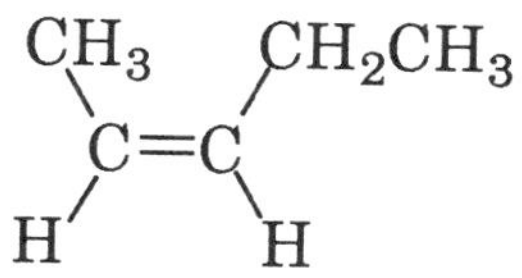

$$\underset{\text{H}}{\overset{\text{CH}_3}{}}\text{C}{=}\text{C}\underset{\text{H}}{\overset{\text{CH}_2\text{CH}_3}{}}$$
cis-2-pentene

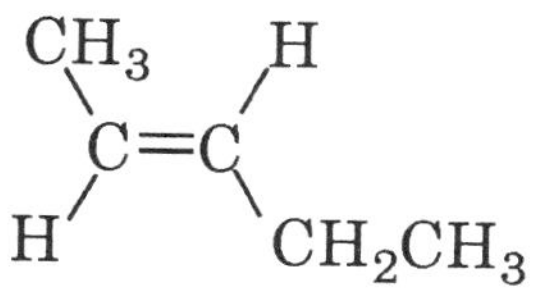

$$\underset{\text{H}}{\overset{\text{CH}_3}{}}\text{C}{=}\text{C}\underset{\text{CH}_2\text{CH}_3}{\overset{\text{H}}{}}$$
trans-2-pentene

$$\underset{\text{CH}_2=\overset{\overset{\displaystyle \text{CH}_3}{|}}{\text{C}}\text{CH}_2\text{CH}_3}{}$$ 2-methyl-1-butene

$$\underset{\text{CH}_2=\text{CH}\overset{\overset{\displaystyle \text{CH}_3}{|}}{\text{C}}\text{H}\text{CH}_3}{}$$ 3-methyl-1-butene

$$\underset{\text{CH}_3\overset{\overset{\displaystyle \text{CH}_3}{|}}{\text{C}}=\text{CHCH}_3}{}$$ 2-methyl-2-butene

2.6

1-methylcyclopentene

3-methylcyclopentene

4-methylcyclopentene

2.7

$HC\equiv CCH_2CH_3$ 1-butyne

$CH_3C\equiv CCH_3$ 2-butyne

2.8

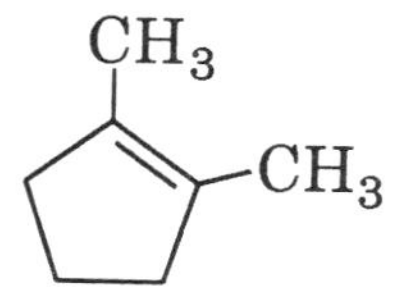

1,2-dimethyl-
cyclopentene

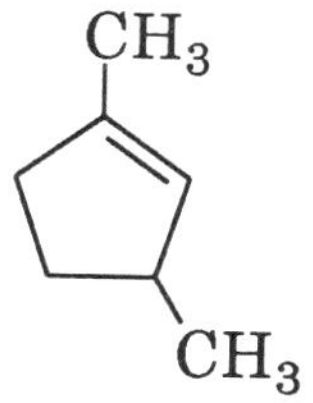

1,3-dimethyl-
cyclopentene

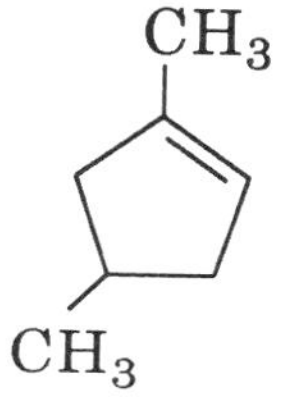

1,4-dimethyl-
cyclopentene

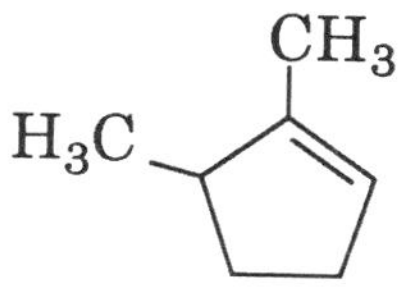

2,3-dimethyl-
cyclopentene

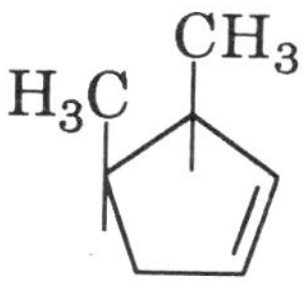

cis-3,4-dimethyl-
cyclopentene

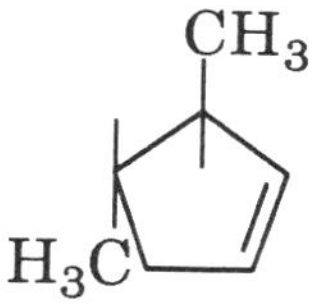

trans-3,4-dimethyl-
cyclopentene

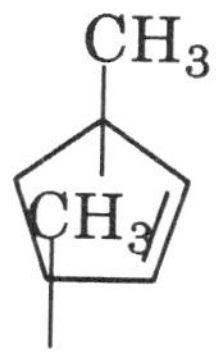

cis-3,5-dimethyl-
cyclopentene

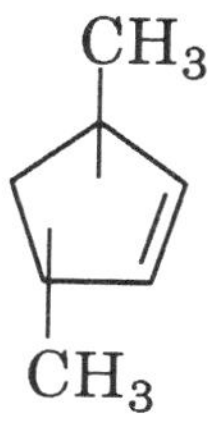

trans-3,5-dimethyl-
cyclopentene

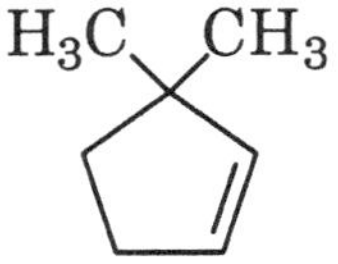

3,3-dimethyl-
cyclopentene

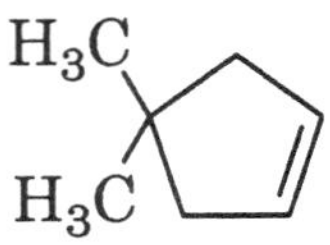

4,4-dimethyl-
cyclopentene

2.9

$CH_2=C=CHCH_2CH_3$ 1,2-pentadiene

$CH_2=CHCH=CHCH_3$ 1,3-pentadiene

$CH_2=CHCH_2CH=CH_2$ 1,4-pentadiene

$CH_3CH=C=CHCH_3$ 2,3-pentadiene

$$\overset{\displaystyle CH_3}{\underset{\displaystyle CH_2=CCH=CH_2}{|}}$$ 2-methyl-1,3-butadiene

2.10   One electron pair is in a sigma bond; the other pair is in a pi bond.
2.11   The sigma bond results from the overlap of two neighboring sp^2 hybrid orbitals.
       The pi bond forms by the overlap of two neighboring p orbitals.
2.12   Free rotation would require that the pi bond break, and this costs too much
       energy. This matters because it gives rise to cis-trans isomers.
2.13   (a)   identical
       (b)   identical
       (c)   isomers
       (d)   identical
       (e)   identical

2.14 (a) 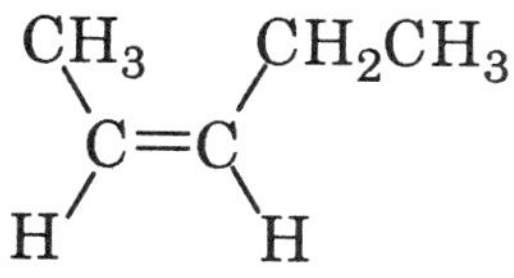 and 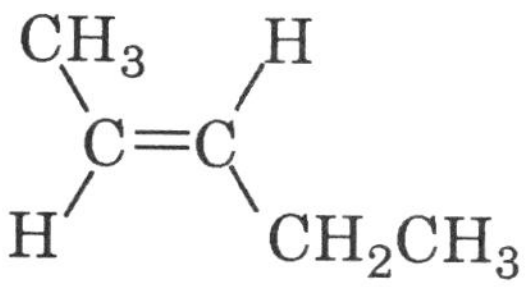

(b) 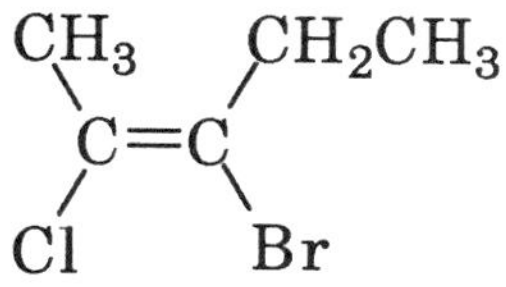 and 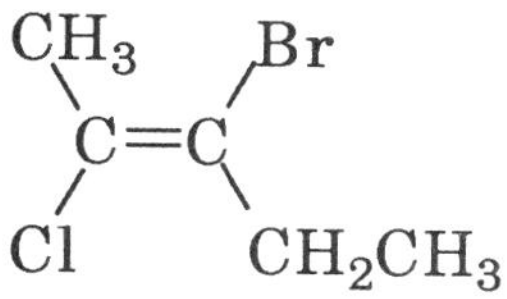

(c) 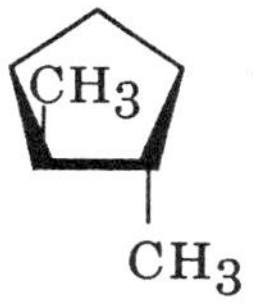 and 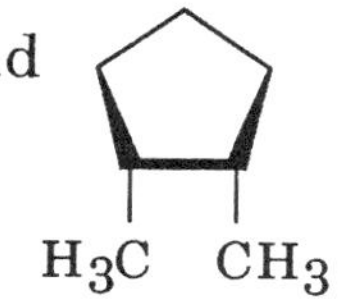

(d) 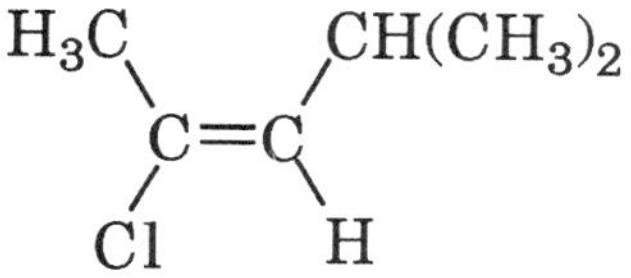 and 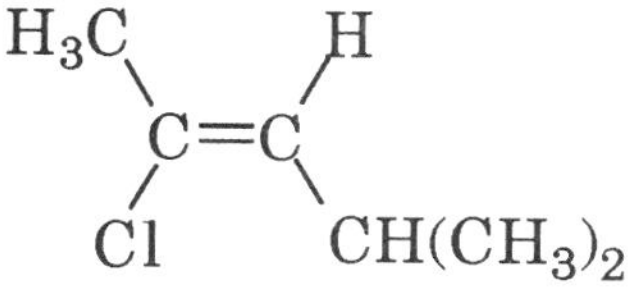

2.15 (a) 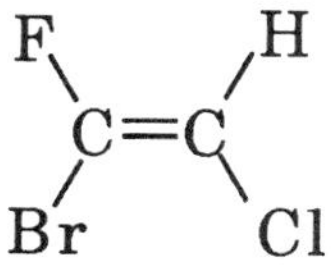 and 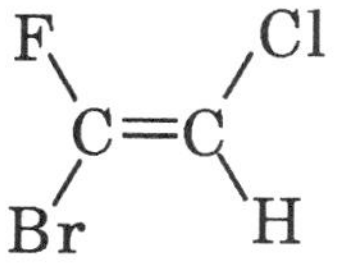

(b) No geometric isomers.

(c) 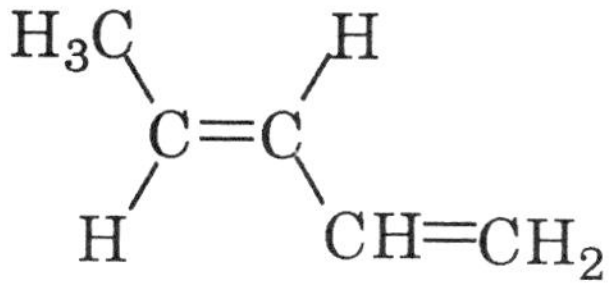 and 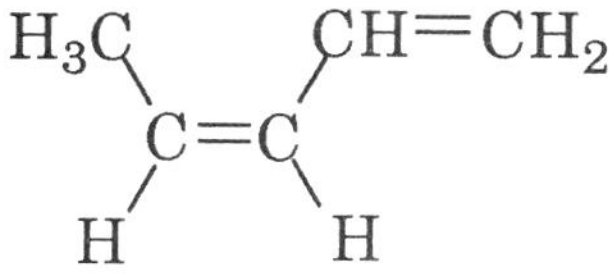

2.16   (a)

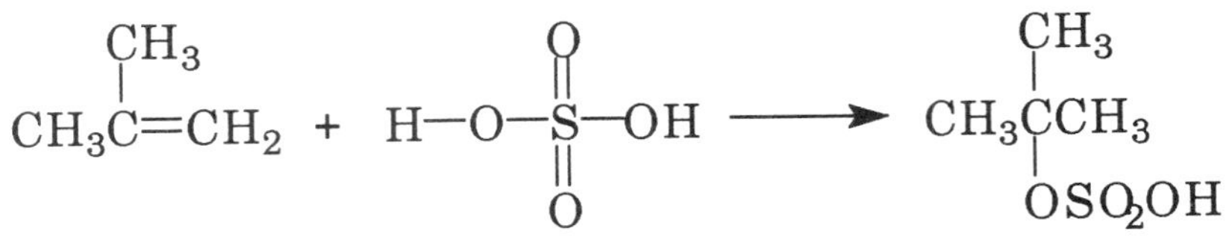

$$CH_3C(CH_3)=CH_2 + H-O-SO_2-OH \longrightarrow CH_3C(CH_3)(OSO_2OH)CH_3$$

(b)

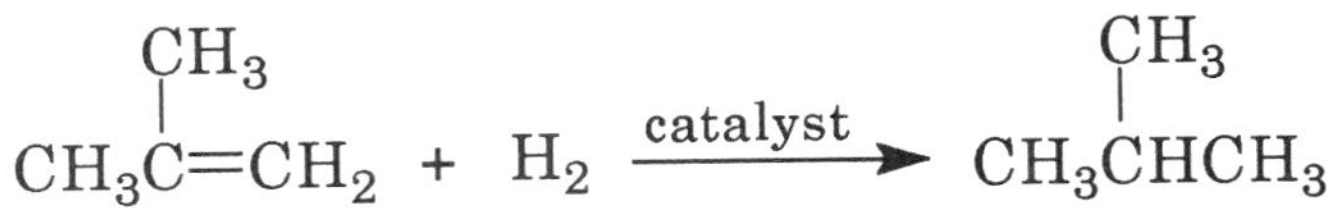

$$CH_3C(CH_3)=CH_2 + H_2 \xrightarrow{\text{catalyst}} CH_3CH(CH_3)CH_3$$

(c)

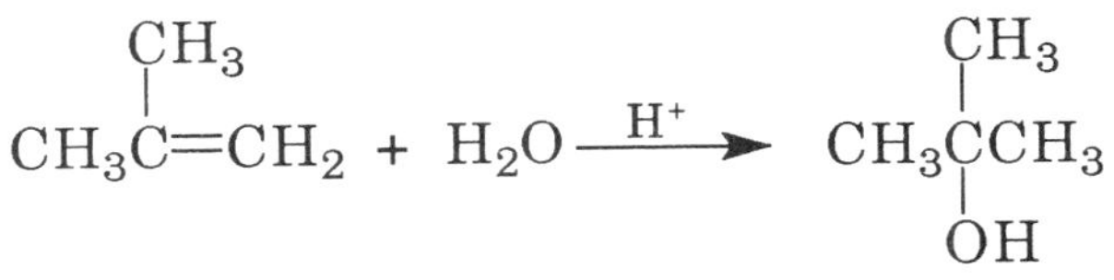

$$CH_3C(CH_3)=CH_2 + H_2O \xrightarrow{\text{H}^+} CH_3C(CH_3)(OH)CH_3$$

(d)

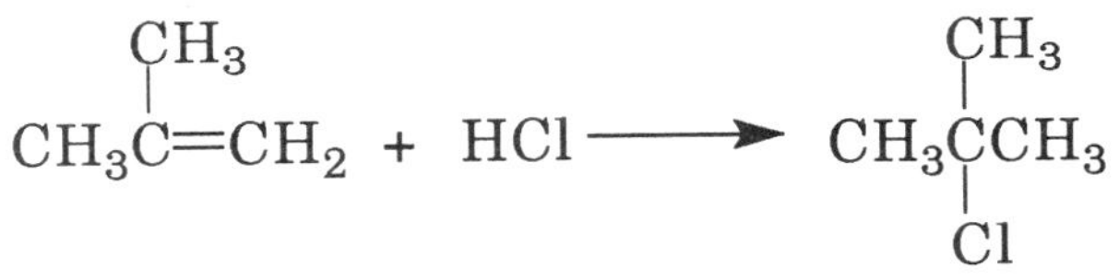

$$CH_3C(CH_3)=CH_2 + HCl \longrightarrow CH_3C(CH_3)(Cl)CH_3$$

(d)

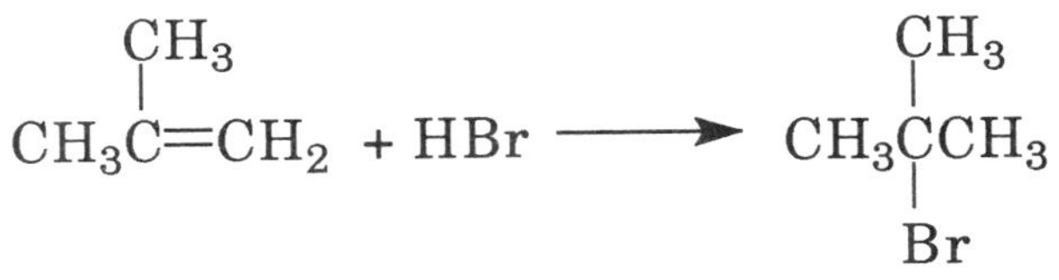

$$CH_3C(CH_3)=CH_2 + HBr \longrightarrow CH_3C(CH_3)(Br)CH_3$$

(e)

$$CH_3C(CH_3)=CH_2 + Br_2 \longrightarrow CH_3C(CH_3)(Br)CH_2Br$$

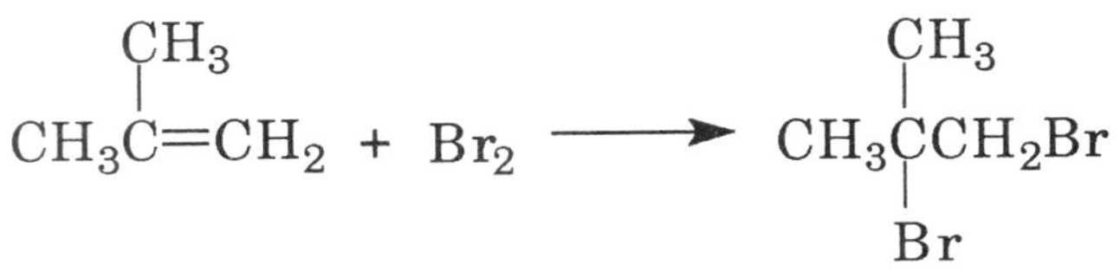

2.17

Compound A could be
either of the following:

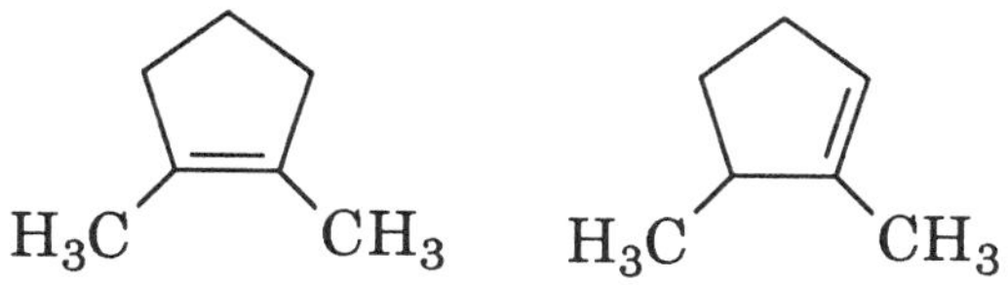

H_3C CH_3 H_3C CH_3

2.18   (a)

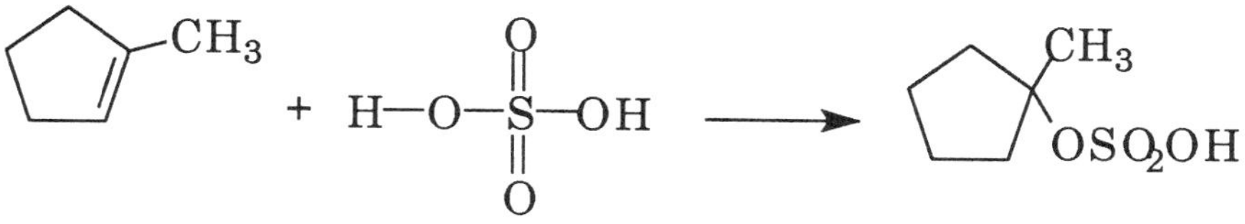

(b)

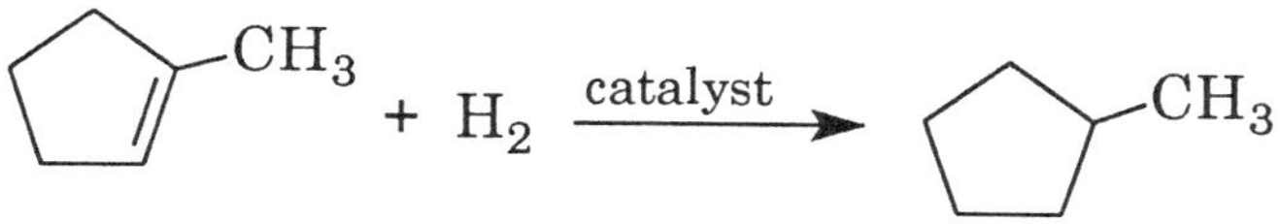

(c)

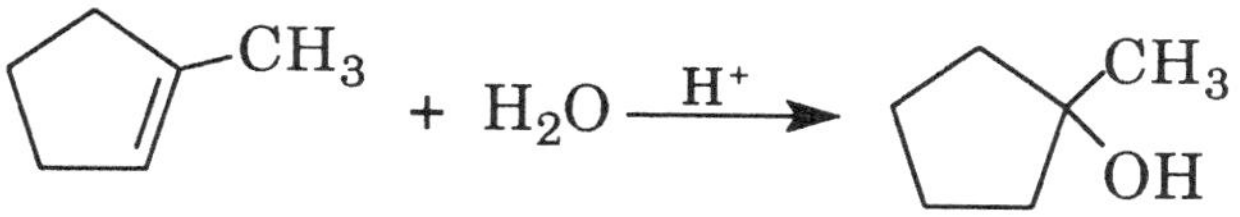

(d)

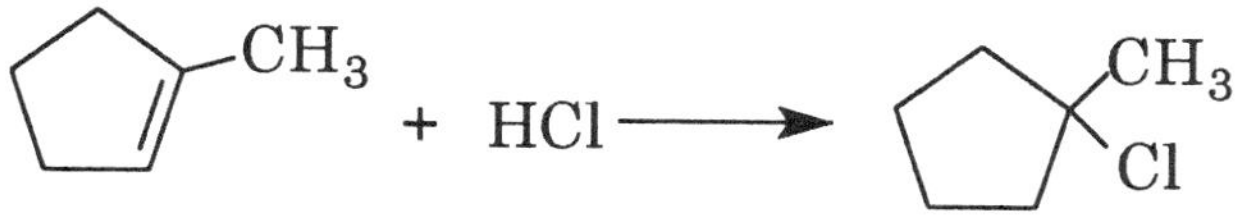

(e)

(e)

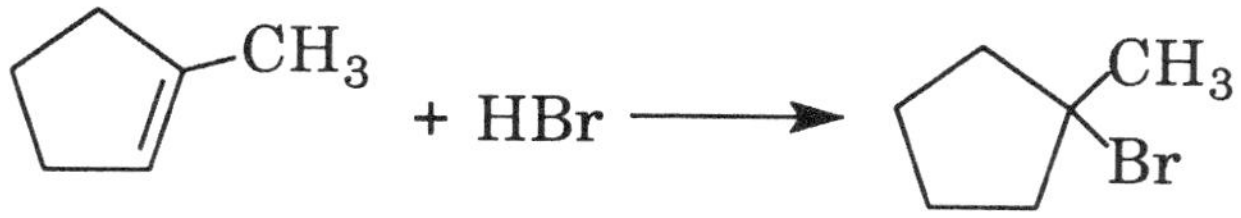

(f)

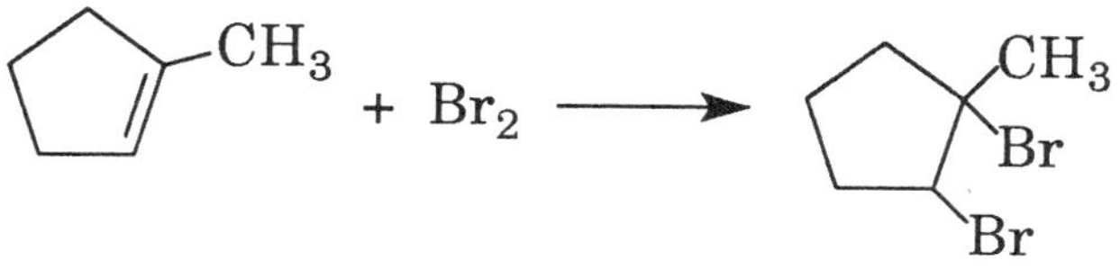

2.19   (a)

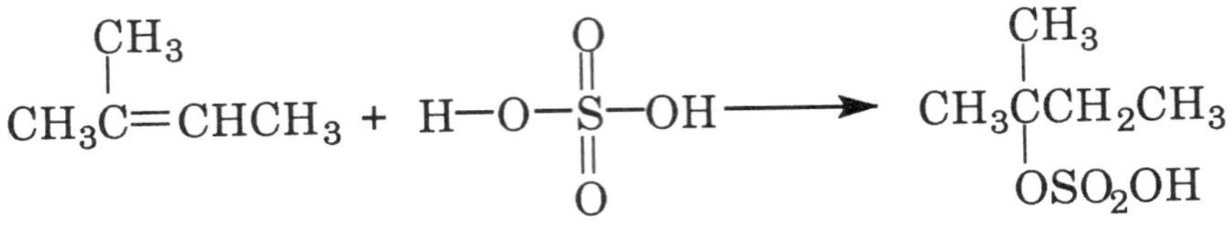

$$CH_3\underset{\underset{CH_3}{|}}{C}=CHCH_3 + H\!-\!O\!-\!\underset{\underset{O}{\|}}{\overset{\overset{O}{\|}}{S}}\!-\!OH \longrightarrow CH_3\underset{\underset{OSO_2OH}{|}}{\overset{\overset{CH_3}{|}}{C}}CH_2CH_3$$

(b)

$$CH_3\underset{\underset{CH_3}{|}}{C}=CHCH_3 + H_2 \xrightarrow{catalyst} CH_3\overset{\overset{CH_3}{|}}{C}HCH_2CH_3$$

(c)

$$CH_3\overset{\overset{CH_3}{|}}{C}=CHCH_3 + H_2O \xrightarrow{H^+} CH_3\underset{\underset{OH}{|}}{\overset{\overset{CH_3}{|}}{C}}CH_2CH_3$$

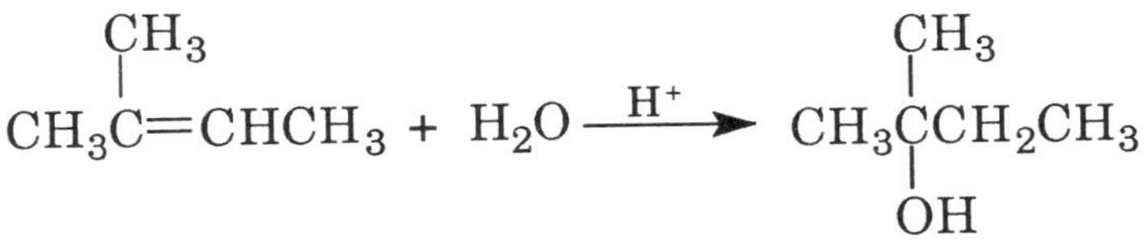

(d)

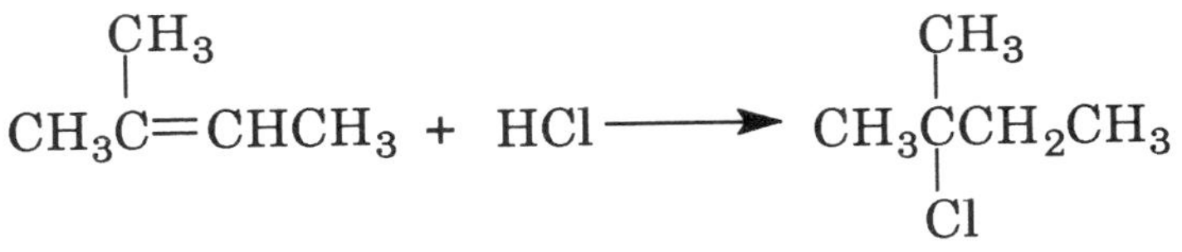

$$CH_3\overset{\overset{CH_3}{|}}{C}=CHCH_3 + HCl \longrightarrow CH_3\underset{\underset{Cl}{|}}{\overset{\overset{CH_3}{|}}{C}}CH_2CH_3$$

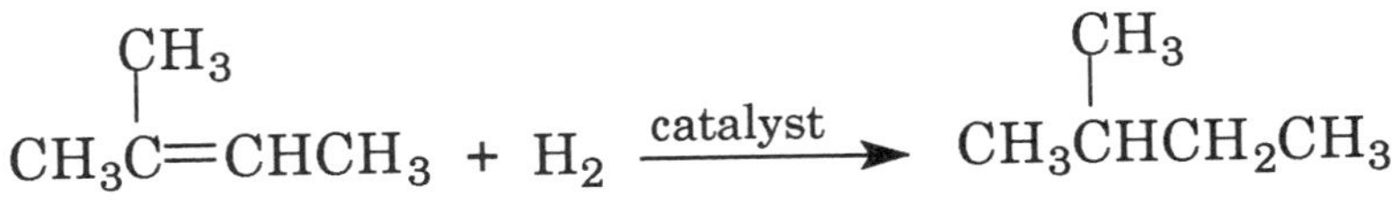

(e)

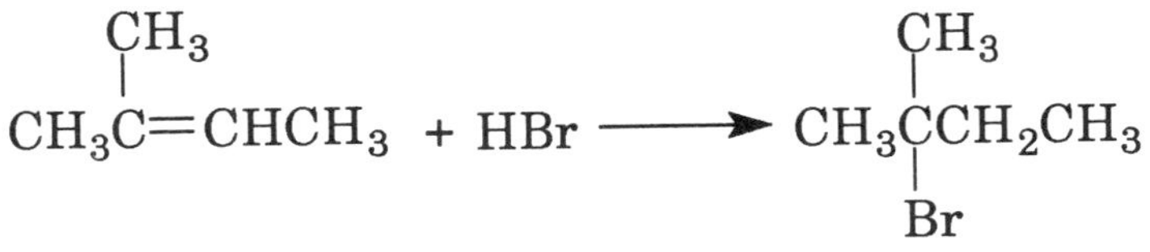

$$CH_3\overset{\overset{CH_3}{|}}{C}=CHCH_3 + HBr \longrightarrow CH_3\underset{\underset{Br}{|}}{\overset{\overset{CH_3}{|}}{C}}CH_2CH_3$$

(f)

$$CH_3\overset{\overset{CH_3}{|}}{C}=CHCH_3 + Br_2 \longrightarrow CH_3\underset{\underset{Br}{|}}{\overset{\overset{CH_3}{|}}{C}}\!-\!\underset{\underset{Br}{|}}{C}HCH_3$$

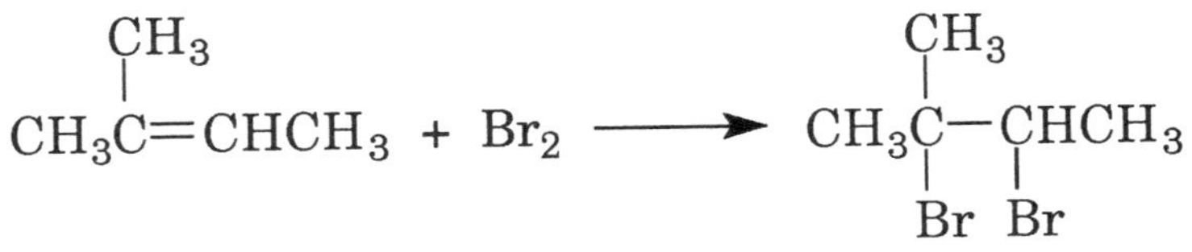

2.20 (a)

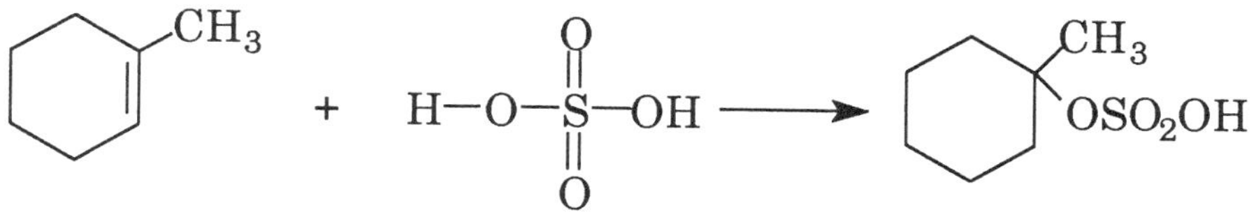

1-methylcyclohexene + H—O—S(=O)(=O)—OH ⟶ 1-methylcyclohexyl (CH₃, OSO₂OH)

(b)

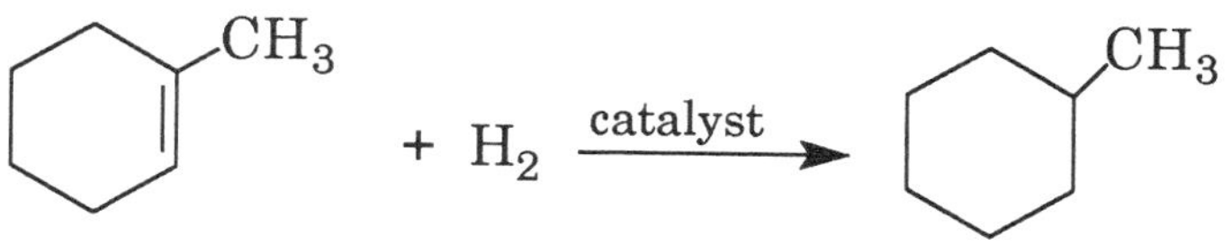

1-methylcyclohexene + H₂ —catalyst→ methylcyclohexane (CH₃)

(c)

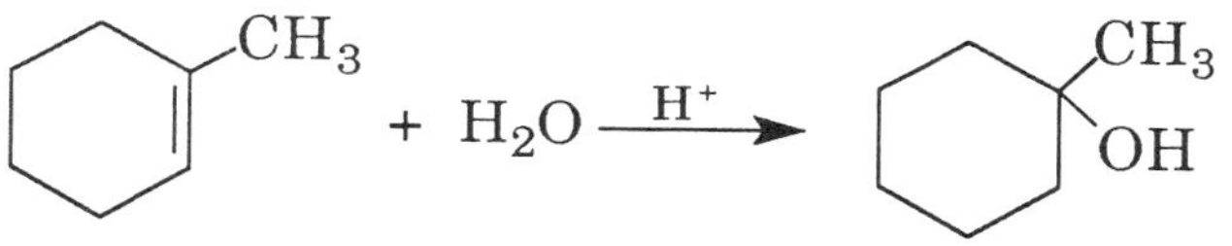

1-methylcyclohexene + H₂O —H⁺→ 1-methylcyclohexanol (CH₃, OH)

(d)

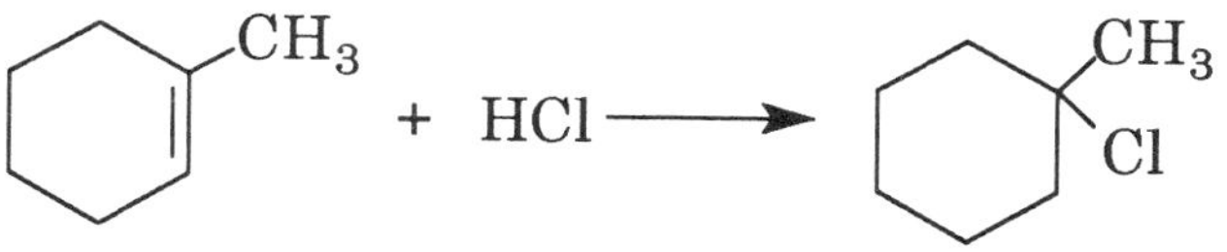

1-methylcyclohexene + HCl ⟶ 1-chloro-1-methylcyclohexane (CH₃, Cl)

(e)

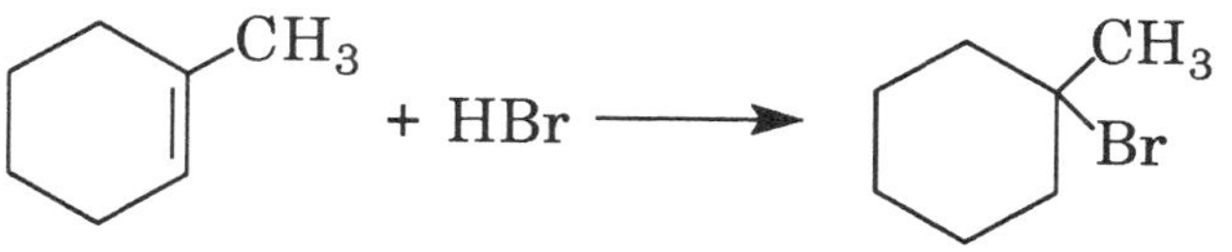

1-methylcyclohexene + HBr ⟶ 1-bromo-1-methylcyclohexane (CH₃, Br)

(f)

1-methylcyclohexene + Br₂ ⟶ 1,2-dibromo-1-methylcyclohexane (CH₃, Br, Br)

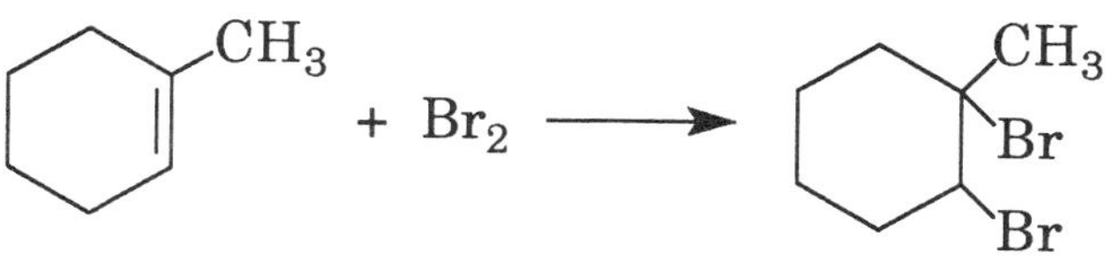

2.21 $CH_2=CH_2 + H_2SO_4 \rightarrow CH_3CH_2OSO_3H$ (ethyl hydrogen sulfate; very polar)

2.22 Cycloalkanes with the formula C_5H_{10} have no alkene groups and so cannot react with the given reactants. For example,

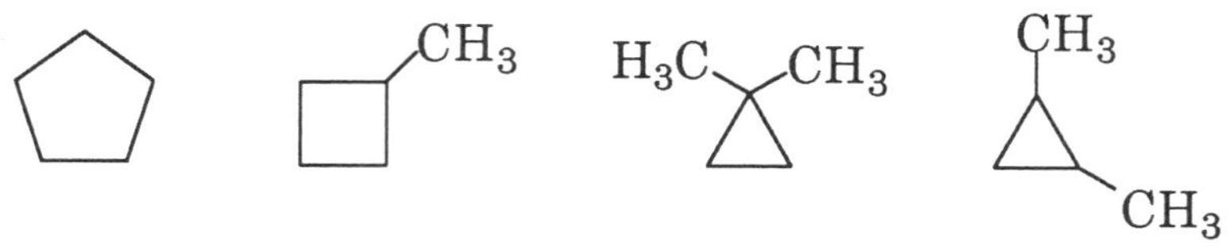

2.23   57.5 g $KMnO_4$

2.24   11.1 g $K_2C_6H_8O_4$

2.25   Of the two possible carbocation intermediates, **A** and **B**, the more stable, 2° carbocation forms (**B**). When Cl^- combines with **B**, the product is 2-chlorobutane.

$$CH_3CH_2CH_2\overset{+}{C}H_2 \qquad\qquad CH_3CH_2\overset{+}{C}HCH_3$$

$$\textbf{A} \qquad\qquad\qquad\qquad \textbf{B}$$

2.26   The two possible carbocation intermediates, **A** and **B**, are both 2° carbocations, so they are equally stable. Some of each forms from 4-methylcyclohexene. When Cl^- ions combine, some react with **A** and some with **B** to give two isomeric products.

2.27   (a)

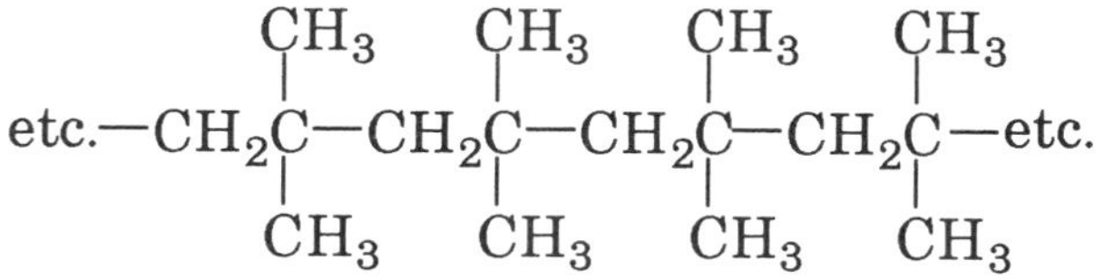

(b)

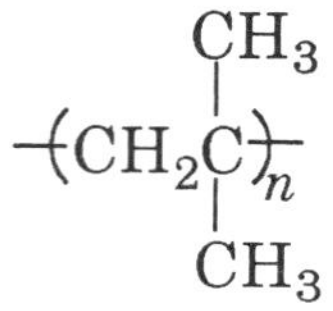

2.28   (a)

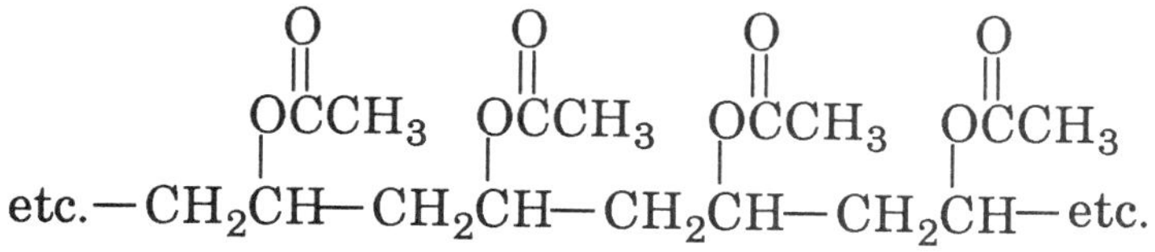

$$etc.-CH_2CH-CH_2CH-CH_2CH-CH_2CH-etc.$$

with $OCCH_3$ (acetate, $\overset{O}{\overset{\|}{OCCH_3}}$) groups on each CH carbon.

(b)

$$-(CH_2CH)_n$$

with an $\overset{O}{\overset{\|}{OCCH_3}}$ group on the CH carbon.

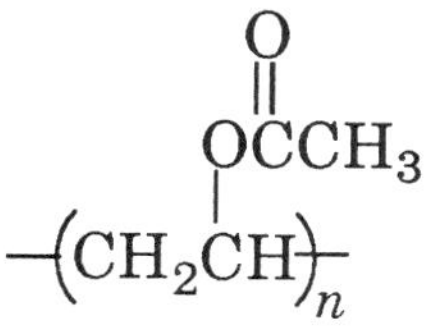

2.29   Some traces of alkenes are also present.

2.30   The traces of alkenes in the fuel are slowly polymerizing to form sticky, low-formula-mass polymers.

2.31   Dipentene has no benzene ring.

2.32   Sulfanilamide has a benzene ring.

2.33   The products of the reactions are the following where C_6H_5 is the phenyl group.
   (a) $C_6H_5SO_2OH$ (or $C_6H_5SO_3H$, benzenesulfonic acid) $+ H_2O$
   (b) $C_6H_5NO_2 + H_2O$
   (c)   No reaction
   (d)   No reaction
   (e)   No reaction (No iron or iron salt catalyst is specified.)
   (f)   No reaction
   (g) $C_6H_5Br + HBr$

2.34   Any monoalkylbenzene, like toluene, $C_6H_5CH_3$, or ethylbenzene, $C_6H_5CH_2CH_3$

2.35

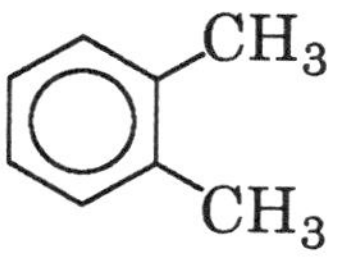

2.36   Your discussion should call attention to the following aspects
   (a)   Between each carbon of the ring there is a sigma bond made from the overlap of sp^2 hybrid orbitals on adjacent carbons.
   (b)   Each C–H bond are made by the overlap of a carbon sp^2 hybrid orbital with an s orbital of H.

(c)   The unhybridized p orbitals of each carbon have axes that are coparallel with each other and perpendicular to the plane of the ring. These p orbitals overlap side to side to form the pi electron network of the benzene ring. Your figure should have the principal features of Figure 2.9.

2.37   With more space in which to spread out, the pi electrons are in a more stable arrangement.

2.38   An addition reaction would break up circular the pi electron network, which costs more energy than it takes to bring about a substitution reaction.

2.39   (a) $C_6H_5CH_3$ (b) $C_6H_5NH_2$ (c) C_6H_5OH (d) $C_6H_5CO_2H$
        (e) C_6H_5CHO (f) $C_6H_5NO_2$

2.40   (a)   (b)   (c)   (d)

2.41   The have alkene groups that are attacked by the oxygen in air, particularly when heated.

2.42   Vegetable oils have more alkene groups per molecule.

2.43   No

2.44   The energy of the photon causes the disengagement of the side-to-side overlap of the p orbitals used to make the pi bond. In this excited state, free rotation can occur with subsequent re-overlapping of the p orbitals.

2.45   Rods enable the seeing of shades of gray, which characterizes very dim light. The owl cannot see in total blackness.

2.46   Oxygen and nitrogen react: $N_2(g) + O_2(g) \rightarrow 2NO(g)$

2.47   The NO in fresh exhaust reacts with the (cooler) O_2 in air.
$$2NO(g) + O_2(g) \rightarrow 2NO_2(g)$$

2.48   Solar radiation provides the energy to decompose some NO_2 into NO and oxygen atoms, O.

$$NO_2(g) \xrightarrow{\text{UV radiation}} NO(g) + O(g)$$

At the surface of a neutral particle, M, oxygen atoms combine with O_2 to give ozone.
$$O(g) + O_2(g) + M \rightarrow O_3(g) + M$$

2.49   It attacks lung tissue.

2.50   NO can destroy ozone ($O_3 + NO \rightarrow NO_2 + O_2$), so anything that destroys NO leads to increasing levels of O_3. Peroxy radicals (ROO) that form from unburned hydrocarbons in smog can destroy NO:

$$ROO + NO \rightarrow RO + NO_2$$

2.51   (a) $CH_3CH=CHCH_3$

       (b) $CH_3\underset{\underset{Cl}{|}}{C}=\underset{\underset{Cl}{|}}{C}CH_3$

       (c) $CH_3\underset{\underset{Br}{|}}{C}=\underset{\underset{Br}{|}}{C}CH_3$

2.52   (a) $CH_3CH_2CH_2CH_3$

       (b) $CH_3\overset{\overset{Cl}{|}}{\underset{\underset{Cl}{|}}{C}}-\overset{\overset{Cl}{|}}{\underset{\underset{Cl}{|}}{C}}CH_3$

       (c) $CH_3\overset{\overset{Br}{|}}{\underset{\underset{Br}{|}}{C}}-\overset{\overset{Br}{|}}{\underset{\underset{Br}{|}}{C}}CH_3$

2.53   The addition would disrupt the pi electron network of the benzene ring, which costs too much energy.

2.54   Because the catalyst aids in generating the reactive species, Cl^+, from Cl_2, not from the catalyst itself.

2.55 $FeBr_3$ is able to react with Br_2 as follows to give $Br^+FeBr_4^-$.

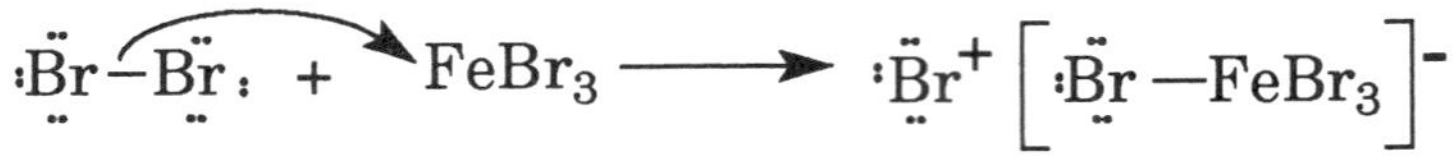

$$\ddot{B}r-\ddot{B}r: \; + \; FeBr_3 \longrightarrow \; :\ddot{B}r^+ \left[:\ddot{B}r-FeBr_3 \right]^-$$

Br^+ then attacks the ring, taking a pair of electrons from the ring's pi electron network and forming a C–Br bond.

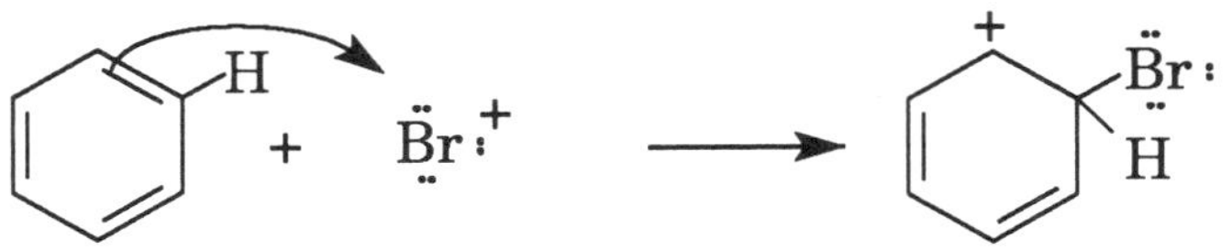

The pair of electrons is restored to the ring as H^+ transfers away, HBr forms, and the catalyst is recovered.

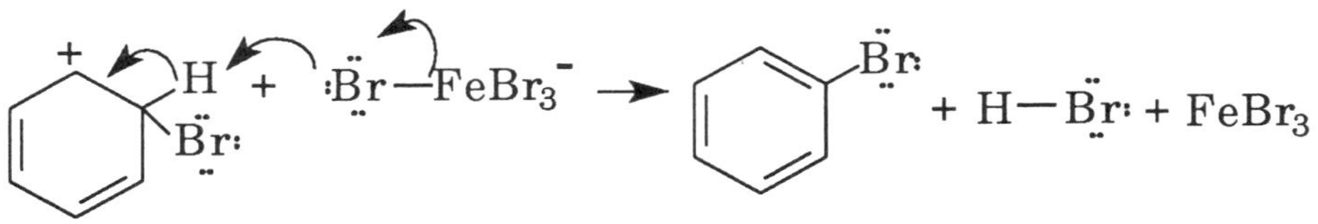

Bromobenzene

2.56   (a)   CH₃CH₂CH₂CHCH₂CH₃
                               |
                              OH

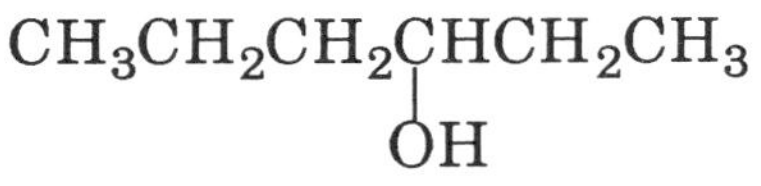

(b)   CH₃
             |
      CH₃CHCH₂CH₃

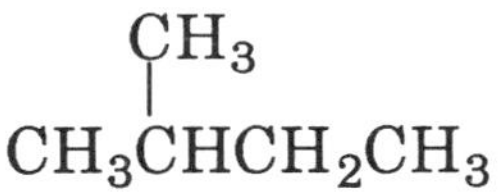

(c)   C₆H₅Br + HBr

(d)   No reaction

(e)

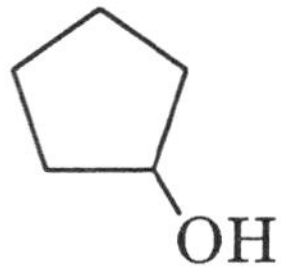

(f) $CH_3CH_2CH_2CH_2CH_3$

(g)   No reaction

(h) 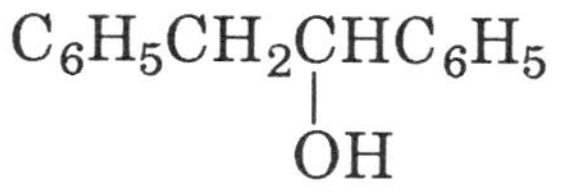 $C_6H_5CH_2CHC_6H_5$
              |
             OH

(i)   No reaction

(j) $C_5H_{12} + 8O_2 \rightarrow 5CO_2 + 6H_2O$

(k)   No reaction

2.57   (a) $CH_3CH_2CH_2CH_2CH_2CH_3$

(b)

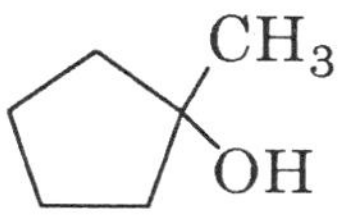

(c) 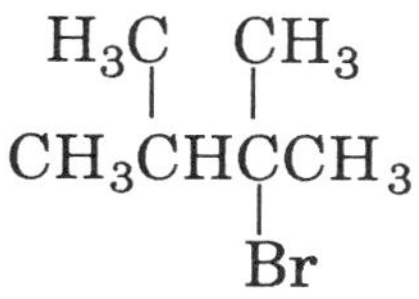 H_3C CH_3
              | |
         CH_3CHCCH_3
                  |
                 Br

(d)   No reaction

(e)

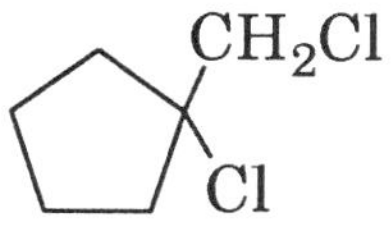

(f) $2C_8H_{10} + 21O_2 \rightarrow 16CO_2 + 10H_2O$

(g) $C_6H_5Cl + HCl$

(h)   No reaction

(i) 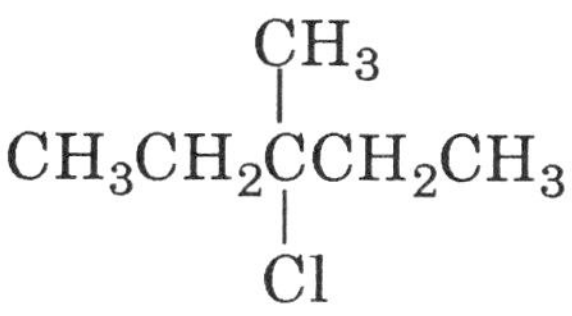 CH_3
                    |
         $CH_3CH_2CCH_2CH_3$
                    |
                   Cl

(j)   No reaction

(j) $CH_3CHCHCH_2CHCH_2Cl$
    | | |
    Cl Cl   Cl

(k) $C_6H_5CHCHCH_3$
    | |
    Br Br

Chapter 3

Practice Exercises, Chapter 3

1. (a)   Monohydric, secondary
   (b)   Monohydric, secondary
   (c)   Dihydric, unstable (two OH groups on the same carbon)
   (d)   Dihydric, both are secondary
   (e)   Monohydric, primary
   (f)   Monohydric, primary
   (g)   Monohydric, tertiary
   (h)   Monohydric, secondary
   (i)   Trihydric, unstable (three OH groups on the same carbon)

2. (a)   Alcohol   (b) Phenol
   (c)   Carboxylic acid   (d) Alcohol (but unstable; an enol)
   (e)   Alcohol   (f) Alcohol

3. (a)   4-Methyl-1-pentanol
   (b)   2-Methyl-2-propanol
   (c)   2-Ethyl-2-methyl-1-pentanol
   (d)   2-Methyl-1,3-propanediol

4. In 1,2-propanediol. Its boiling point is 189 °C, much higher than that of 1-butanol (b.p. 117 °C). 1,2-Propanediol is more soluble in water.

5. (a) $CH_3CH=CH_2$ (b) $CH_3CH=CH_2$

   (b) CH_3 (c)
             |
       $CH_2=CCH_3$

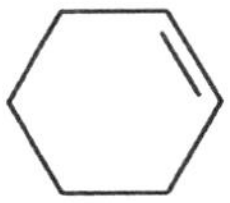

6.

(a) $\overset{CH_3}{\underset{|}{CH_3CHCH{=}O}}$ then $\overset{CH_3}{\underset{|}{CH_3CHCO_2H}}$

(b) $C_6H_5CH{=}O$ then $C_6H_5CO_2H$

7.

(a) $\overset{O}{\overset{\|}{CH_3CCH_2CH_3}}$ (b) $\overset{O}{\overset{\|}{C_6H_5CCH_3}}$ (c)

8.

(a) $\overset{O}{\overset{\|}{CH_3CCH_2CHO}}$ and $\overset{O}{\overset{\|}{CH_3CCH_2CO_2H}}$

(b) No reaction

(c) $\overset{CH_3}{\underset{\underset{CH_3}{|}}{CH_3CCH{=}O}}$ and $\overset{CH_3}{\underset{\underset{CH_3}{|}}{CH_3CCO_2H}}$

(d) $\overset{O}{\underset{\underset{CH_3}{|}}{CH_3CHCCH_3}}$

9. (a) CH_3OCH_3 (b) $CH_3CH_2CH_2OCH_2CH_2CH_3$

(c)

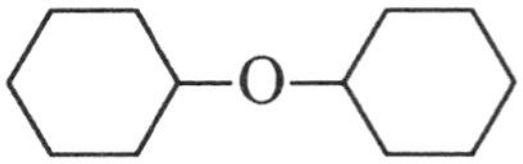

10.

(a) $2\,CH_3SH$

(b)

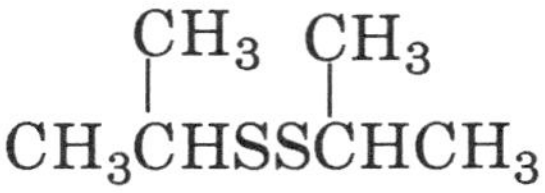

$$\underset{CH_3CHSSCHCH_3}{\overset{CH_3 \quad CH_3}{|\qquad\quad|}}$$

(c)

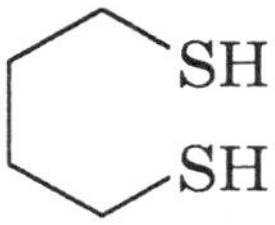

(d)

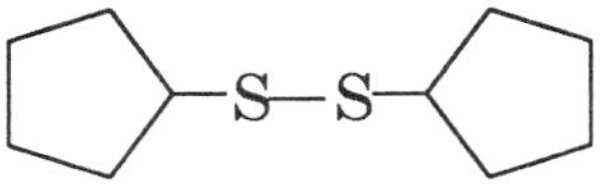

Review Exercises, Chapter 3

3.1   1, 1°alcohol; 2, ketone; 3, 2°alcohol; 4, ketone; 5, alkene
3.2   1, ketone; 2, carboxylic acid; 3, 2°alcohol; 4, alkene; 5, 2°alcohol
3.3

(a)

$$\underset{CH_3CHCH_2OH}{\overset{CH_3}{|}}$$

(b)

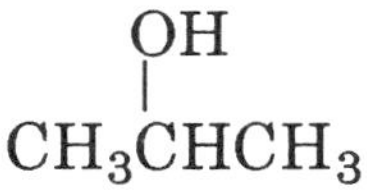

$$\underset{CH_3CHCH_3}{\overset{OH}{|}}$$

(c) $CH_3CH_2CH_2OH$

(d) 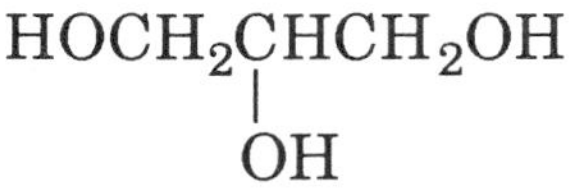 $\underset{\qquad\;\;OH}{\overset{}{HOCH_2CHCH_2OH}}$

3.4

(a) CH_3OH

(b)

$$\underset{CH_3}{\overset{CH_3}{\underset{|}{\overset{|}{CH_3COH}}}}$$

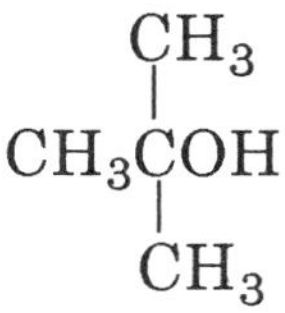

(c) CH_3CH_2OH

(d) $CH_3CH_2CH_2CH_2OH$

3.5   (a)   Propyl alcohol   (b)   Butyl alcohol
        (c)   *t*-Butyl alcohol   (d)   Isobutyl alcohol
3.6   HOCH$_2$CH$_2$OH, 1,2-ethanediol. (Remember, no carbon may hold two
        or more OH groups.)
3.7   1,2,3-Propanetriol

$$HOCH_2\underset{\underset{\displaystyle OH}{|}}{C}HCH_2OH$$

3.8   (a)   2-Methyl-1-propanol   (b) 2-Propanol
        (c)   1-Propanol   (d) 1,2,3-Propanetriol
3.9   (a)   Methanol   (b) 2-Methyl-2-propanol
        (c)   Ethanol   (d) 1-Butanol
3.10   (a)   1-Propanol   (b) 1-Butanol
        (c)   2-Methyl-2-propanol   (d) 2-Methyl-1-propanol
3.11   2-Ethyl-1-pentanol
3.12   Hydrogen bonds of the following type form.

3.13

B < D < A < C

3.14

(a)
$$CH_3\underset{|}{\overset{|}{C}}H_3C\overset{CH_3}{\underset{}{C}}{=}CH_2$$

(b) CH$_3$CH=CHCH$_3$ + CH$_2$=CHCH$_2$CH$_3$
        mostly   some

(c)

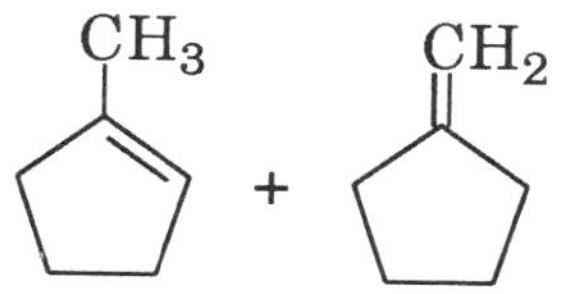

     mostly   some

(d) C$_6$H$_5$CH=C(CH$_3$)$_2$

(e)
$$CH_3CH_2CH{=}\overset{CH_3}{\underset{}{C}}CH_3$$
        mostly

$$+\ CH_3CH_2CH_2\overset{CH_3}{\underset{}{C}}{=}CH_2$$
        some

(f) 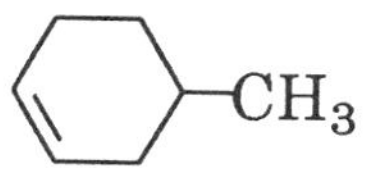
—CH$_3$

3.15

(a)

$$CH_3\overset{\overset{\displaystyle O}{\|}}{C}\underset{\underset{\displaystyle CH_3}{|}}{H}\overset{}{C}H \qquad CH_3\underset{\underset{\displaystyle CH_3}{|}}{C}H\overset{\overset{\displaystyle O}{\|}}{C}OH$$

(b)

$$CH_3\overset{\overset{\displaystyle O}{\|}}{C}CH_2CH_3$$

(c)   No reaction

(d)

$$C_6H_5\overset{\overset{\displaystyle O}{\|}}{C}CH(CH_3)_2$$

(e)   No reaction

(f)

$O{=}\!\!\bigcirc\!\!{-}CH_3$

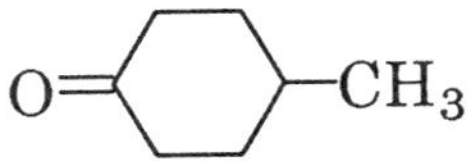

3.16

(a) $HOCH_2CH_2CH_3$ or

$$CH_3\underset{\underset{\displaystyle OH}{|}}{C}HCH_3$$

(b) $-OH$

(c) $\underset{\underset{\displaystyle \quad}{}}{CH_3}\overset{\overset{\displaystyle CH_2OH}{|}}{C}HCH_3$ or $CH_3\underset{\underset{\displaystyle OH}{|}}{\overset{\overset{\displaystyle CH_3}{|}}{C}}CH_3$

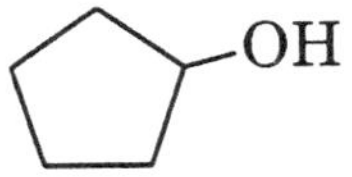

(d) $\underset{CH_3}{OH}$ or $\overset{OH}{}{-}CH_3$

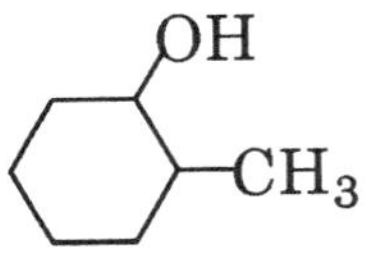

3.17

(a) $HOCH_2CH_2CH_2CH_3$

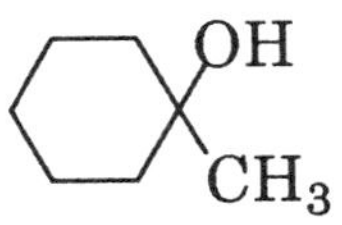

(b)

$$CH_3CH_2CH_2\underset{\underset{\displaystyle OH}{|}}{C}HCH(CH_3)_2$$

(c) $HOCH_2{-}$

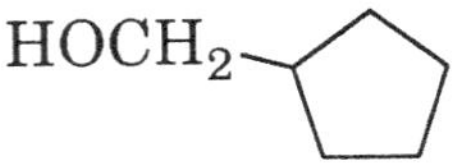

(d) $C_6H_5CH_2OH$

3.18

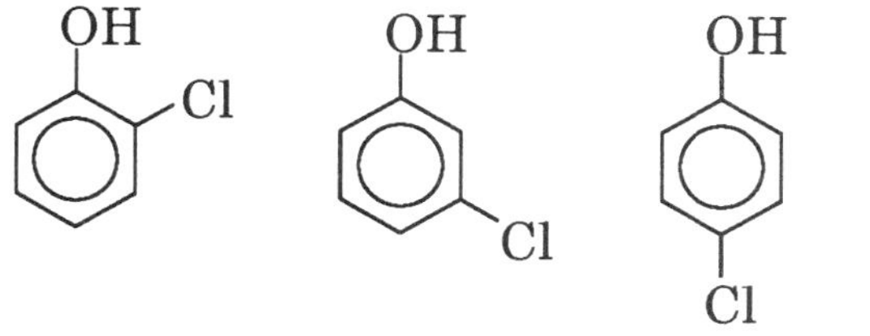

o-chloro-   m-chloro-   p-chloro-
phenol   phenol   phenol

3.19   **A** reacts with aqueous NaOH; **B** does not. **B** can be dehydrated to an alkene;
       A cannot. (Both react with oxidizing agents, but in different ways.)

3.20   Compound **A**. It is a phenol that changes to a water-soluble salt:.

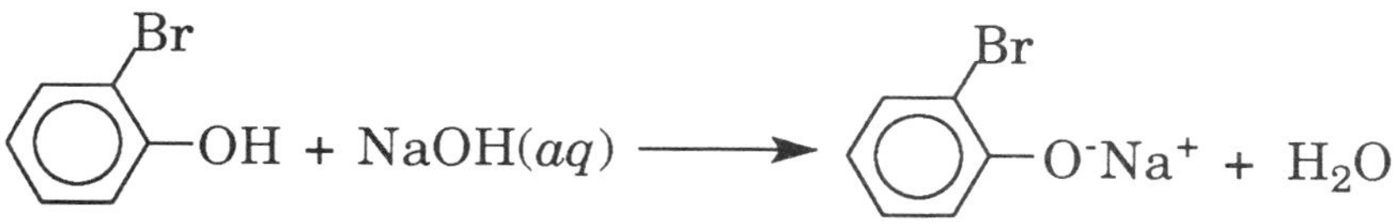

3.21

(a) $CH_3CH_2OCH_2CH_3$ (b)

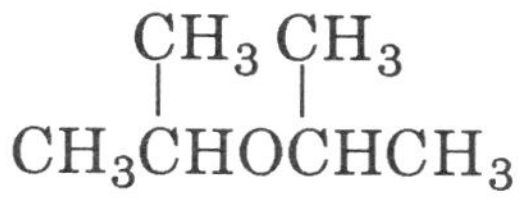

(c) $CH_3OCH_2CH_2OCH_2CH_2OCH_3$ (d)

3.22

(a) $CH_3CH_2CH_2OH$ (b)

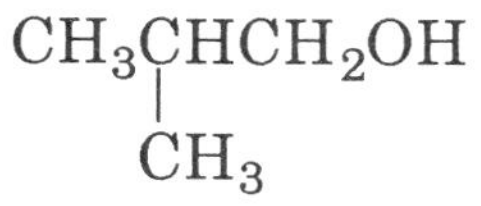

(c)  (d)

3.23 $CH_3OCH_3 + CH_3OCH_2CH_3 + CH_3CH_2OCH_2CH_3$

3.24   No reaction occurs. Ethers are stable in base.

3.25

(a) $CH_3CH_2SSCH_2CH_3$ (b) $HSCH_2\underset{\underset{SH}{|}}{C}HCH_3$

(a) $CH_3CH_2CH_2SSCH_2CH_2CH_3$ (b) $(CH_3)_2CHCH_2SH$

3.26

(c)   (d) $(CH_3)_2CHSSCH(CH_3)_2$

3.27   Hydrogen bonding in the thioalcohol family does exist, but the hydrogen bonds are not as strong as those in the alcohol family.

3.28   (a)   Ethanol
          (b)   Ethanol, 2-propanol
          (c)   1,2,3-Propanetriol (glycerol)
          (d)   Methanol
          (e)   1,2-Ethanediol and 1,2-propanediol
          (f)   1,2,3-Propanetriol (glycerol)

3.29   (a)   1,2,3-Propanetriol (glycerol)
          (b)   Ethanol
          (c)   Methanol

3.30 $C_5H_9OH + H_2SO_4 \rightleftharpoons C_5H_9OH_2^+ + HSO_4^-$

3.31

3.32   It must lose H⁺. The most abundant and so the most likely proton acceptor is a molecule of cyclopentanol.

3.33   Carbocations have a carbon atom lacking an outer octet, but the sodium ion has such an octet.

3.34   In dilute sulfuric acid, the likeliest proton donor is H_3O^+.

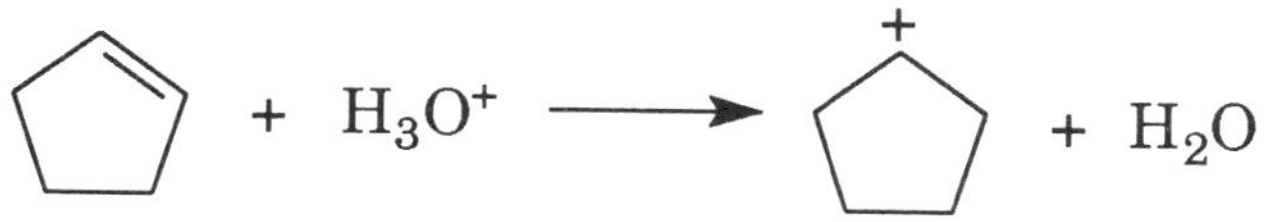

The carbocation reacts with water to give the protonated form of the alcohol.

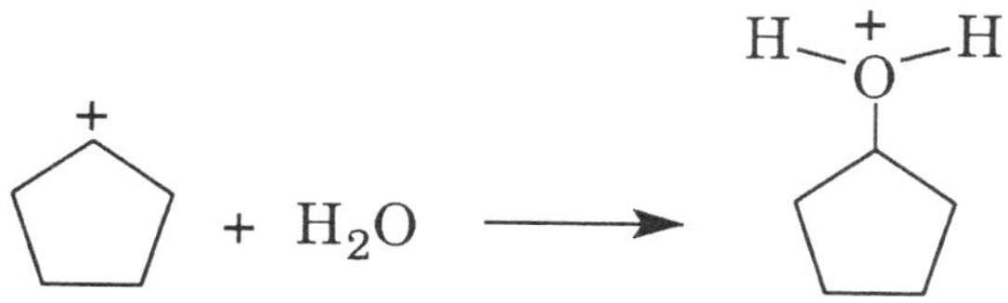

A proton transfers to an acceptor (not shown) and the alcohol forms.

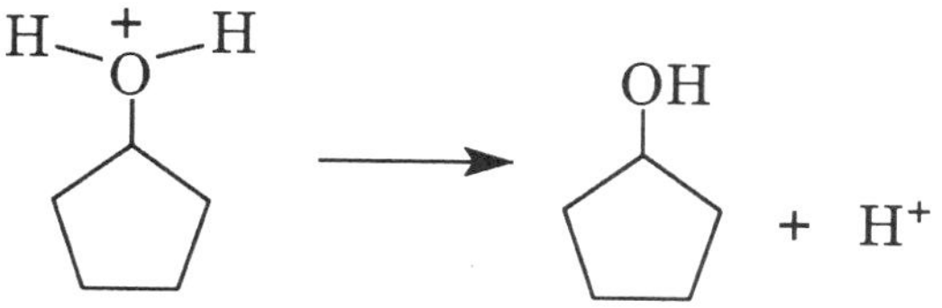

3.35   (a)   vanillin
        (b)   phenol
        (c)   eugenol
        (d)   members of the urushiol family

3.36   Their greater ability to react with oxidizing agents (and thus food-spoiling agents) than key constituents of the food itself.

3.37   It is a general protoplasmic poison and too harmful to healthy tissue.

3.38   Its volatility; it exists in the gas phase at body temperature.

3.39   Its ability to react so rapidly with oxygen in air (given a spark) as to cause an explosion.

3.40   Like cholesterol, methyl *t*-butyl ether is largely hydrocarbon-like and so this ether is able to dissolve cholesterol.

3.41   The dioxins.

3.42   The operation of municipal incinerators.
3.43

(a)

(b) $CH_3CH_2OCH_2CH_3$

(c) $\overset{\overset{\textstyle O}{\textstyle \|}}{CH_3CCH_2CH_3}$

(d)   ⬠—CH_3

(e)   No reaction

(f)   No reaction

(g)   No reaction

(h) $CH_3CH_2\underset{\underset{\textstyle CH_3}{|}}{C}HCH_3$

(i)   ⬡$\overset{CH_3}{\underset{OH}{}}$

(j) $O{=}$⬠$—CH_3$

3.44

(a)   No reaction

(b) $CH_3CH_2\overset{\overset{\textstyle Cl}{|}}{\underset{\underset{\textstyle CH_3}{|}}{C}}CH_3$

(c) $CH_3\overset{\overset{\textstyle CH_3}{|}}{\underset{\underset{\textstyle O}{\|}}{C}}CHCH_3$

(d) $CH_3\overset{}{\underset{\underset{\textstyle CH_3}{|}}{C}}HO\overset{}{\underset{\underset{\textstyle CH_3}{|}}{C}}HCH_3$

(e) 
$$\underset{\text{HCCH}_2\text{CH}_3}{\overset{\text{O}}{\|}}$$

(f)

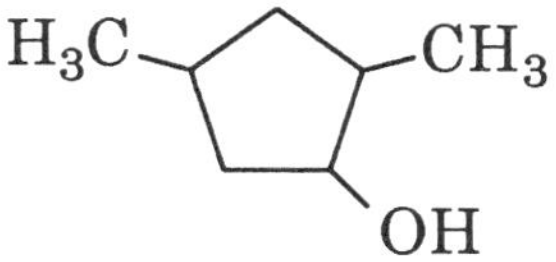

(g)   No reaction

(h)

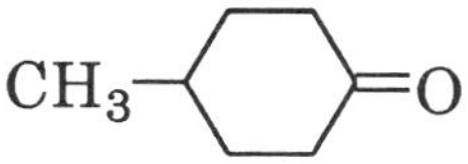

(i) $CH_3CH{=}CH_2$

(j) 
$$\underset{\text{CH}_3\text{CCH}_2\text{OCH}_3}{\overset{\text{O}}{\|}}$$

3.45   (a)   2.50 mol acetone
     (b)   0.120 mol $KMnO_4$
     (c)   22.1 g $KMnO_4$
     (d)   12.2 g acetone; 12.1 g MnO_2

3.46   (a) $3C_3H_8O + 2Cr_2O_7{}^{2-} + 16H^+ \rightarrow 3C_3H_6O_2 + 4Cr^{3+} + 11H_2O$
     (b) $3C_3H_8O + 2K_2Cr_2O_7 + 16HCl \rightarrow 3C_3H_6O_2 + 4CrCl_3 + 4KCl + 11H_2O$
     (c)   2/3 mol $K_2Cr_2O_7$
     (d) $K_2Cr_2O_7 \cdot 2H_2O$; 2/3 mol
     (e)   15.3 g propanoic acid
     (f)   53.1 g $K_2Cr_2O_7 \cdot 2H_2O$

3.47 $3C_5H_{10}O + Cr_2O_7{}^{2-} + 8H^+ \rightarrow 3C_5H_8O + 2Cr^{3+} + 7H_2O$
     20.5 g cyclopentanol needed; 20.8 g $Na_2Cr_2O_7$ required.

Chapter 4

Practice Exercises, Chapter 4

1.   (a)   2-Methylpropanal
     (b)   3-Bromobutanal
     (c)   4-Ethyl-2,4,6-trimethylheptanal

2.   2-Isopropylpropanal would have the structure:

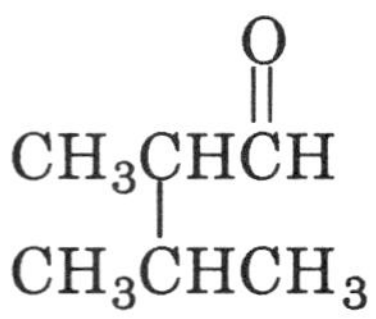

$$CH_3\underset{\underset{\displaystyle CH_3CHCH_3}{|}}{C}H\overset{\displaystyle O}{\overset{\|}{C}}H$$

and it should be named 2,3-dimethylbutanal.

3.   (a)   2-Butanone
      (b)   6-Methyl-2-heptanone
      (c)   2-Methylcyclohexanone

4.

(a) 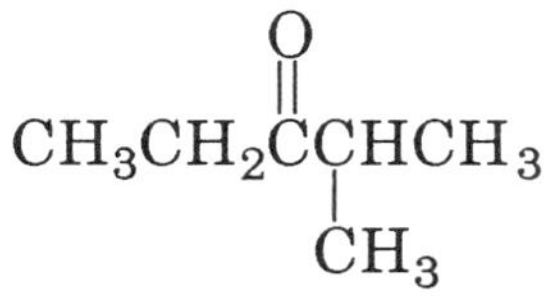 $CH_3CH_2\underset{\underset{\displaystyle CH_3}{|}}{\overset{\displaystyle O}{\overset{\|}{C}}}CHCH_3$

(b) $CH_3\overset{\displaystyle O}{\overset{\|}{C}}C_6H_5$

(c) $CH_3CH_2CH_2\overset{\displaystyle O}{\overset{\|}{C}}CH_2CH_2CH_3$

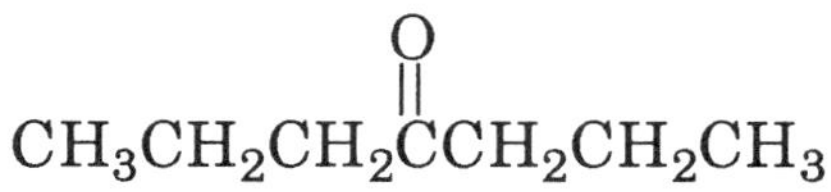

(d) 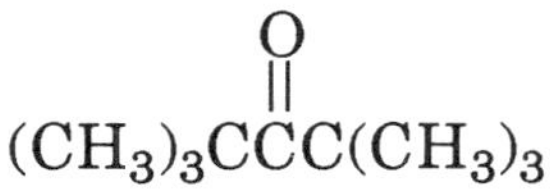 $(CH_3)_3C\overset{\displaystyle O}{\overset{\|}{C}}C(CH_3)_3$

5.

(a) 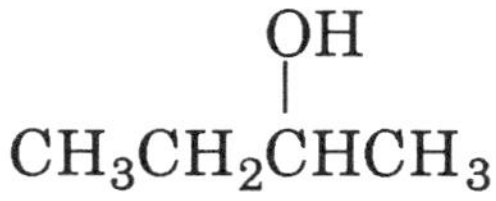 $CH_3CH_2\underset{\underset{}{}}{\overset{\displaystyle OH}{\overset{|}{C}}}HCH_3$

(b) $CH_3\underset{\underset{\displaystyle CH_3}{|}}{C}HCH_2CH_2OH$

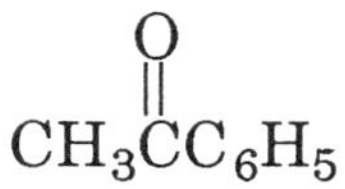

(c)   —OH

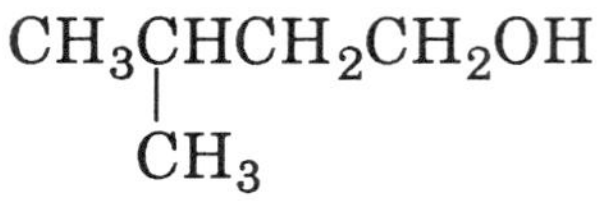

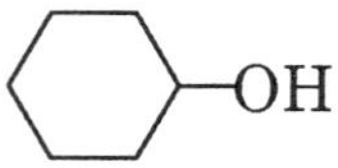

6.

(a) Not a hemiacetal

(b) Not a hemiacetal

(c) $HOCH_2OCH_2CH_3$

    ↑
    hemiacetal
    position

(d)

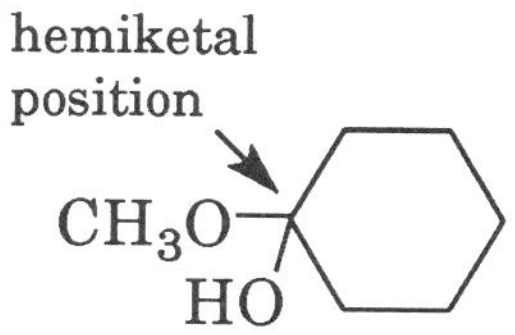

hemiketal
position
↘

CH_3O-
HO

7.

(a)

$\overset{\displaystyle OH}{\underset{\displaystyle |}{}}$
CH_3CHOCH_3

(b)

$\overset{\displaystyle OH}{\underset{\displaystyle |}{}}$
$CH_3CH_2CH_2CHOCH_2CH_3$

(c)

$\overset{\displaystyle OH}{\underset{\displaystyle |}{}}$
$C_6H_5CHOCH_2CH_2CH_3$

(d) $HOCH_2OCH_3$

8.

(a)

$\overset{\displaystyle O}{\underset{\displaystyle ||}{}}$
$CH_3CH_2CH + HOCH_3$

(b)

$\overset{\displaystyle O}{\underset{\displaystyle ||}{}}$
$CH_3CH_2OH + HCCH_2CH_3$

9.   (a) Neither an acetal nor a ketal

(b) A ketal. Its carbon atom that holds two oxygen atoms
was a keto group carbon. The breakdown products are

$\overset{\displaystyle O}{\underset{\displaystyle ||}{}}$
$2CH_3CH_2OH + CH_3CCH_3$

10.

(a)

$$2CH_3OH + \overset{O}{\underset{\|}{HCH}}$$

(b) No reaction

(c)

$$2CH_3OH + \overset{H_3C}{\underset{|}{CH_3CH}}\overset{O}{\underset{\|}{CCH_3}}$$

Review Exercises, Chapter 4

4.1 (a) Ketone (b) Aldehyde
(c) Ketone (d) Carboxylic acid
(e) Aldehyde (f) Ether + ketone

4.2

$$CH_3CH_2\overset{O}{\underset{\|}{CH}} \qquad CH_3\overset{O}{\underset{\|}{C}}CH_3 \qquad CH_3CH_2\overset{O}{\underset{\|}{C}}OH \qquad CH_3\overset{O}{\underset{\|}{C}}OCH_3$$

aldehyde   ketone   carboxylic acid   ester

4.3

(a) $CH_3CH_2\underset{\underset{CH_3}{|}}{CH}CHO$

(b)

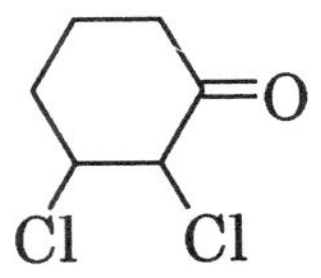

(c) $C_6H_5\overset{O}{\underset{\|}{C}}CH_3$

(d) $(CH_3)_2CHCH_2\overset{O}{\underset{\|}{C}}CH_2CH(CH_3)_2$

(e) $CH_3\overset{O}{\underset{\|}{C}}CH_2CH_2\overset{O}{\underset{\|}{C}}CH_3$

4.4

(a)

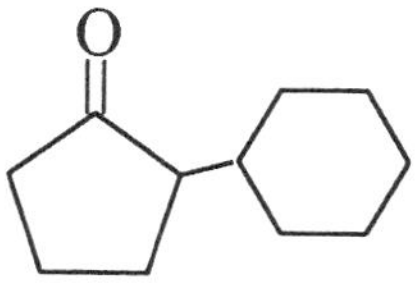

(b) 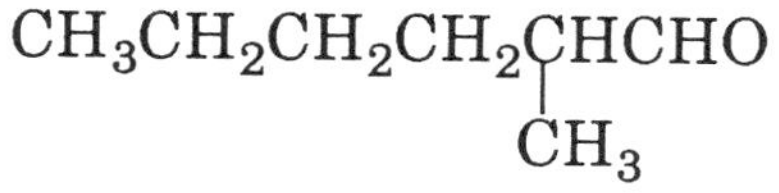 $CH_3CH_2CH_2CH_2\underset{\underset{CH_3}{|}}{C}HCHO$

(c)

$CH_3CH_2\underset{\underset{CH_3}{|}}{C}H\overset{\overset{O}{\|}}{C}\underset{\underset{CH_3}{|}}{C}HCH_2CH_3$

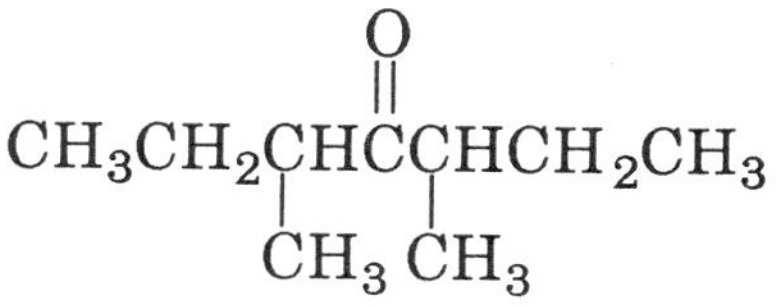

(d)

$C_6H_5CH_2\overset{\overset{O}{\|}}{C}CH_2C_6H_5$

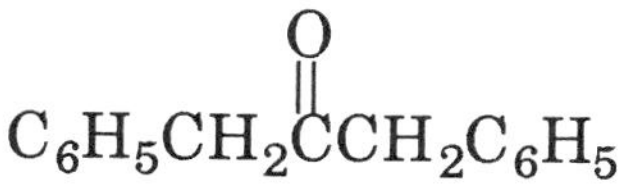

(e)

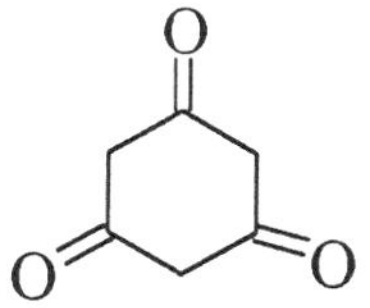

4.5   (a)   2-Methylcyclohexanone   (b) Propionic acid
        (c)   2-Pentanone   (d) Propanal
        (e)   2-Methylpentanal
4.6   (a)   2-Ethyl-3-methylpentanal   (b) 1-Phenylethanone
        (c)   2,3-dimethylcyclopentanone   (d) Acetic acid
        (e)   Butanal
4.7   5-Ketohexanal
4.8   2-Formylbenzoic acid
4.9   (a)   3-Methylbutanal
        (b)   3-Methylpentanal
        (c)   3-Ethyl-5-methylhexanal
        (d)   4-Methyl-2-pentanone
        (e)   3-Propylcyclopentanone
4.10   (a)   3-Isobutyl-5-methyl-2-heptanone
        (b)   4-t-Butyl-6-methyloctanal
        (c)   4-Phenyl-2-butanone
        (d)   3,4-Dimethylcyclohexanone
        (e)   2,2-Dimethylcyclooctanone
4.11   Valeraldehyde

4.12   Glyceric acid

4.13

$$B < A < D < C$$

4.14

$$C < B < D < A$$

4.15

$$B < A < D < C$$

4.16

$$C < B < D < A$$

4.17

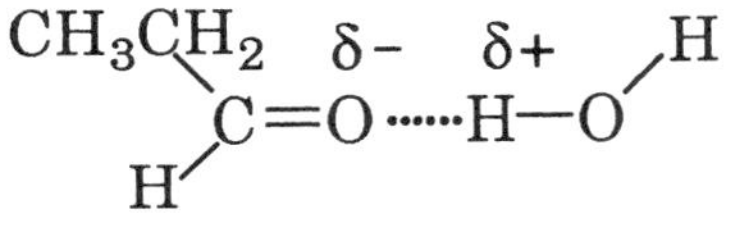

4.18

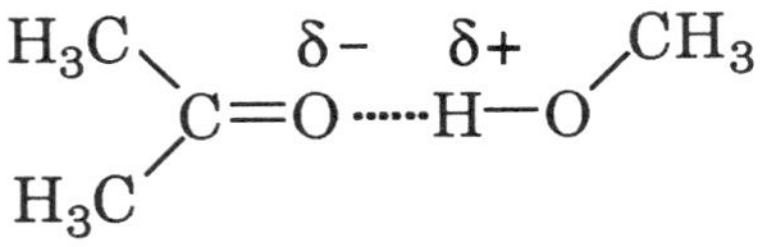

4.19

(a)

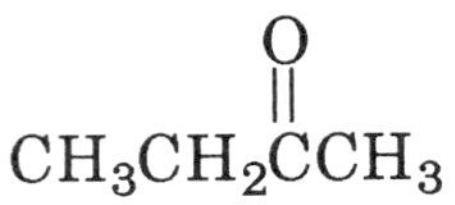

2-Butanone

(b)

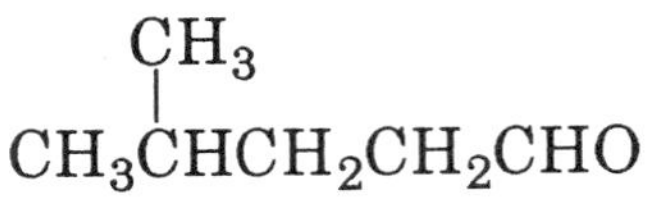

4-Methylpentanal

(c)

3-Methylcyclopentanone

(d) $C_6H_5CH_2\underset{\underset{\textstyle CH_3}{|}}{C}HCHO$

2-Methyl-3-phenylpropanal

4.20

(a)

$$CH_3\overset{\overset{\textstyle O}{||}}{C}CH_2CH_3$$

(b)

$$H\overset{\overset{\textstyle O}{||}}{C}\underset{\underset{\textstyle }{|}}{\overset{\overset{\textstyle CH_3}{|}}{C}}HCH_3$$

(c)   No oxidation product

(d)   No oxidation product

(e)

$$CH_3OCH_2CH_2\overset{\overset{\textstyle O}{||}}{C}CH_3$$

(f)

4.21 C_3H_6O is $CH_3CH_2CH=O$; $C_3H_6O_2$ is $CH_3CH_2CO_2H$

4.22

$$CH_3\overset{\overset{\textstyle O}{||}}{C}CH_2OH \qquad CH_3\overset{\overset{\textstyle O}{||}}{C}CO_2H$$

$$C_3H_6O_2 \qquad\qquad C_3H_4O_3$$

4.23   Positive Benedict's tests are given by a and b.
4.24   Benedict's tests are given by b and d.
4.25   (a)   The cations of transition metals.
     (b)   Electron-rich particles, whether electrically neutral or negatively charged.
     (c) F^-, Cl^-, Br^-, and I^- are four examples in the same family.
     (d) H_2O, which forms $Cu(H_2O)_4{}^{2+}$
          NH_3, which forms $Cu(NH_3)_4{}^{2+}$
4.26   (a) $Ag(NH_3)_2{}^+$. Hydroxide ions are persent, which react with Ag^+ to give Ag_2O, an insoluble compound. Hydroxide ions do not react with $Ag(NH_3)_2{}^+$.
     (b)   It forms a complex ion with Cu^{2+} which retains Cu^{2+} in solution in the presence of base. (Cu^{2+} is otherwise insoluble in base; it forms CuO.)

4.27 Cu_2O

4.28   To test for glucose in urine specimens.

4.29   Manufacture of silvered mirrors.

4.30

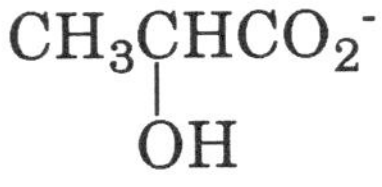

$$CH_3\underset{\underset{OH}{|}}{C}HCO_2^-$$

4.31

$$CH_3\overset{\overset{O}{\parallel}}{C}CH_2CO_2^-$$

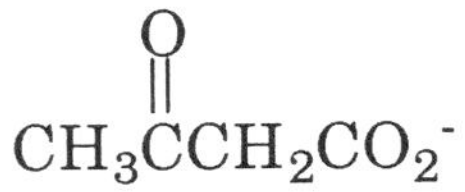

4.32   **B** is oxidized to a compound with the keto group.

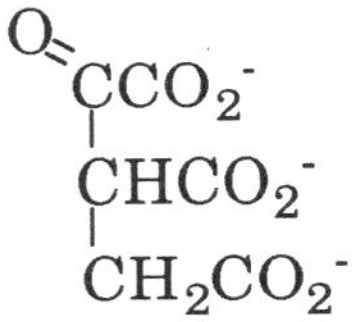

$$\underset{\underset{CH_2CO_2^-}{|}}{\underset{\underset{CHCO_2^-}{|}}{\overset{\overset{O}{\diagdown}}{C}CO_2^-}}$$

4.33 $H{:}^- + CH_3CH_2OH \rightarrow H_2 + CH_3CH_2O^-$

4.34   (a) $CH_3CH_2O^-$
       (b) $CH_3CH_2O^- + H_2O \rightarrow CH_3CH_2OH + OH^-$
       (c)   Ethanol

4.35

(a) $$CH_3\underset{\underset{}{|}}{\overset{\overset{O^-}{|}}{C}}HCH_3$$

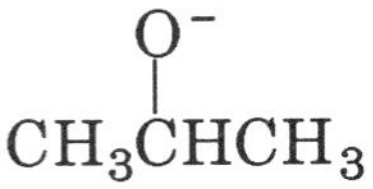

(b) $$CH_3\overset{\overset{O^-}{|}}{C}HCH_3 + H_2O \longrightarrow CH_3\overset{\overset{OH}{|}}{C}HCH_3 + OH^-$$

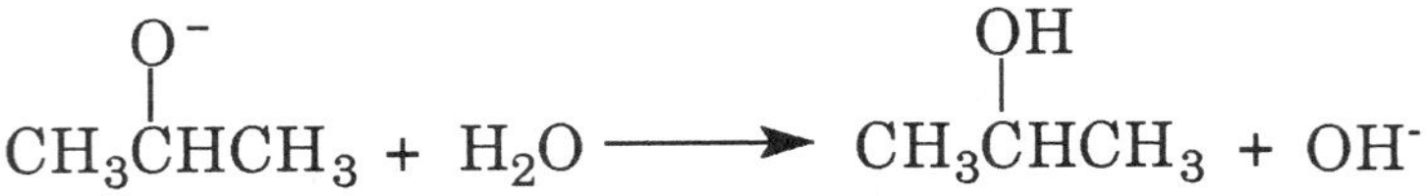

(c)   2-Propanol

4.36

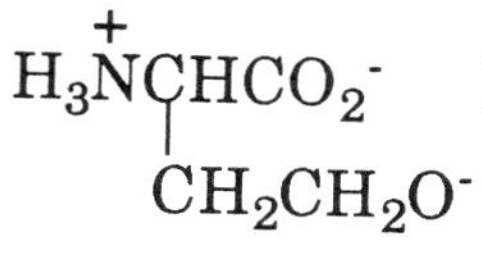

$$\underset{\textbf{A}}{\overset{+}{H_3}N\underset{\underset{\textstyle CH_2CH_2O^-}{|}}{C}HCO_2^-} \qquad \underset{\textbf{B}}{\overset{+}{H_3}N\underset{\underset{\textstyle CH_2CH_2OH}{|}}{C}HCO_2^-}$$

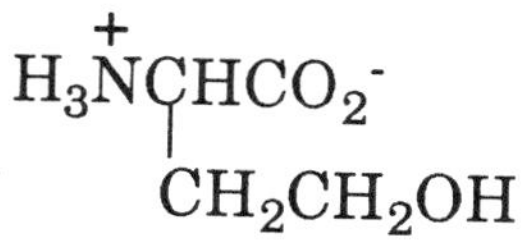

4.37

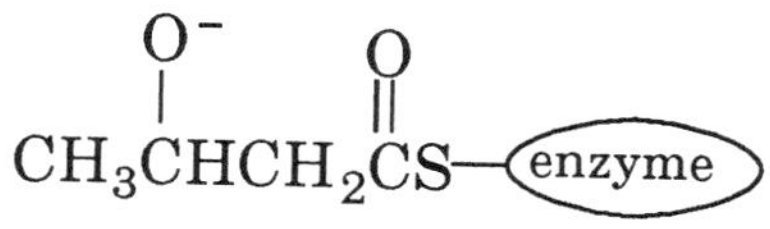

$$\underset{\textbf{A}}{CH_3\underset{\underset{\textstyle O^-}{|}}{C}HCH_2\overset{\overset{\textstyle O}{\|}}{C}S-\boxed{enzyme}} \qquad \underset{\textbf{B}}{CH_3\underset{\underset{\textstyle OH}{|}}{C}HCH_2\overset{\overset{\textstyle O}{\|}}{C}S-\boxed{enzyme}}$$

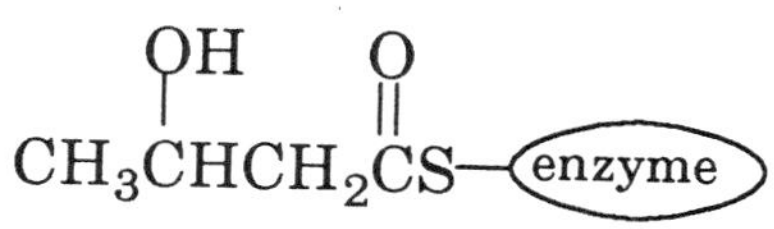

4.38

(a) $CH_3\overset{\overset{\textstyle O}{\|}}{C}CH_2CH_3$

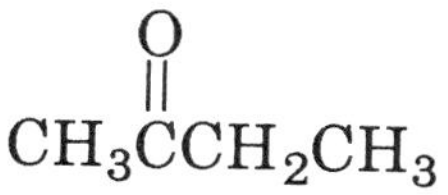

(b) $H\overset{\overset{\textstyle O}{\|}}{C}\underset{\underset{\textstyle OCH_3}{|}}{C}HCH_2CH_3$

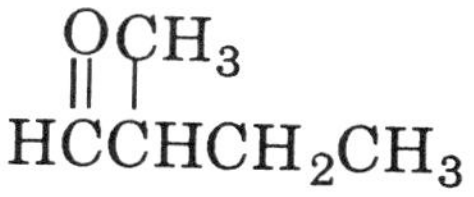

(c)

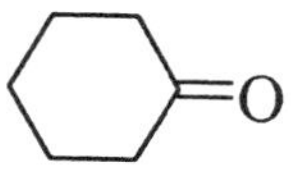

(d) $CH_3-\langle\bigcirc\rangle-\overset{\overset{\textstyle O}{\|}}{C}H$

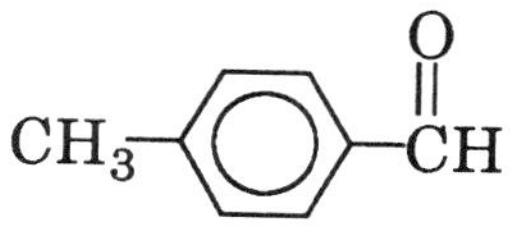

4.39

(a) $H\overset{\overset{\textstyle O}{\|}}{C}CH_2OCH_3$

(b) $CH_3O\underset{\underset{\textstyle CH_3}{|}}{C}H\overset{\overset{\textstyle O}{\|}}{C}H$

(c) $CH_3\overset{\overset{\textstyle O}{\|}}{C}CH_2\underset{\underset{\textstyle CH_3}{|}}{\overset{\overset{\textstyle OH}{|}}{C}}CH_2$

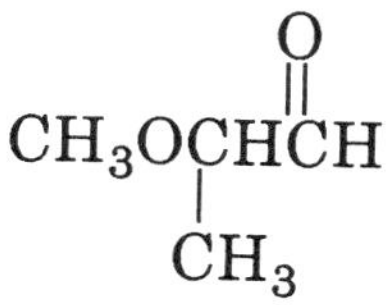

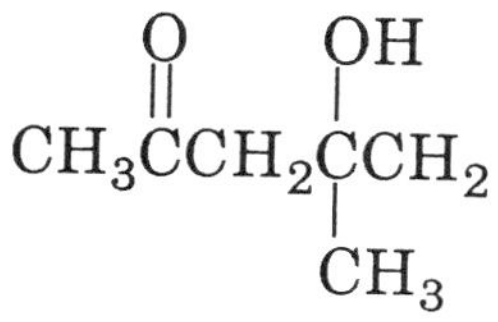

(d) $CH_3CH_2O-\langle\bigcirc\rangle-\overset{\overset{\textstyle O}{\|}}{C}CH_3$

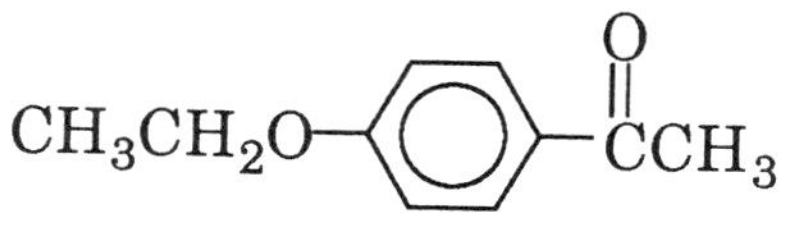

4.40   (a)   Hemiacetal   (b) Acetal
         (c)   Something else (a 1,2-di-ether)   (d) Ketal

4.41   (a)   Acetal   (b) Hemiacetal
         (c)   Hemiacetal   (d) Something else (an ether-
                                  alcohol)

4.42

(a)
$$\underset{CH_3CH_2\overset{\displaystyle OH}{\underset{|}{C}}HOCH_3}{}$$

$$\underset{CH_3CH_2\overset{\displaystyle OCH_3}{\underset{|}{C}}HOCH_3}{}$$

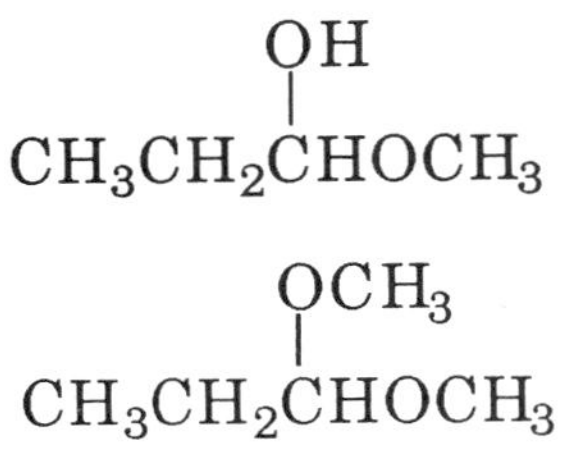

(b)
$$\underset{CH_3CH_2\overset{\displaystyle OH}{\underset{|}{C}}HOCH_2CH_3}{}$$

$$\underset{CH_3CH_2\overset{\displaystyle OCH_2CH_3}{\underset{|}{C}}HOCH_2CH_3}{}$$

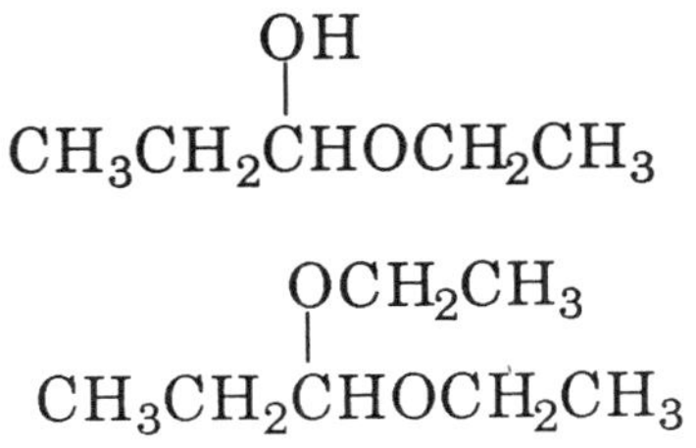

4.43

(a)
$$CH_3\overset{\displaystyle OCH_3}{\underset{\displaystyle CH_3}{\overset{|}{\underset{|}{C}}}}OCH_3$$

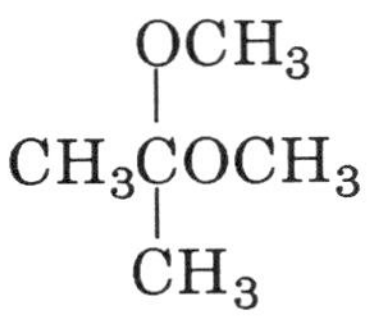

(b)
$$CH_3\overset{\displaystyle OCH_2CH_3}{\underset{\displaystyle CH_3}{\overset{|}{\underset{|}{C}}}}OCH_2CH_3$$

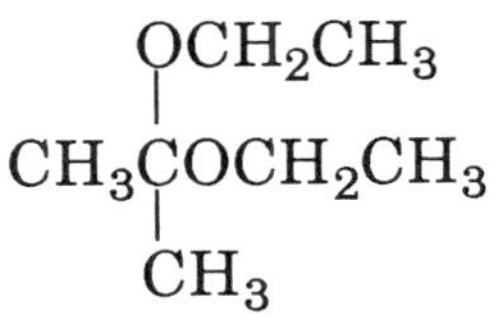

4.44

$$
\begin{array}{c}
CH_3 \\
\diagdown \\
CH-OH \\
\diagup \\
CH_2 \qquad CH{=}O \\
\diagdown \qquad \diagup \\
CH_2{-}CH_2
\end{array}
$$

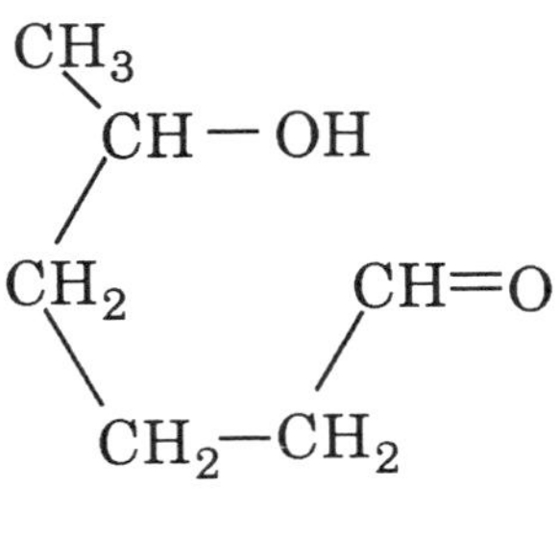

4.45

(a) 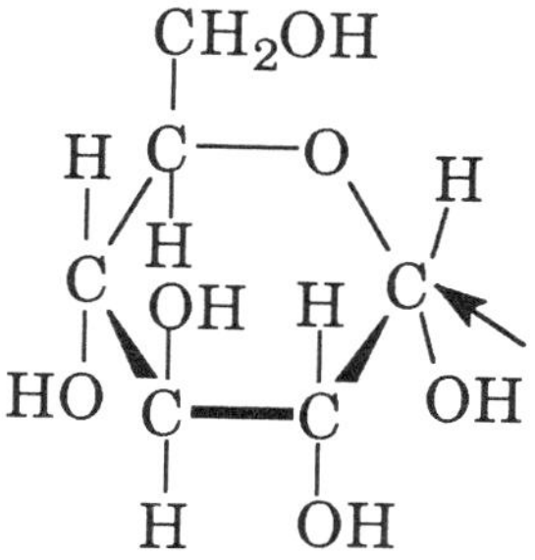 (b) 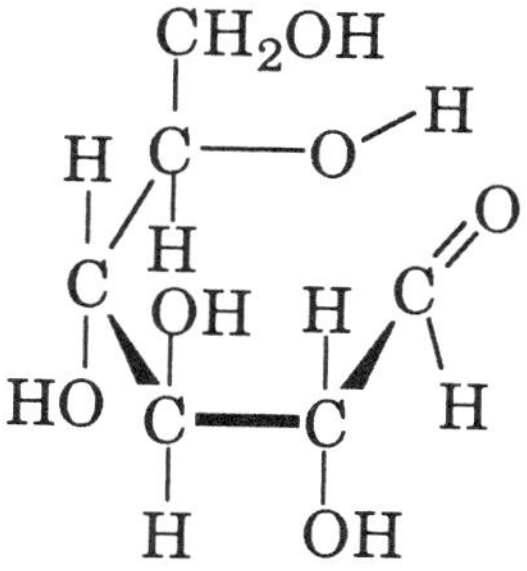

4.46

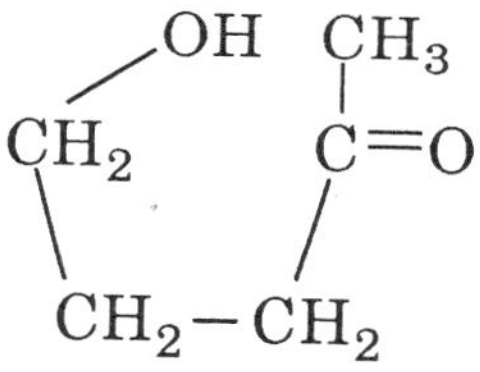

4.47

(a) 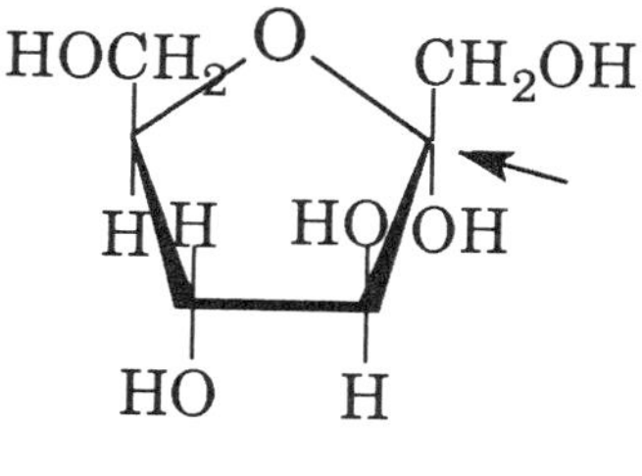 (b) 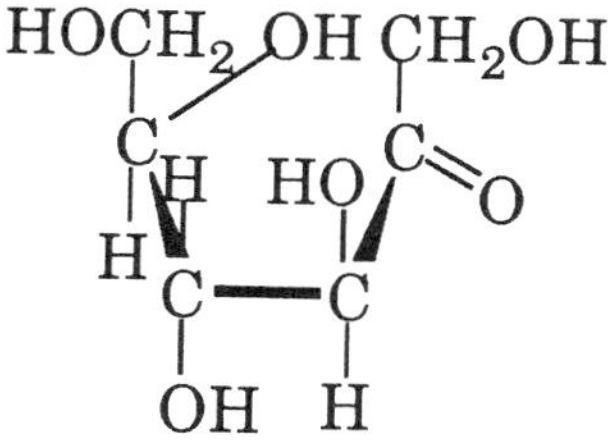

4.48

(a) $CH_3CH_2CHO + 2CH_3OH$ (b) No reaction

(c) $CH_3\overset{\overset{\displaystyle O}{\|}}{C}CH_3 + 2CH_3CH_2OH$ (d) 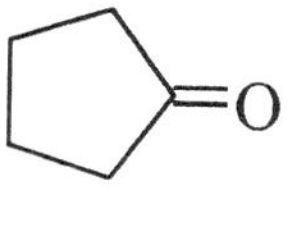 + $2CH_3OH$

4.49

(a) CH_3CH_2CHO

  $+ HOCH_2CH_3$

  $+ HOCH(CH_3)_2$

(b) No reaction

(c) 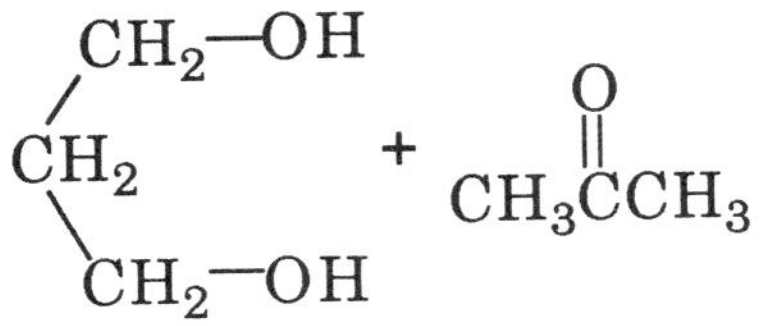
$$\begin{array}{l} CH_2\text{—}OH \\ / \\ CH_2 \\ \backslash \\ CH_2\text{—}OH \end{array} + CH_3\overset{O}{\overset{\|}{C}}CH_3$$

(d) $CH_3OCH_2CHO + HOCH_2CH_3$

  $+ HOCH_3$

4.50 (a) Estrone
(b) Acetone
(c) Formaldehyde

4.51 (a) sp^2 on C overlapping with s on H
(b) sp^2 on C overlapping with sp^2 on O
(c) $2p_z$ on C overlapping with $2p_z$ on O

4.52 120°. Yes

4.53

(a) 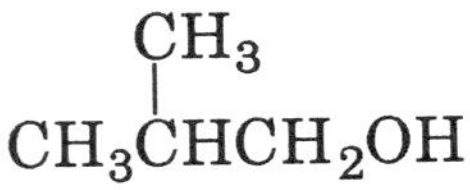
$$\begin{array}{l} CH_3 \\ | \\ CH_3CHCH_2OH \end{array}$$

(b) 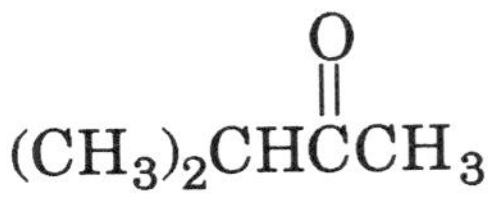
$$(CH_3)_2CH\overset{O}{\overset{\|}{C}}CH_3$$

(c) No reaction

(d) $CH_3CH_2CH_2CH_2CH_3$

(e) 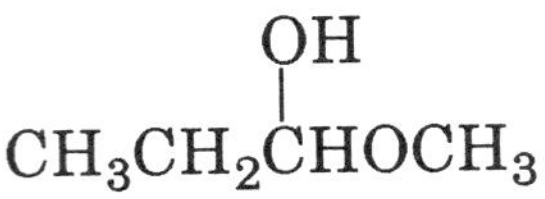
$$\begin{array}{l} OH \\ | \\ CH_3CH_2CHOCH_3 \end{array}$$

(f) $CH_3CH_2OH + Mtb^+$

(g)

$$OCH_2CH_3$$
$$CH_3CHOCH_2CH_3$$

(h) $CH_3CHO + 2CH_3OH$

(i)

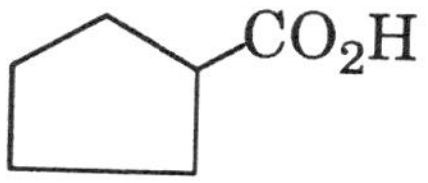

CO_2H

(j)   No reaction

4.54

(a)   No reaction

(b)   No reaction

(c)

$$OH$$
$$CH_3CHOCH_2CH_2CH_3$$

(d)   No reaction

(e)

$$O$$
$$CH_3CCH_3 + HOCH_2CH_3$$

(f) $CH_3OCH_2CH_2CH_2OH + Mtb^+$

(g)

$$O$$
$$CH_3CH_2CCH_3$$
$$+ 2HOCH_2CH_3$$

(h)

$$O$$
$$CH_3CC_6H_5$$

(i)

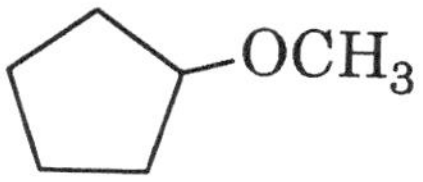

OCH_3

(j)

$$OCH_3$$
$$C_6H_5CHOCH_3$$

4.55

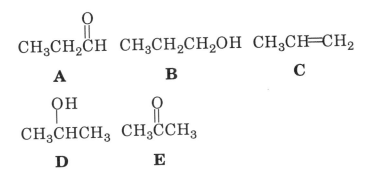

CH₃CH₂CH (A) — CH₃CH₂CH₂OH (B) — CH₃CH=CH₂ (C)

CH₃CHCH₃ (D, with OH) — CH₃CCH₃ (E, with =O)

4.56

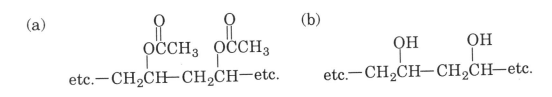

(CH₃)₂CH₂OH (F) — (CH₃)₂CHCH (G) — (CH₃)₂CHCOH (H)

(CH₃)₂C=CH₂ (I) — (CH₃)₃COH (J)

4.57 (a) 0.173 mol butanal
(b) 1.23 mol CH_3OH
(c) Yes, 11.1 g CH_3OH needed but 39.4 g taken
(d) 3.11 g H_2O obtained
(e) To ensure that the equilibria involved in the reaction are all shifted as much as possible to the right, in favor of the products.

4.58

(a) etc.—CH₂CH—CH₂CH—etc. (with OCCH₃ groups)

(b) etc.—CH₂CH—CH₂CH—etc. (with OH groups)

(c)

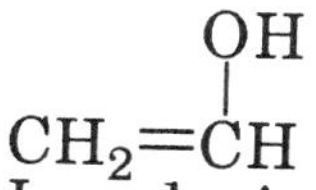

OH
|
CH₂=CH   The "alcohol" is actually an enol and is unstable. In polyvinyl alcohol, the OH groups are proper alcohol group because all are bound to saturated carbon atoms; all OH groups are 2° alcohol groups.

(d)

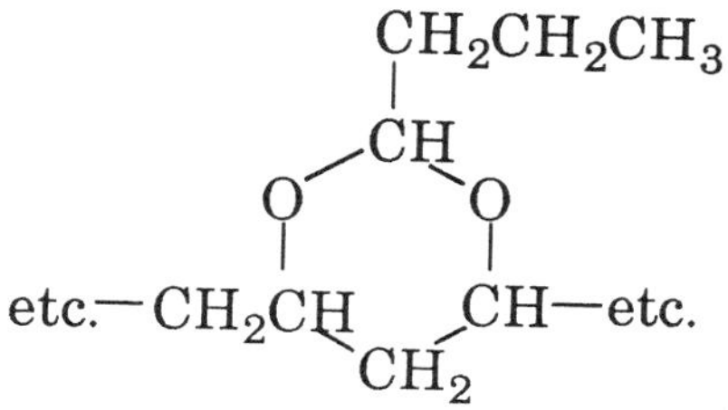

$$CH_2CH_2CH_3$$
|
CH
O   O
| |
etc.—CH₂CH   CH—etc.
        CH₂

Chapter 5

Practice Exercises, Chapter 5

1.
    (a)   2,2-Dimethylpropanoic acid
    (b)   5-Ethyl-5-isopropyl-3-methyloctanoic acid
    (c)   Sodium ethanoate
    (d)   5-Chloro-3-methylheptanoic acid
2.   Pentanedioic acid
3.   9-Octadecenoic acid
4.

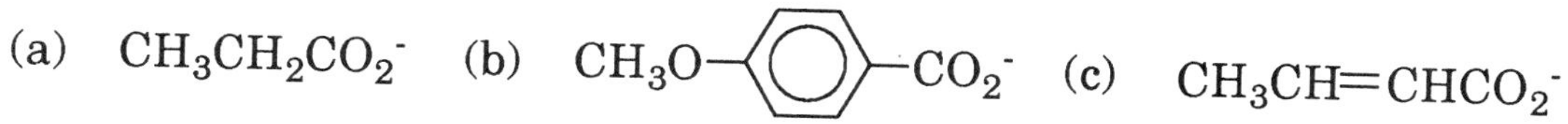

(a) $CH_3CH_2CO_2^-$ (b) $CH_3O-\bigcirc-CO_2^-$ (c) $CH_3CH=CHCO_2^-$

5.

(a) $CH_3O-\bigcirc-CO_2H$ (b) $CH_3CH_2CO_2H$ (c) $CH_3CH=CHCO_2H$

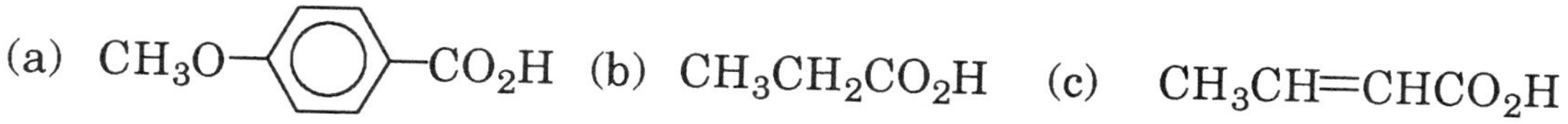

6.

(a)

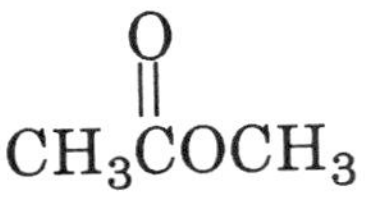

$$CH_3\overset{O}{\overset{\|}{C}}OCH_3$$

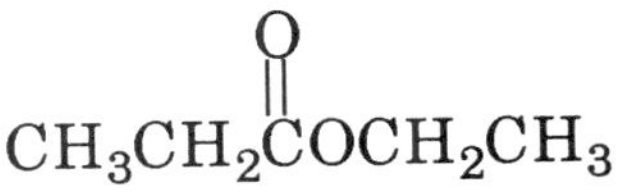

(b)

$$CH_3\overset{O}{\overset{\|}{C}}OCH_2CH_2CH_3$$

(c)

$$CH_3\overset{O}{\overset{\|}{C}}O\overset{CH_3}{\overset{|}{C}}HCH_3$$

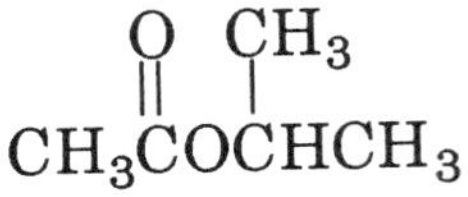

7.

(a)

$$H\overset{O}{\overset{\|}{C}}OCH_2CH_3$$

(b)

$$CH_3CH_2\overset{O}{\overset{\|}{C}}OCH_2CH_3$$

(c)

$$C_6H_5\overset{O}{\overset{\|}{C}}OCH_2CH_3$$

8. (a) Methyl propanoate   (b) Propyl 3-methylpentanoate
9. (a) t-Butyl acetate   (b) Ethyl butyrate
10. (a) $CH_3OH + CH_3CO_2H$ (b) $(CH_3)_2CHOH + CH_3CH_2CO_2H$
    (c) $CH_3CH_2CH_2OH + (CH_3)_2CHCO_2H$

11. (a) $C_6H_5OH + CH_3CO_2^-$ (b) 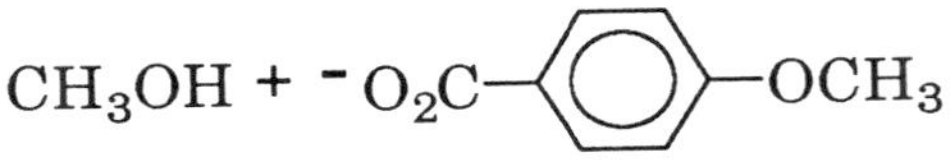 $CH_3OH + {}^-O_2C\text{—}\langle\bigcirc\rangle\text{—}OCH_3$

Review Exercises, Chapter 5

5.1 (a) **B**   (b) **A** (c) **B**   (d) **C**
5.2 Fatty acids
5.3 Acetic acid; lactic acid
5.4 (a) $CH_3CH_2CO_2H$ (b) $C_6H_5CO_2H$
     (c) CH_3CO_2H (d) HCO_2H

5.5

(a)

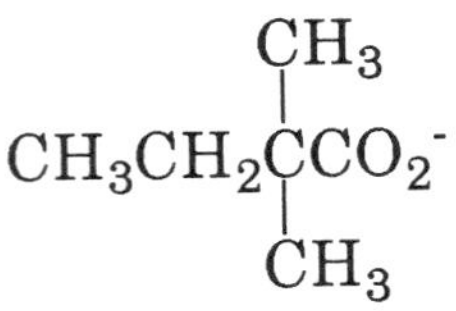

$$CH_3CH_2\overset{CH_3}{\underset{CH_3}{\overset{|}{\underset{|}{C}}}}CO_2^-$$

(b) $CH_3CH_2CH_2\overset{}{\underset{Cl}{\overset{|}{C}}}HCH_2\overset{}{\underset{CH_3}{\overset{|}{C}}}HCO_2H$

(c)

$$HO\overset{O}{\overset{\|}{C}}CH_2CH_2\overset{O}{\overset{\|}{C}}OH$$

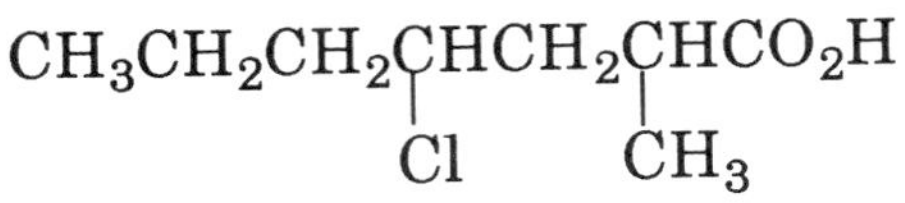

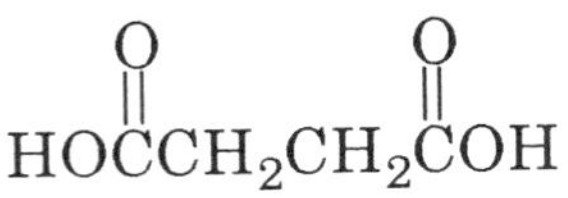

(b) $C_6H_5CO_2^-$

5.6 (a) 2-Methylpropanoic acid (b) 2,2-Dimethylbutanoic acid
     (c) Sodium 3-chloropentanoate (d) Potassium benzoate

5.7 (a) 2,4-Dimethylpentanoic acid (b) 3,4-Dimethylpentanoic acid
     (c) Sodium decanoate (d) Potassium 2-bromopropanoate

5.8 sodium 3-hydroxybutanoate (b) Sodium β-hydroxybutyrate

5.9 *trans*-Butenedioic acid

5.10

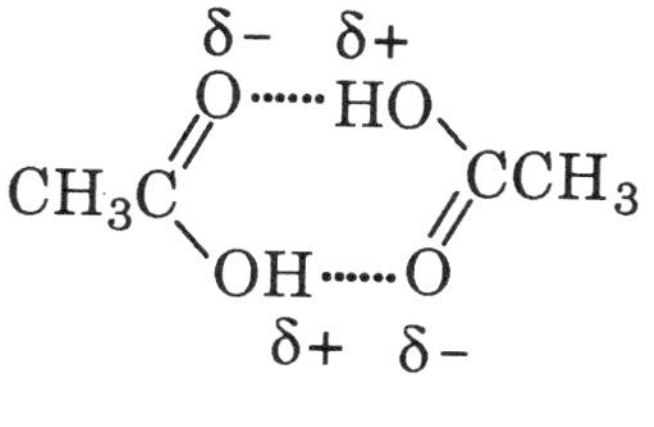

5.11

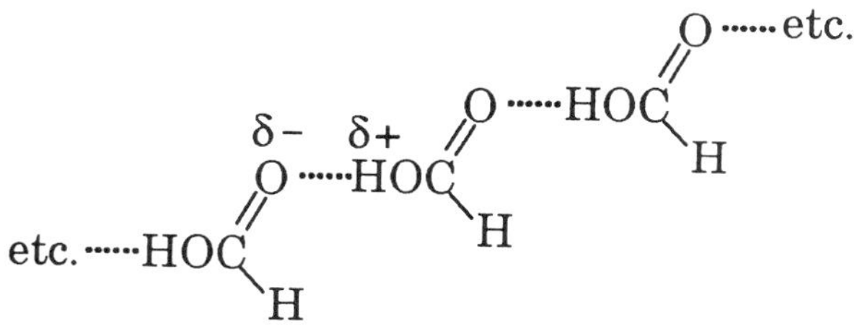

5.12

C < A < B

5.13

B < C < D < A

5.14 (a) $CH_3CO_2H + H_2O \rightleftharpoons CH_3CO_2^- + H_3O^+$
     (b) Toward acetic acid. The addition of $HCl(aq)$ makes the concentration of H_3O^+ increase. This puts a stress on the equilibrium, which shifts to the left to relieve the stress.

     (c) $K_a = \dfrac{[CH_3CO_2^-][H^+]}{[CH_3CO_2H]}$

     (d) Weaker

5.15   (a) $HCO_2H + H_2O \rightleftharpoons HCO_2^- + H_3O^+$

(b)   Toward the formate ion. The added OH^- (from NaOH) neutralizes H_3O^+ and so reduces the concentration of H_3O^+ in the equilibrium. The equilibrium thus must shift to the right to replace the lost H_3O^+.

(c) $K_a = \dfrac{[HCO_2^-][H^+]}{[HCO_2H]}$

(d)   Stronger

5.16

$$A < C < D < B$$

5.17

$$B < A < C < D$$

5.18

(a)  $CH_3-\langle\bigcirc\rangle-OH + NaOH \longrightarrow CH_3-\langle\bigcirc\rangle-O^-Na^+ + H_2O$

(b)   No reaction

(c)  $CH_3-\langle\bigcirc\rangle-CO_2H + NaOH \longrightarrow CH_3-\langle\bigcirc\rangle-CO_2^-Na^+ + H_2O$

(d) $HNO_3 + NaOH \longrightarrow NaNO_3 + H_2O$

5.19

(a) $HO_2CCH_2CH_2CO_2H + 2OH^- \longrightarrow {}^-O_2CCH_2CH_2CO_2^- + 2H_2O$

(b) $HOCH_2CH_2CH_2CO_2H + OH^- \longrightarrow HOCH_2CH_2CH_2CO_2^- + H_2O$

(c)

$$\underset{\overset{\displaystyle O}{\displaystyle \|}}{H\overset{}{C}}CH_2CH_2CH_2CO_2H \;+\; OH^- \;\longrightarrow\; \underset{\overset{\displaystyle O}{\displaystyle \|}}{H\overset{}{C}}CH_2CH_2CH_2CO_2^- \;+\; H_2O$$

(d)

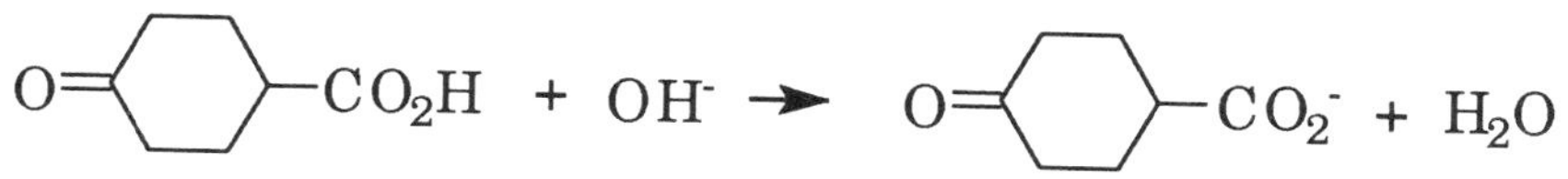

$$O{=}\!\!\left\langle\text{cyclohexane}\right\rangle\!\!{-}CO_2H \;+\; OH^- \;\longrightarrow\; O{=}\!\!\left\langle\text{cyclohexane}\right\rangle\!\!{-}CO_2^- \;+\; H_2O$$

5.20

5.21

5.22

B. The ionic compound, **A**, is very insoluble in a nonpolar solvent. **B** is not ionic. **A**, an ionic compound, is much more soluble in water than the molecular compound, **B**.

(a) $CH_3CH_2CO_2^- + H^+ \longrightarrow CH_3CH_2CO_2H$

(b) $^-O_2CCH_2CH_2CH_2CO_2^- + H^+ \longrightarrow HO_2CH_2CH_2CH_2CO_2^-$

(c) $NH_3 + H^+ \longrightarrow NH_4^+$

5.23

(a) $HOCH_2CH_2CO_2^- + H^+ \longrightarrow HOCH_2CH_2CO_2H$

(b) No reaction

(c) $C_6H_5O^- + H^+ \longrightarrow C_6H_5OH$

5.24

(a)
$$CH_3CH_2\overset{O}{\overset{\|}{C}}Cl$$

(b)
$$CH_3CH_2\overset{O}{\overset{\|}{C}}O\overset{O}{\overset{\|}{C}}CH_2CH_3$$

(c)
$$CH_3CH_2\overset{O}{\overset{\|}{C}}OH$$

5.25

(a)
$$C_6H_5\overset{O}{\overset{\|}{C}}OH + HOCH_3$$

(b)
$$C_6H_5\overset{O}{\overset{\|}{C}}Cl + HOCH_3$$

(c)
$$C_6H_5\overset{O}{\overset{\|}{C}}O\overset{O}{\overset{\|}{C}}C_6H_5 + HOCH_3$$

5.26

(a)
$$CH_3\overset{O}{\overset{\|}{C}}OCH_2CH_3$$

(b)   The electronegativities of O and Cl place a relatively large δ+ charge on the carbon atom of the carbonyl group in acetyl chloride. This charge is able quite strongly to attract the δ− charge on the O atom of the alcohol molecule. In addition, the Cl⁻ ion is a very stable leaving group and so quite readily leaves the carbonyl carbon atom of the acetyl chloride molecule when the alcohol molecule attacks.

5.27   The electronegativies of the O atoms of the anhydride create a sizeable δ+ charge on the carbonyl carbon atoms. This charge is able quite strongly to attract the δ− charge on the O atom of the alcohol. In addition, the acetate ion is a stable leaving group and so quite readily leaves the carbonyl carbon atom when the alcohol molecule attacks.

5.28

(a)
$$CH_3CH_2\overset{O}{\overset{\|}{C}}OCH_2CH_3 + H_2O$$

(b)

$$(CH_3)_2CHCOCH_2CH_3 + H_2O$$

with C=O above

(c)

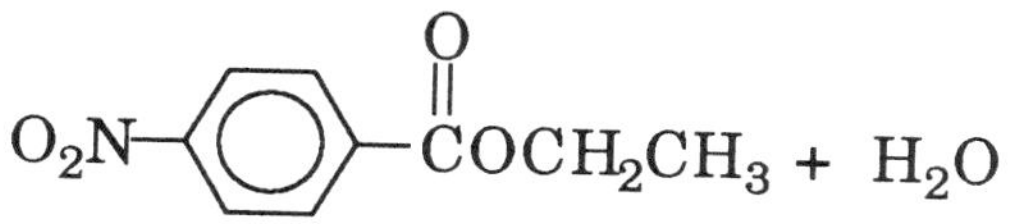

$$O_2N-\!\!\bigcirc\!\!-COCH_2CH_3 + H_2O$$

(d)

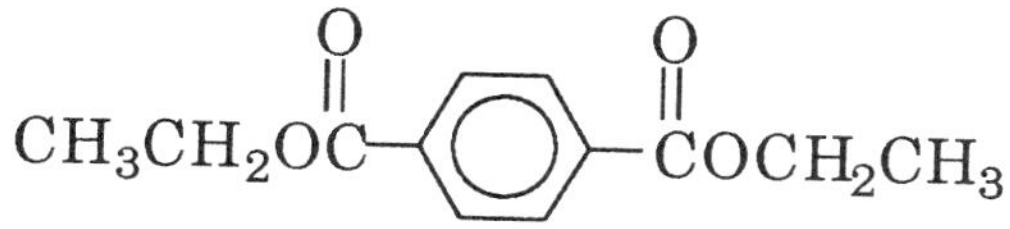

$$CH_3CH_2OC-\!\!\bigcirc\!\!-COCH_2CH_3$$

5.29

(a)

$$CH_3COCH_3$$

(b)

$$CH_3COCH_2CH(CH_3)_2$$

(c)

$$CH_3COC_6H_5$$

(d)

$$CH_3COCH_2CH_2OCCH_3$$

5.30   In the first step, H+ transfers from the hydronium ion to form a bond to oxygen in **I** to make **II**.

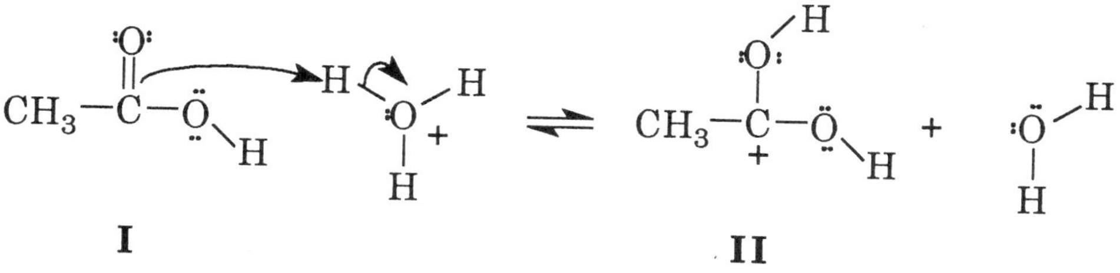

I   **II**

II now attracts a molecule of ethyl alcohol to form **III**. Then a proton transfers from one O atom to another in **III** to create a stable leaving group, H_2O, in the product, **IV**.

In the next step, **IV** drops off a water molecule and **V** forms.

Finally, **V** drops off the proton and electrons relocate to reestablish an outer octet for the carbonyl carbon atom as the ester forms.

5.31   A much stronger attraction can exist between the OH⁻ ion and the δ+ charge on the carbonyl carbon atom of the ester, the specific site attacked in both hydrolysis and saponification, than between this δ+ site and the δ− charge on a water molecule.

5.32   Shift it to the right, because the stress in the equilibrium is the loss of a *product*. In accordance with Le Châtelier's principle, the equilibrium must shift in the direction that tries to replace this loss.

5.33   The acid chloride has a more stable leaving group (the weakly basic Cl^- ion) than the ester, for which the leaving group is a very strongly basic anion of an alcohol

5.34

(a)

$$\overset{\overset{\displaystyle O}{\|}}{H}COCH_2CH_3$$

(b)

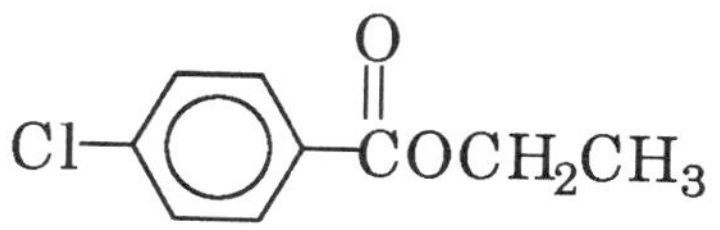

Cl—⟨benzene ring⟩—$\overset{\overset{\displaystyle O}{\|}}{C}OCH_2CH_3$

5.35

(a)

$$CH_3CH_2\overset{\overset{\displaystyle O}{\|}}{C}OC(CH_3)_3$$

(b)

$$CH_3CH_2\overset{\overset{\displaystyle H_3C}{|}}{C}H\overset{\overset{\displaystyle O}{\|}}{C}OCH(CH_3)_2$$

5.36

C < A < B < D

5.37

D < B < C < A

5.38

(a)

$$CH_3\overset{\overset{\displaystyle O}{\|}}{C}OCH_2\overset{\overset{\displaystyle CH_3}{|}}{C}HCH_3 + H_2O \xrightarrow{H^+} CH_3\overset{\overset{\displaystyle O}{\|}}{C}OH + HOCH_2\overset{\overset{\displaystyle CH_3}{|}}{C}HCH_3$$

(b)

$$CH_3CH_2O\overset{\overset{\displaystyle O}{\|}}{C}-\langle\text{cyclohexane}\rangle + H_2O \xrightarrow{H^+} HO\overset{\overset{\displaystyle O}{\|}}{C}-\langle\text{cyclohexane}\rangle + HOCH_2CH_3$$

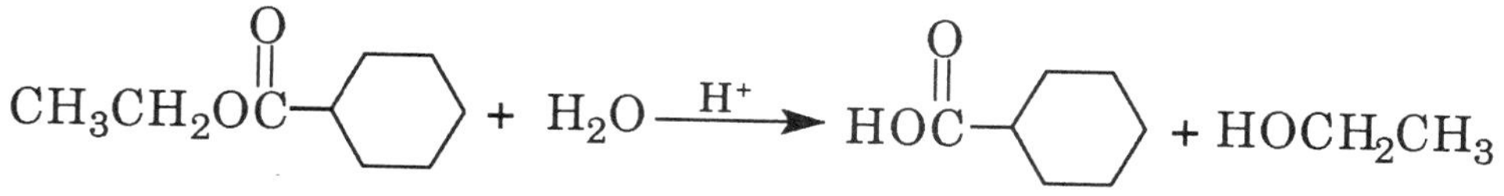

(c) No reaction   (d) No reaction

5.39

(a)

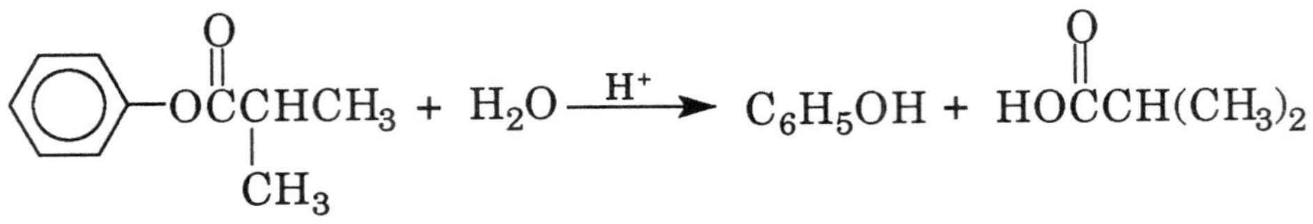

$$
\text{C}_6\text{H}_5\text{-O}\overset{\text{O}}{\overset{\|}{\text{C}}}\text{CHCH}_3 \;(\text{CH}_3) + \text{H}_2\text{O} \xrightarrow{\text{H}^+} \text{C}_6\text{H}_5\text{OH} + \text{HO}\overset{\text{O}}{\overset{\|}{\text{C}}}\text{CH(CH}_3)_2
$$

(b)

$$
\text{C}_6\text{H}_5\text{-}\overset{\text{O}}{\overset{\|}{\text{C}}}\text{OCHCH}_3\;(\text{CH}_3) + \text{H}_2\text{O} \xrightarrow{\text{H}^+} \text{C}_6\text{H}_5\overset{\text{O}}{\overset{\|}{\text{C}}}\text{OH} + \text{HOCH(CH}_3)_2
$$

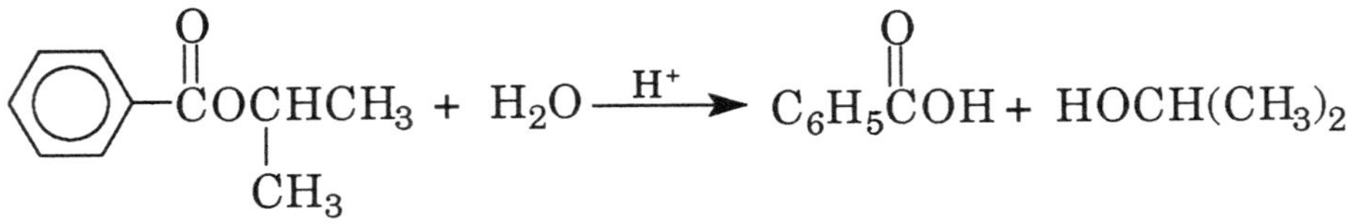

(c)   No reaction

(d)

$$
\text{CH}_3\text{CH}_2\text{O}\overset{\text{O}}{\overset{\|}{\text{C}}}\text{CH}_2\text{CH}_2\overset{\text{O}}{\overset{\|}{\text{C}}}\text{OCH}_2\text{CH}_3 + 2\text{H}_2\text{O} \xrightarrow{\text{H}^+}
$$

$$
\text{HO}\overset{\text{O}}{\overset{\|}{\text{C}}}\text{CH}_2\text{CH}_2\overset{\text{O}}{\overset{\|}{\text{C}}}\text{OH} + 2\text{CH}_3\text{CH}_2\text{OH}
$$

5.40

$$
\underset{\underset{\text{OH}}{|}}{\text{HOCH}_2\text{CHCH}_2\text{OH}} + \text{CH}_3(\text{CH}_2)_{12}\text{CO}_2\text{H} + \text{CH}_3(\text{CH}_2)_{14}\text{CO}_2\text{H}
$$

$$
+ \text{CH}_3(\text{CH}_2)_{10}\text{CO}_2\text{H}
$$

5.41 $\text{HOCH}_2\text{CH}_2\text{CH}_2\text{CO}_2\text{H}$

5.42

(a)

$$
\text{CH}_3\overset{\text{O}}{\overset{\|}{\text{C}}}\text{O}^-\text{Na}^+ + \underset{\underset{\text{CH}_3}{|}}{\text{HOCH}_2\text{CHCH}_3}
$$

(b)

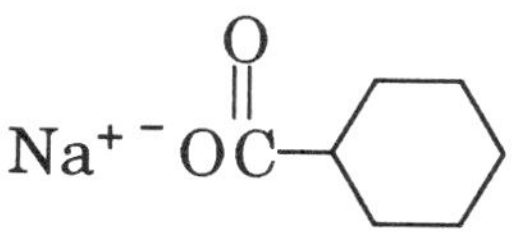

$$Na^+ \ ^-OC-\hexagon + HOCH_2CH_3$$

(c) No reaction   (d) No reaction

5.43

(a) $C_6H_5O^-K^+ + K^+ \ ^-O_2CCH(CH_3)_2$

(b)

$$C_6H_5CO^- \ K^+ + HOCH(CH_3)_2$$

(c)   No reaction

(d)

$$K^+ \ ^-OCCH_2CH_2CO^- K^+ + 2CH_3CH_2OH$$

5.44

$$HOCH_2CHCH_2OH + CH_3(CH_2)_{12}CO_2^- \ Na^+ + CH_3(CH_2)_{14}CO_2^- \ Na^+$$
$$OH \qquad\qquad + CH_3(CH_2)_{10}CO_2^- \ Na^+$$

5.45 $HOCH_2CH_2CH_2CO_2^-$

5.46 If a large mole excess of ethyl alcohol is used, the following equilibrium will lie so much on the side of the products that essentially all of the expensive acid will be converted to the ester.

$$RCO_2H + CH_3CH_2OH \rightleftharpoons RCO_2CH_2CH_3 + H_2O$$

5.47 The ester, **I**, accepts a proton from the acid catalyst.

The protonated ester, **II**, reacts with a water molecule to give **III**, which undergoes an internal proton transfer to give **IV**.

IV drops off a water molecule to give **V**.

The proton is transferred from **V** as an internal shift of an electron pair occurs to give the carbonyl group of the carboxylic acid.

5.48

(a)

$$CH_3O-\overset{\overset{\textstyle O}{\|}}{\underset{\underset{\textstyle OH}{|}}{P}}-OH$$

(b)

$$CH_3CH_2O-\overset{\overset{\textstyle O}{\|}}{\underset{\underset{\textstyle OH}{|}}{P}}-O-\overset{\overset{\textstyle O}{\|}}{\underset{\underset{\textstyle OH}{|}}{P}}-OH$$

(c)

$$CH_3CH_2CH_2O-\overset{\overset{\textstyle O}{\|}}{\underset{\underset{\textstyle OH}{|}}{P}}-O-\overset{\overset{\textstyle O}{\|}}{\underset{\underset{\textstyle OH}{|}}{P}}-O-\overset{\overset{\textstyle O}{\|}}{\underset{\underset{\textstyle OH}{|}}{P}}-OH$$

5.49   The phosphate esters are more soluble in water.

5.50   The phosphoric anhydride system. The breakup of this system releases considerable energy, but the breakup reaction is very slow in the absence of the appropriate enzyme.

5.51   In acetyl chloride, the attack by water occurs at the carbonyl carbon atom, which is very open and exposed. The site of attack in the phosphoric anhydride group is P, which is surrounded by electron-rich O atoms that tend to repel a water molecule.

5.52   (a)   Acetic acid
       (b)   Acetic acid
       (c)   Sorbic acid and sorbate salts

5.53   Parabens. As mold-inhibiting additives in cosmetics, pharmaceuticals, and food.

5.54   Two (or more) monomers are combined to give the polymer.

5.55   Dacron

5.56   An analgesic is a pain suppressant; an antipyretic is a fever reducer.

5.57   Salicylic acid is a stomach irritant.

5.58   (a)   Phenol group and the carboxylic acid group.
       (b)   The phenol group.
       (c)   The carboxylic acid group.

5.59   (a)   **A**
       (b) $C_6H_5CO_2^-$ and $C_6H_5O^-$
       (c) $C_6H_5O^-$

5.60   (a)   6.29 g methyl benzoate
       (b)   1.48 g CH_3OH; 1.88 ml CH_3OH
       (c)   The reaction involves an equilibrium. By using a large excess of methyl alcohol, the equilibrium shifts in accordance with Le Châtelier's principle so that essentially all of the benzoic acid is converted to the ester.

5.61

   (a) $CH_3\overset{|}{C}HCO_2^-\ Na^+$
         CH_3O

   (b) $CH_3CH_2CO_2H$ + CH_3OH

   (c) CH_3CO_2H

   (d)

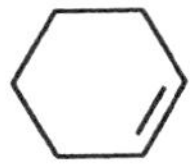

   (e) $(CH_3)_2CHCH_2CO_2^-\ Na^+$ + CH_3OH

   (f) $\overset{\displaystyle O}{\overset{\|}{}}$
         $(CH_3)_2CHCCH_3$

   (g) $\overset{\displaystyle O}{\overset{\|}{}}$
         $CH_3CH_2COCH_2CH_3$

   (h) CH_3CH_2CHO + $2CH_3OH$

(i)   No reaction

(j)

$$C_6H_5\overset{\overset{\displaystyle O}{\|}}{C}OCH_2CH_3$$

(k)   CH$_3$CHCH$_2$CH$_3$
      |
      Cl

(l)   No reaction

5.62

(a)

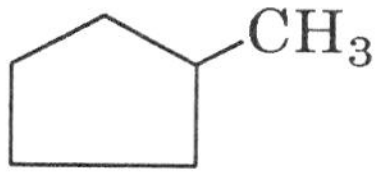

(b)

$$CH_3\overset{\overset{\displaystyle O}{\|}}{C}OCH_2CH_2CH_3 + CH_3\overset{\overset{\displaystyle O}{\|}}{C}OH$$

(c) $(CH_3)_2CHCH_2CO_2H$

(d)   No reaction

(e)

$$CH_3CH_2OH + Na^+\ ^-O\overset{\overset{\displaystyle O}{\|}}{C}CH_2CH_2\overset{\overset{\displaystyle O}{\|}}{C}O^-\ Na^+ + HOCH_3$$

(f)

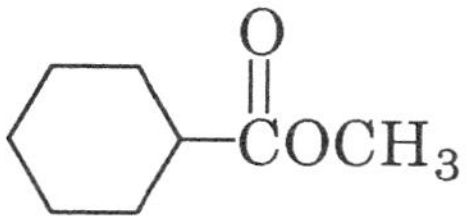

(g)

$$CH_3CH_2OH + Na^+\ ^-O\overset{\overset{\displaystyle O}{\|}}{C}CH_2CH_2OH + Na^+\ ^-O\overset{\overset{\displaystyle O}{\|}}{C}CH_2CH_3$$

(h)

$$C_6H_5\overset{\overset{\displaystyle O}{\|}}{C}H \ + \ 2CH_3OH$$

(i) $Na^+ \ ^-O_2CCH_2CH_2CH_2CH_3$

(j)

$$CH_3CH_2CH_2OH \ + \ HO\overset{\overset{\displaystyle O}{\|}}{C}CH_2CH_2OH \ + \ HO\overset{\overset{\displaystyle O}{\|}}{C}CH_3$$

(k)

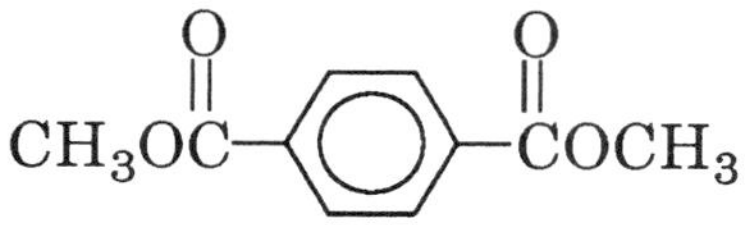

$$CH_3O\overset{\overset{\displaystyle O}{\|}}{C}{-}\!\!\bigcirc\!\!{-}\overset{\overset{\displaystyle O}{\|}}{C}OCH_3$$

(l) $CH_3OCH_2CH_2CO_2H$

Chapter 6

Practice Exercises, Chapter 6

1.   (a)   Isopropyldimethylamine
     (b)   Cyclohexylamine
     (c)   t-Butylisobutylamine

2.

(a) $(CH_3)_3CNH\underset{\overset{\displaystyle |}{CH_3}}{CH}CH_2CH_3$

(b)

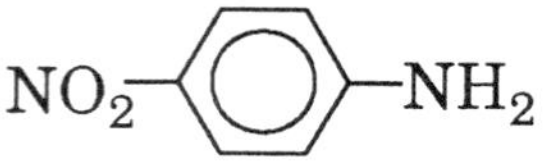

$$NO_2{-}\!\!\bigcirc\!\!{-}NH_2$$

(c)

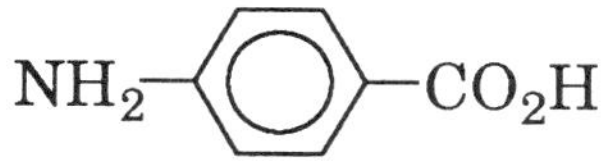

$$NH_2 - \text{⟨benzene ring⟩} - CO_2H$$

3.

(a) $C_6H_5NH_3^+$ (b) $(CH_3)_3NH^+$ (c) $^+NH_3CH_2CH_2NH_3^+$

4.

(a)

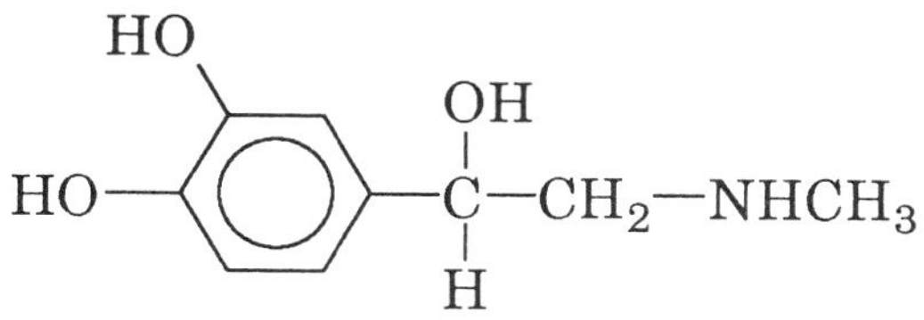

$$HO - \text{⟨benzene ring with HO substituent⟩} - \overset{\displaystyle OH}{\underset{\displaystyle H}{C}} - CH_2 - NHCH_3$$

(b)

$$CH_3O - \text{⟨benzene ring with } OCH_3 \text{ top and } OCH_3 \text{ bottom⟩} - CH_2CH_2NH_2$$

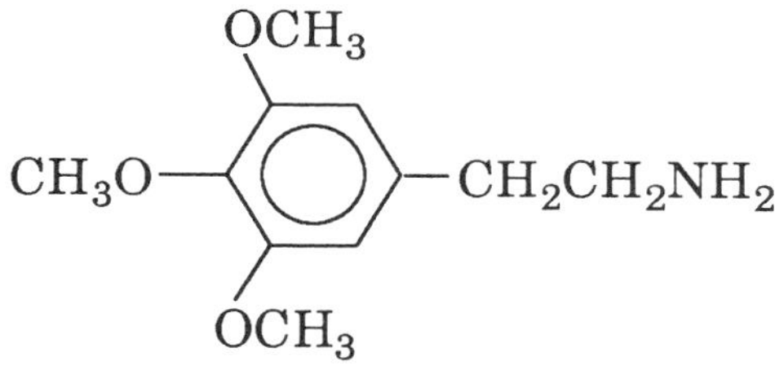

5.   (a) 4-Methylhexanamide
         (b) 2-Ethylbutanamide

6.

(a)

$$\underset{\displaystyle (CH_3)_2CH\overset{\displaystyle O}{\overset{\|}{C}}NHCH_3}{}$$

(b)

$$\underset{\displaystyle CH_3\overset{\displaystyle O}{\overset{\|}{C}}NHC_6H_5}{}$$

(c) No amide forms.   (d) No amide forms.

7.

(a) $C_6H_5CO_2H + NH_2CH_3$ (b) No hydrolysis occurs.

(c) $C_6H_5NH_2 + HO_2CCH_3$ (d) $NH_2CH_2CH_2NH_2 + 2CH_3CO_2H$

8.

$$NH_2CH_2CO_2H + \underset{\underset{CH_3}{|}}{NH_2CHCO_2H} + \underset{\underset{\underset{CH_3}{|}}{CH_3CH}}{NH_2CHCO_2H} + \underset{\underset{CH_2SH}{|}}{NH_2CHCO_2H}$$

Review Exercises, Chapter 6

6.1   (a) Aliphatic amide + ether group
        (b) Aliphatic amine + ester group
        (c) Aliphatic, heterocyclic amide
        (d) Aromatic *compound* overall because of the benzene ring, but the 2° amine is an *aliphatic* amine. (The amino group is not attached directly to the ring.)

6.2   (a) Aliphatic, heterocyclic amine + keto group
        (b) Aliphatic, heterocyclic amide
        (c) Aliphatic, heterocyclic amide
        (d) Aliphatic, heterocyclic amine + keto group

6.3   (a) 2° Amine and heterocyclic
        (b) 1, 3° Amine; 2, ester; 3, 1° amine (aromatic)
        (c) 1, Heterocyclic amine; 2, 2° (heterocyclic) amine
        (d) 1, 2° Alcohol; 2, 2° amine

6.4   (a) 1, Alkene; 2, ester; 3, 3° amine
        (b) 1, 3° Amine; 2, 1° alcohol; 3, ester
        (c) 1, Alkene; 2, 2° alcohol; 3, ether; 4, heterocyclic amine
        (d) 1, Amide; 2, 3° amine; 3, alkene; 4, 2° (aliphatic, heterocyclic) amine

6.5   (a) Isopropylpropylamine   (b) Ethylmethylpropylamine
        (c) *p*-Bromoaniline   (d) Dipropylamine

6.6   (a) Trimethylammonium chloride (b) Cyclohexylmethylamine
        (c) 3,5-Dichloroaniline   (d) Triisopropylamine

6.7   (a) $CH_3CH_2CH_2NH_3^+$ (b) $CH_3CH_2CH_2NH_2$
        (c) No reaction   (d) No reaction

6.8   (a)

(b)

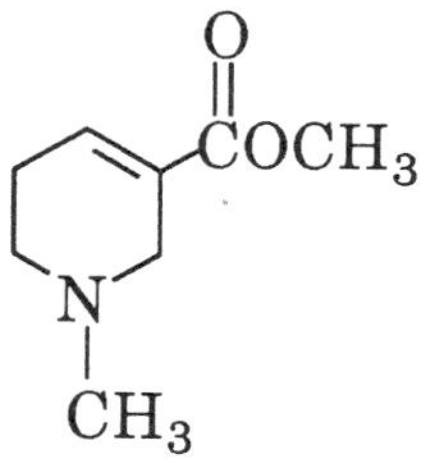

(c)

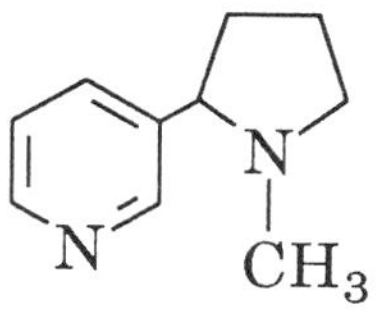

(d)

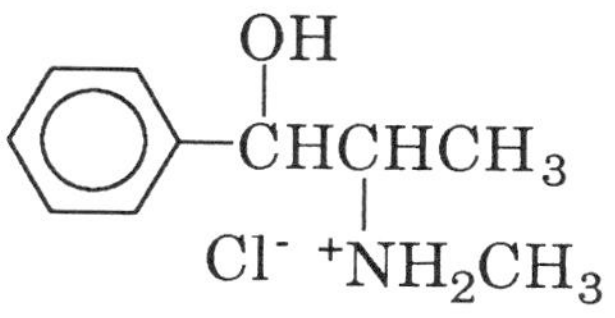

(e)

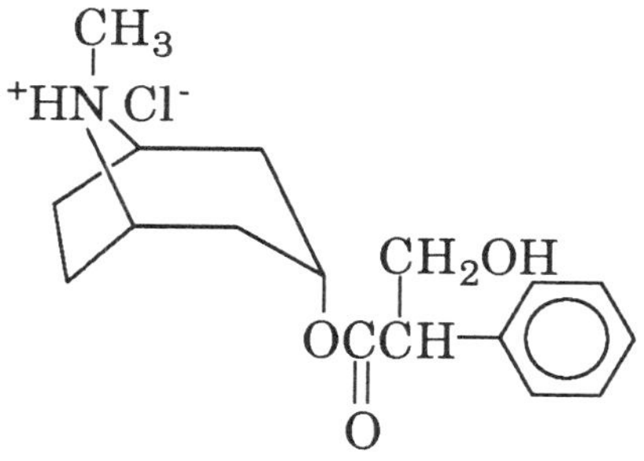

6.9   **A** is the stronger base; it is an amine (plus a ketone) **B** is an amide.
6.10   **B** is the stronger proton acceptor; it's a 3° amine. **A** has no space on N to accept a proton.
6.11   (a)   Butanamide
       (b)   3-Methylbutanamide
6.12   Caproamide
6.13 $C_6H_5CON(CH_3)_2$

6.14

$$\underset{\displaystyle \text{O}}{\overset{\displaystyle \|}{\text{NH}_2\text{C}}}\text{CH}_2\text{CH}_2\underset{\displaystyle \text{O}}{\overset{\displaystyle \|}{\text{C}}}\text{NH}_2$$

6.15

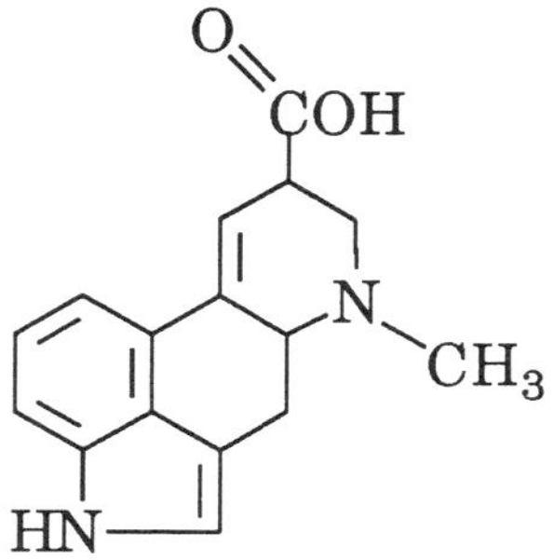

6.16

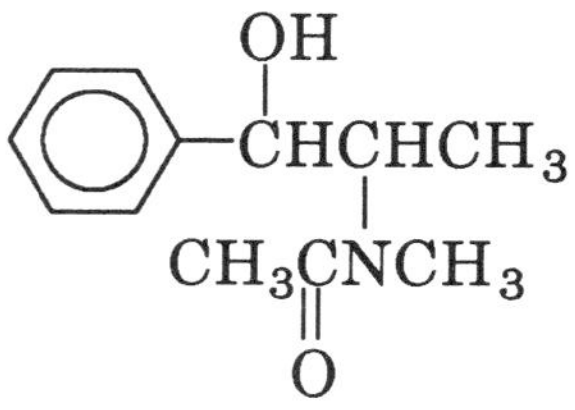

6.17

$$\underset{\displaystyle \text{O}}{\overset{\displaystyle \|}{\text{CH}_3\text{C}}}\text{Cl} + 2\,\text{NH}_3 \longrightarrow \underset{\displaystyle \text{O}}{\overset{\displaystyle \|}{\text{CH}_3\text{C}}}\text{NH}_2 + \text{NH}_4^+\text{Cl}^-$$

$$\text{CH}_3\overset{\displaystyle \text{O}}{\overset{\displaystyle \|}{\text{C}}}\text{O}\overset{\displaystyle \text{O}}{\overset{\displaystyle \|}{\text{C}}}\text{CH}_3 + 2\,\text{NH}_3 \longrightarrow \text{CH}_3\overset{\displaystyle \text{O}}{\overset{\displaystyle \|}{\text{C}}}\text{NH}_2 + \text{NH}_4^+ \ ^-\text{O}\overset{\displaystyle \text{O}}{\overset{\displaystyle \|}{\text{C}}}\text{CH}_3$$

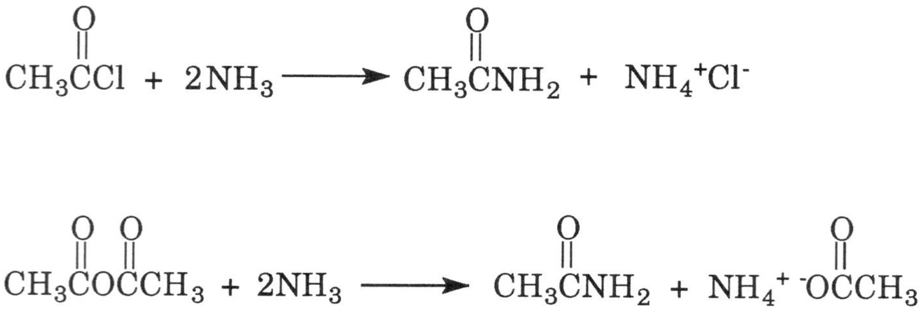

6.18

$$\underset{\overset{\displaystyle O}{\|}}{CH_3CCl} + 2NH_2CH_3 \longrightarrow \underset{\overset{\displaystyle O}{\|}}{CH_3CNHCH_3} + CH_3NH_3{}^+Cl^-$$

$$\underset{\overset{\displaystyle O\ \ O}{\|\ \ \|}}{CH_3COCCH_3} + 2NH_2CH_3 \longrightarrow \underset{\overset{\displaystyle O}{\|}}{CH_3CNHCH_3} + CH_3NH_3{}^+ \ \underset{\overset{\displaystyle O}{\|}}{{}^-OCCH_3}$$

6.19

(a)

$$\underset{\overset{\displaystyle \ CH_3}{|}}{NH_2CH_2\underset{\overset{\displaystyle O}{\|}}{C}NHCH\underset{\overset{\displaystyle O}{\|}}{C}-}$$

(b) Two

6.20

$$CH_3CH_2\underset{\overset{\displaystyle O}{\|}}{C}NH_2 + CH_3\underset{\overset{\displaystyle O}{\|}}{C}NH_2$$

6.21 (a) $CH_3CH_2NH_2 + CH_3CO_2H$
  (b) $(CH_3)_2CHNH_2 + CH_3CH_2CO_2H$
  (c) $CH_3NH_2 + (CH_3)_2CHCO_2H$
  (d) Does not hydrolyze.

6.22

(a)
$$\underset{\overset{\displaystyle \ \ \ CH_3}{|}}{NH_2CHCH_2CO_2H} + NH_2CH_2CO_2H$$

(b)
$$2NH_3 \ + \ \underset{\overset{\displaystyle \ \ \ \ \ \ \ \ \ \ \ \ \ \ \ \ CH_3}{|}}{HO_2CCH_2CHCO_2H}$$

(c) $CH_3CHCH_2CH_2CH_2CO_2H$
          |
          NH_2

(d) $2NH_3 + (H_2CO_3)$. The latter breaks up into $CO_2 + H_2O$.

6.23   Compounds that the body makes in special glands to serve as chemical messengers.

6.24   Adrenergic agents

6.25   Epinephrine and norepinephrine

6.26   Adrenergic drugs

6.27   The 1,2-hydroxybenzene ring system (the catechol system).

6.28   Yes

6.29   The β-phenylethylamines

6.30   (a) Dexidrin   (b) Amphetamines

6.31   The amide group

6.32   The hydrogen bond

6.33   (a) Water reacts quantitatively with alkenes, acetals or ketals, esters, and amides. (Not shown are the hydrolyses of acid chlorides, acid anhydrides, and esters of phosphoric acid.) The R groups can be alike or different. (H)R means that the group can be H or R.

$$\ce{>C=C<} + H_2O \xrightarrow{H^+} \quad \underset{\substack{| \quad | \\ H \quad OH}}{-\overset{|}{C}-\overset{|}{C}-}$$

Alkene   Alcohol

$$\underset{\substack{\text{Acetal or} \\ \text{ketal}}}{(H)R\overset{\displaystyle (H)R}{\underset{|}{C}}(OR)_2} + H_2O \xrightarrow{H^+} \underset{\substack{\text{Aldehyde} \\ \text{or ketone}}}{(H)R\overset{O}{\overset{||}{C}}R(H)} + \underset{\text{Alcohol}}{2HOR}$$

$$\underset{\text{Ester}}{(H)R\overset{O}{\overset{||}{C}}OR} + H_2O \xrightarrow{H^+} \underset{\substack{\text{Carboxylic} \\ \text{acid}}}{(H)R\overset{O}{\overset{||}{C}}OH} + \underset{\text{Alcohol}}{HOR}$$

$$\text{(H)RCNH}_2 + \text{H}_2\text{O} \xrightarrow{\text{H}^+} \text{(H)RCOH} + \text{NH}_3$$

Amide   Carboxylic   Ammonia
                    acid

(The Hs on N of the amide can be replaced by one or two alkyl groups.

(b)   The groups that can be hydrogenated are alkenes, aldehydes and ketones, and disulfides.

$$\text{C=C} + \text{H}_2 \xrightarrow{\text{catalyst}} -\overset{|}{\underset{\text{H}}{\text{C}}}-\overset{|}{\underset{\text{H}}{\text{C}}}-$$

Alkene   Alkane

$$\text{(H)RCR(H)} + \text{H}_2 \xrightarrow{\text{catalyst}} \text{(H)R}\overset{\text{OH}}{\text{C}}\text{HR(H)}$$

Aldehyde   Alcohol
or ketone

$$\text{RSSR} + \text{H}_2 \xrightarrow{\text{catalyst}} \text{2RSH}$$

Disul-   Thioalcohol
fide

(c)   Oxidizable groups in our study are 1° and 2° alcohols, aldehydes, a n d thioalcohols.

$$\text{RCH}_2\text{OH} + \text{(O)} \longrightarrow \text{RCH} \overset{\text{O}}{\|}$$

1° Alcohol   Aldehyde

$$\underset{\text{2° Alcohol}}{\overset{\overset{\displaystyle OH}{|}}{RCHR}} + (O) \longrightarrow \underset{\text{Ketone}}{\overset{\overset{\displaystyle O}{\|}}{RCR}}$$

$$\underset{\text{Thioalcohol}}{2RSH} + (O) \longrightarrow \underset{\substack{\text{Disul-}\\\text{fide}}}{RSSR}$$

6.34   Acetal (or ketal) group in carbohydrates; ester group in fats and oils; and the amide group in proteins.

6.35   (a)   1.66 g benzoic acid
　　　 (b)   28.3 mL 0.482 M HCl

6.36

(a) $CH_3CO_2H + CH_3OH$

(b) $$\underset{CH_3CHCH_2CH_3}{\overset{\overset{\displaystyle OH}{|}}{}}$$

(c)   No reaction

(d) $$\underset{CH_3CH_2CNH_2}{\overset{\overset{\displaystyle O}{\|}}{}}$$

(e)   No reaction

(f)   No reaction

(g) $CH_3CHO + 2HOCH_2CH_3$

(h) $CH_3CH_2CO_2H$

(i)   No reaction

(j)
$$CH_3CH_2\overset{\overset{\textstyle O}{\|}}{C}OCH_3$$

(k) $CH_3CH_2CO_2^-\,Na^+\ +\ NH_3$

(l)
$$\underset{CH_3CH_2\overset{|}{C}HOCH_3}{\overset{OCH_3}{}}$$

(m) $CH_3CH_2SSCH_2CH_3$

(n) $C_6H_5CO_2^-\,Na^+\ +\ HOCH_2CH(CH_3)_2$

(o) $Cl^-\ {}^+NH_3CH_2CH_2CH(CH_3)_2$

(p) $CH_3CH_2CH_2CH_2OCH_3$

6.37

(a) $Na^+\ {}^-O_2CCH_2CH_2CH_3$

(b)   No reaction

(c) $CH_3CH_2OCH_2CH_2CO_2H$

(d) $Cl^-\ {}^+NH_3CH_2CH_2NH_3{}^+\ Cl^-$

(e)
$$\langle\!\!\!\bigcirc\!\!\!\rangle\!-\!\underset{CH_3CH_2}{\overset{OH}{\overset{|}{C}H}}CH_2CH_3$$

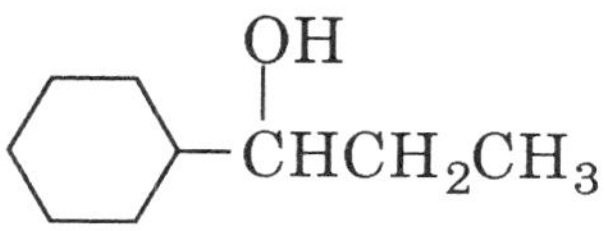

(f)   No reaction

(g)

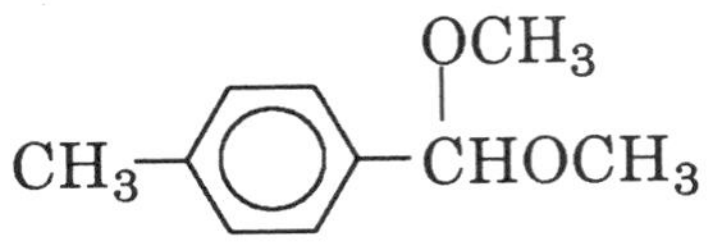

$$CH_3-\!\!\left\langle\bigcirc\right\rangle\!\!-\overset{\overset{\displaystyle OCH_3}{|}}{C}HOCH_3$$

(h)

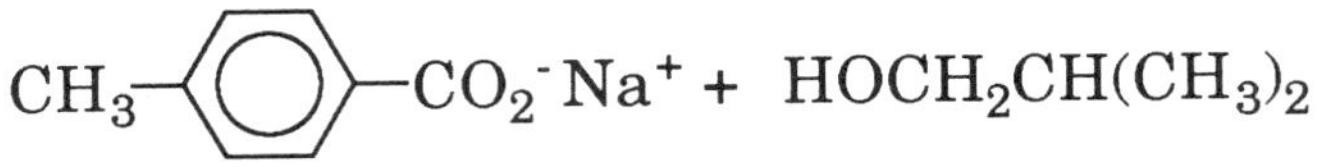

$$CH_3-\!\!\left\langle\bigcirc\right\rangle\!\!-CO_2^-\,Na^+ + \ HOCH_2CH(CH_3)_2$$

(i)

$$CH_3\overset{\overset{\displaystyle O}{\|}}{C}CH_2CH_3$$

(j)

$$CH_3\overset{\overset{\displaystyle O}{\|}}{C}CH_3 + \ 2HOCH_2CH_3$$

(k) $HO_2C(CH_3)_3CO_2H + 2\,CH_3CH_2OH$

(l) $CH_3CH_2CHO + 2CH_3OH$

(m)

$$CH_3\overset{\overset{\displaystyle O}{\|}}{C}NH_2 \ + \ NH_4^+\ {}^-\overset{\overset{\displaystyle O}{\|}}{O}CCH_3$$

(n) $2CH_3SH$

(o) $CH_3(CH_2)_5CO_2CH_3$

(p) $CH_3NH_2 + HO_2CCH_2CH_2NH_2 + HO_2CCH_2CH_3$

6.38   (a)   methyl butanoate
       (b)   1-bromo-2-methylpropane
       (c)   3-methylbutanal
       (d)   2-pentene
       (e)   2,2,4-trimethylpentane
       (f)   2-methyl-3-hexanone
       (g)   2-methyl-2-propanol
       (h)   pentanoic acid
       (i)   sodium ethanoate
6.39   (a)   sodium benzoate
       (b)   propylamine
       (c)   aniline
       (d)   propionic acid
       (e)   butyraldehyde
       (f)   isobutyl alcohol
       (g)   phenol
       (h)   diethyl ether
       (i)   ethyl butyrate
       (j)   acetamide
       (k)   acetone
6.40   **B**. It is a carboxylic acid that will become an anion at the basic pH and so more
       soluble in water. (**A** is an ester and **C** is an amine.)
6.41   **B**. It is an amine (plus a ketone), and so will form a water-soluble salt in acid.
       (**A** is an amide; **C** is a carboxylic acid)

Chapter 7

Practice Exercises, Chapter 7

1.

(a) 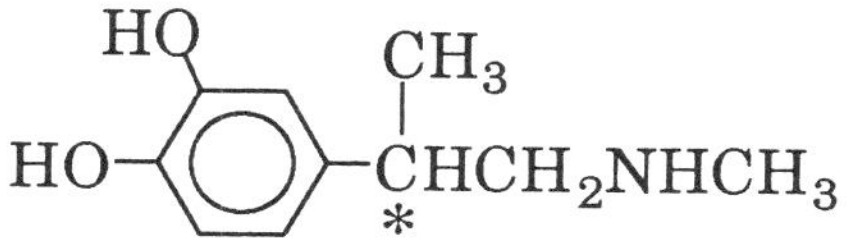   HO

      HO—⟨◯⟩—CHCH$_2$NHCH$_3$ with CH$_3$ on the CH carbon, marked *

(b) 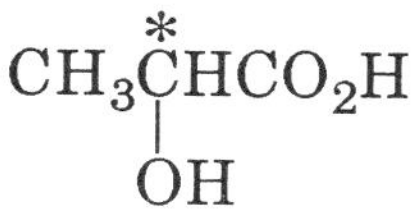   CH$_3$*CHCO$_2$H
                |
                OH

(c) $CH_3\overset{*}{C}H\overset{*}{C}HCO_2^-$
$\quad\quad\;\; |\quad\;\; |$
$\quad\quad\; HO\;\; \overset{+}{N}H_3$

(d)

$$HOCH_2\overset{*}{C}H-\overset{*}{C}H-\overset{*}{C}H\overset{O}{\overset{||}{C}}H$$
$$\;\;\;\;\;\;\;\;\;\;\;\; |\;\;\;\;\;\;\;\; |\;\;\; |$$
$$\;\;\;\;\;\;\;\;\;\;\;\; HO\;\;\; OH\;\; OH$$

2.   (a) 3   (b) 8   (c) 4
3.

$$CH_3\overset{*}{C}H\overset{*}{C}HCH_3$$
$$\;\;\;\;\;\; |\;\;\; |$$
$$\;\;\;\;\;\; HO\;\; OH$$

The two tetrahedral stereocenters are identical; they hold identical sets of four different groups, CH_3, OH, H, and $CH_3CH(OH)$.

Review Exercises

7.1   Their molecular structures must have different atom-to-atom sequences.
7.2 $CH_3CH_2CH_2OH$ and $(CH_3)_2CHOH$
7.3   The must have identical constitutions—identical molecular formulas, functional groups, and heavy-atom skeletons—but display different geometries.
7.4   *Enantiomers* are substances whose molecules are related as object to mirror image that cannot be superimposed. *Diastereomers* are stereoisomers that are not enantiomers.
7.5   (a) Stereoisomers   (b) Constitutional
7.6   Each unique substance must have a unique molecular structure.
7.7

(a) $HOCH_2-\overset{*}{C}H-\overset{*}{C}H-\overset{*}{C}H-\overset{*}{C}H-CH=O$
$$\; |\;\;\;\; |\;\;\;\; |\;\;\;\; |$$
$$\; OH\;\; OH\;\; OH\;\; OH$$

(b)   All are different.
(c)   16
(d)   8

7.8   No; glycine has no tetrahedral stereocenter.

7.9   (a)   No; citric acid has no tetrahedral stereocenter.

(b) $CH_2CO_2CH_3$
$HO\overset{*}{C}CO_2H$
CH_2CO_2H

(The bottom carboxyl group, but not the middle one, might also have been chosen for showing it as a methyl ester.)

7.10   148.5 °C. The designations (+) and (−) placed before otherwise identical names tell us that the two compounds are enantiomers, and enantiomers have identical physical properties.

7.11   They have identical constitutions, bond lengths, and bond angles and so must have identical polarities and identical London forces between molecules. Because the factors that determine physical properties are identical, the properties must be identical.

7.12   The ability of the molecules of enantiomers to be approached by and fit to an *achiral* molecule must be identical in the same way that either hand can "react" with an achiral water glass or broom handle.

7.13   In the same way that one hand fits best to its matching glove, both hand and glove being chiral, so must one enantiomer interact differently than the other with a chiral ion or molecule.

7.14   −0.375°

7.15   −21°

7.16   The amount of rotation is directly proportional to the actual population of chiral molecules encountered by the polarized light. This population increases directlyi with concentration or with path length.

7.17   The optical activity is lost. The effect on the polarized light by the chiral molecules of one enantiomer is cancelled by the opposite effect exerted by the molecules of the other enantiomer.

7.18   3.01 g/100 mL

7.19   4.86 g/100 mL

7.20   Strychnine. The calculated specific rotation for the sample is −139°, which corresponds to the value for strychnine, not for brucine.

7.21   Cortisone. The calculated specific rotation for the sample is +209, which corresponds to cortisone, not corticosterone.

7.22   (a)   They are not related as object to mirror image.
(b)   Because they are stereoisomers of each other.
(c)   What makes them different is not a lack of free rotation.
(d)   Diastereomers

7.23   Methane lacks a tetrahedral stereocenter and is not a member of a *set* of stereoisomers.

7.24   It has tetrahedral stereocenters and is a member of a set of stereoisomers some of which are chiral and optically active.

7.25   A 50:50 mixture of enantiomers consists of chiral molecules but the mixture as a whole is optically inactive.

7.26   (a)   2-Butanol
       (b)   3-Methylhexane

7.27   The cis isomer exists as a pair of enantiomers involving the tetrahedral stereocenter (with the OH group), and the trans isomer likewise exists a a pair of enantiomers.

Chapter 8

Practice Exercises, Chapter 8

1.   (a)   a = b
            c = e
     (b)   Compounds a and d are enantiomers.
     (c)   Compound c (or e) is a meso compound.

2.

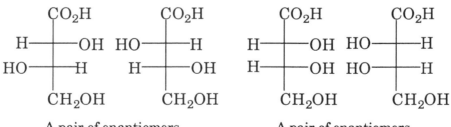

A pair of enantiomers   A pair of enantiomers

Review Exercises, Chapter 8

8.1   Materials, energy, and information.
8.2   Carbohydrates, lipids, and proteins.
8.3   Nucleic acids.
8.4   Organization.
8.5   (a)   C   (b)   D
      (c)   A   (d)   B and D

8.6

(a) 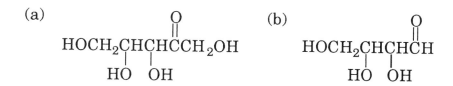 (b)

$$HOCH_2CHCHCCH_2OH$$
$$\quad\quad\; HO\;\; OH$$
with O double bonded

$$HOCH_2CHCHCH$$
$$\quad\quad HO\;\; OH$$
with O double bonded

8.7

$$HOCH_2CHCH$$
$$\quad\quad OH$$
with O double bonded

Glyceraldehyde

8.8

$$HOCH_2CCH_2OH$$
with O double bonded

Dihydroxyacetone

8.9   Polysaccharide.
8.10   Reducing carbohydrate.
8.11   (a) 2   (b) Trisaccharide
8.12   Because the formula mass of CH_2O is 30, the unknown must have 5 CH_2O units
       and be a pentose, $C_5H_{10}O_5$. The change to $C_{13}H_{18}O_9$ signifies a gain of 4 O atoms;
       thus 4OH groups have been acetylated. Gentle oxidation *without loss of carbon*
       signifies an aldehyde; the unknown is an aldopentose. Its reduction to pentane
       means a *straight chain*. So, write the structure of an aldopentose with a
       straight chain.

$$HOCH_2CHCHCHCH{=}O$$
$$\quad\quad\; HO\;\; OHOH$$

8.13

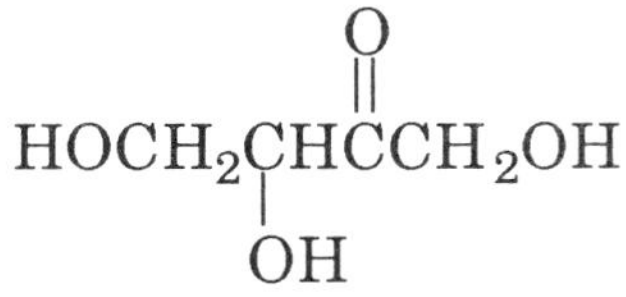

$$\text{HOCH}_2\underset{\underset{\text{OH}}{|}}{\text{CH}}\overset{\overset{\text{O}}{||}}{\text{C}}\text{CH}_2\text{OH}$$

8.14   **A** is ruled out because it has one carbon holding two OH groups, an unstable system.

8.15   Glucose

8.16   (a)   2

(b)

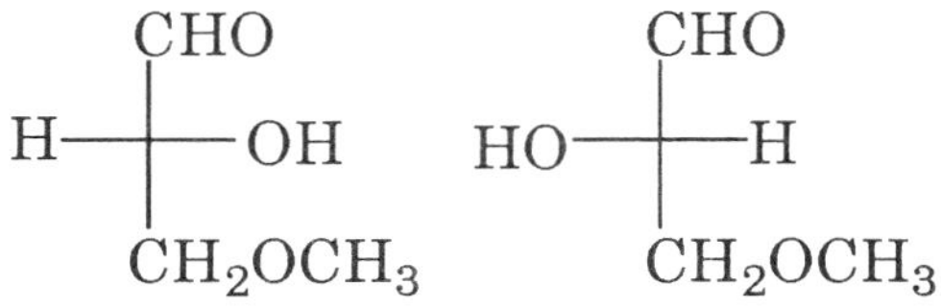

$$\begin{array}{c} \text{CHO} \\ \text{H} \underline{\qquad} \text{OH} \\ \text{CH}_2\text{OCH}_3 \end{array} \qquad \begin{array}{c} \text{CHO} \\ \text{HO} \underline{\qquad} \text{H} \\ \text{CH}_2\text{OCH}_3 \end{array}$$

(c)   **D**   **L**

8.17   It has no tetrahedral stereocenter

8.18

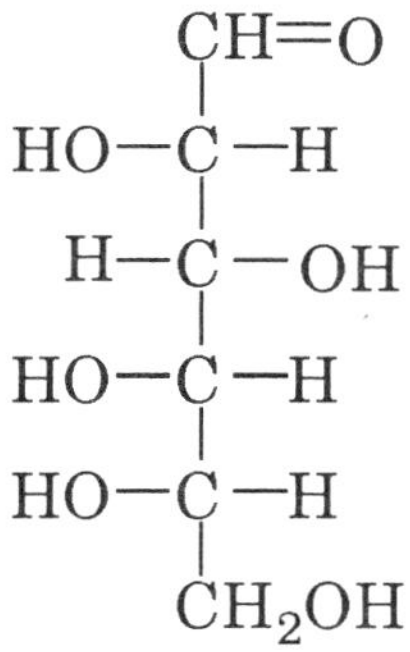

$$\begin{array}{c} \text{CH}{=}\text{O} \\ \text{HO}-\text{C}-\text{H} \\ \text{H}-\text{C}-\text{OH} \\ \text{HO}-\text{C}-\text{H} \\ \text{HO}-\text{C}-\text{H} \\ \text{CH}_2\text{OH} \end{array}$$

8.19

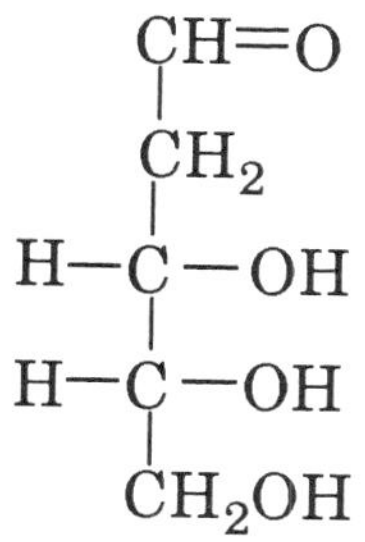

```
        CH=O
         |
        CH₂
         |
   H—C—OH
         |
   H—C—OH
         |
        CH₂OH
```

8.20

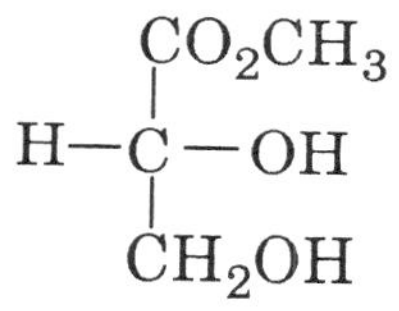

```
        CO₂CH₃
         |
   H—C—OH
         |
        CH₂OH
         D
```

It must be in the same configurational family as **D**-glyceraldehyde because no bonds to the tetrahedral stereocenter are broken during its synthesis from **D**-glyceraldehyde.

8.21   (a)   **L** family

(b)

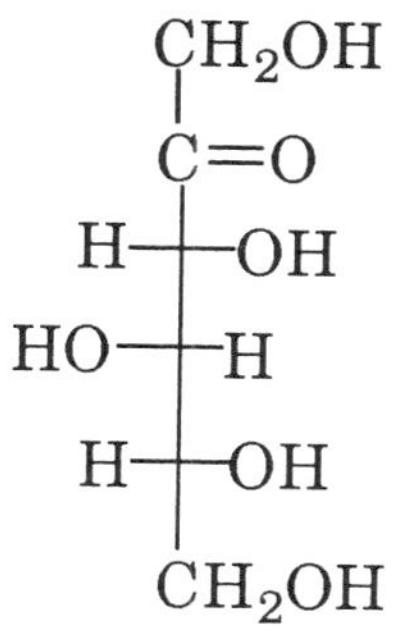

```
        CH₂OH
         |
        C=O
   H——OH
  HO——H
   H——OH
         |
        CH₂OH
```

(c)

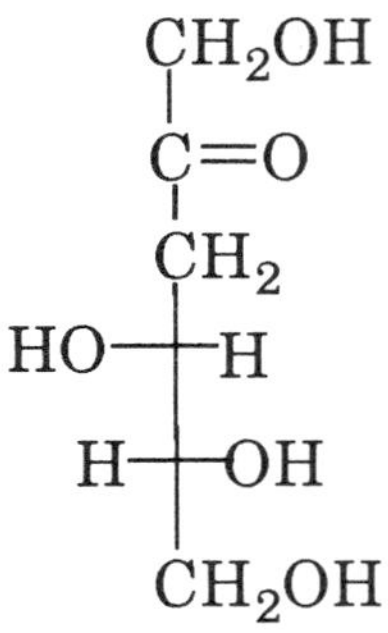

$$
\begin{array}{c}
\text{CH}_2\text{OH} \\
| \\
\text{C}=\text{O} \\
| \\
\text{CH}_2 \\
\text{HO}-\!\!-\text{H} \\
\text{H}-\!\!-\text{OH} \\
\text{CH}_2\text{OH}
\end{array}
$$

8.22

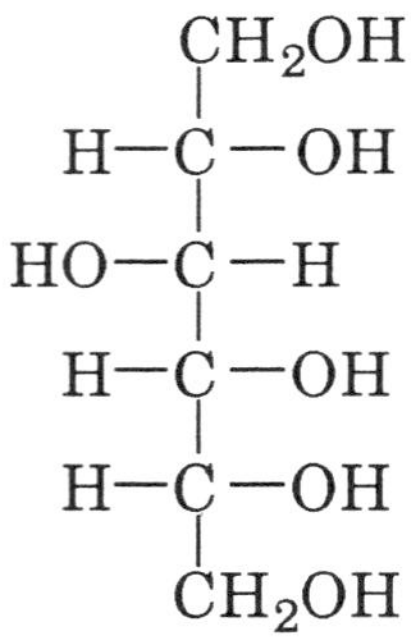

$$
\begin{array}{c}
\text{CH}_2\text{OH} \\
| \\
\text{H}-\text{C}-\text{OH} \\
| \\
\text{HO}-\text{C}-\text{H} \\
| \\
\text{H}-\text{C}-\text{OH} \\
| \\
\text{H}-\text{C}-\text{OH} \\
| \\
\text{CH}_2\text{OH}
\end{array}
$$

8.23

$$
\begin{array}{c}
\text{CO}_2\text{H} \\
| \\
\text{H}-\text{C}-\text{OH} \\
| \\
\text{HO}-\text{C}-\text{H} \\
| \\
\text{H}-\text{C}-\text{OH} \\
| \\
\text{H}-\text{C}-\text{OH} \\
| \\
\text{CH}_2\text{OH}
\end{array}
$$

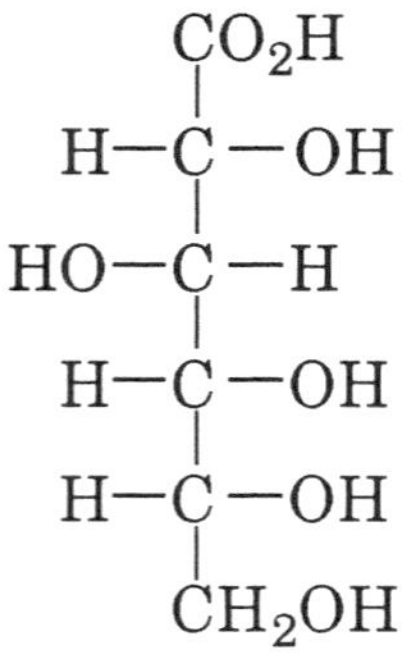

8.24

(a)

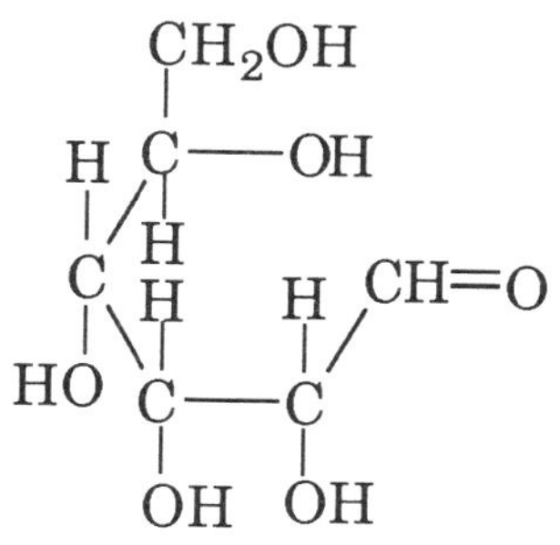

(b)   At carbon 3.

(c)   **D**-family. The relative positions of the CH_2OH group and the O atom of the ring tell us that the compound is in the **D**-family.

(d)   **D**-allose

(e)   An epimer

8.25

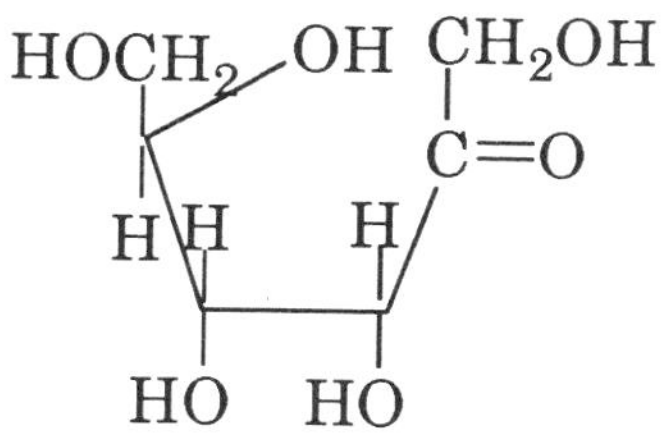

8.26

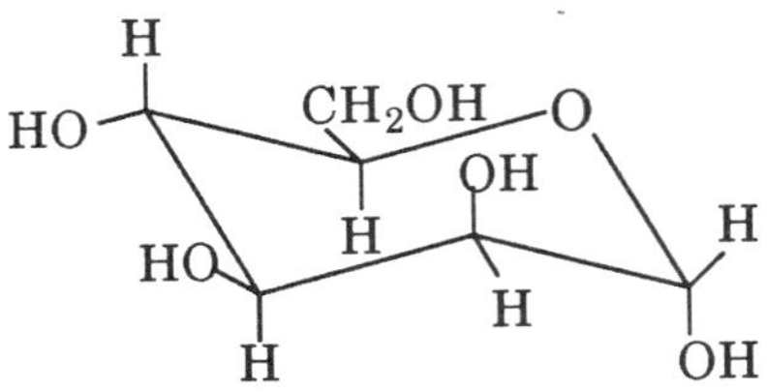

α -Mannose

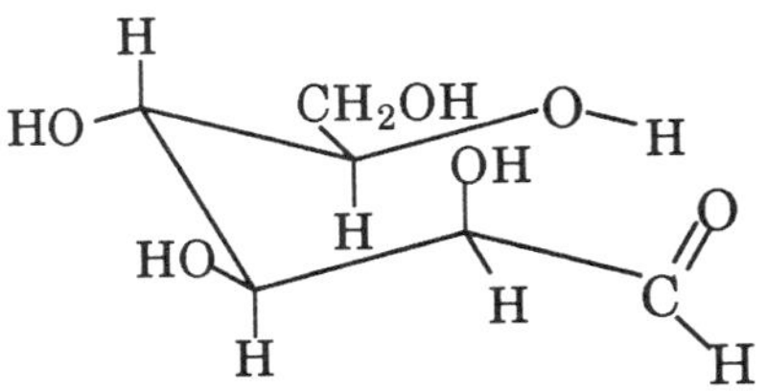

Open form of mannose

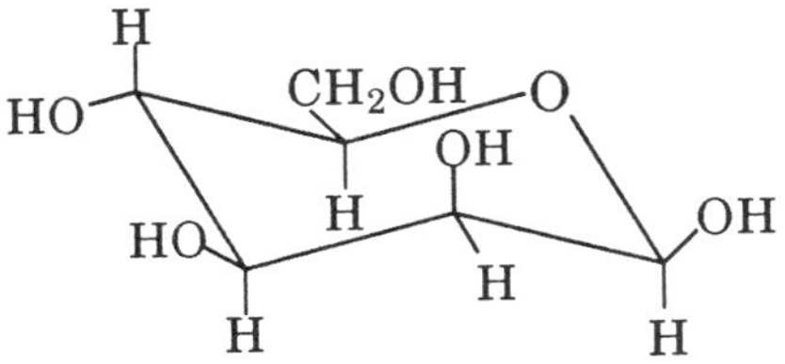

β-Mannose

8.27

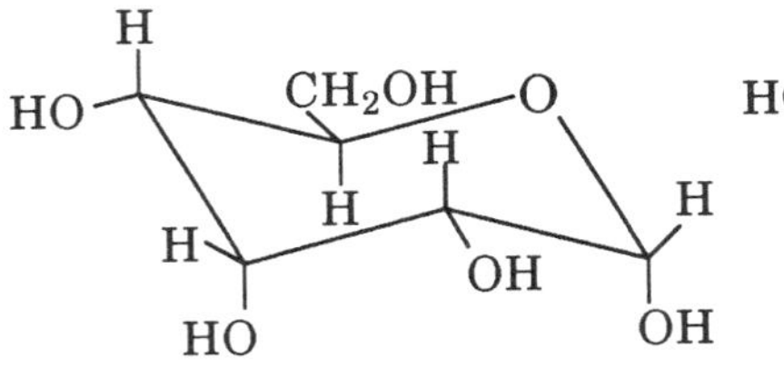

α -Allose

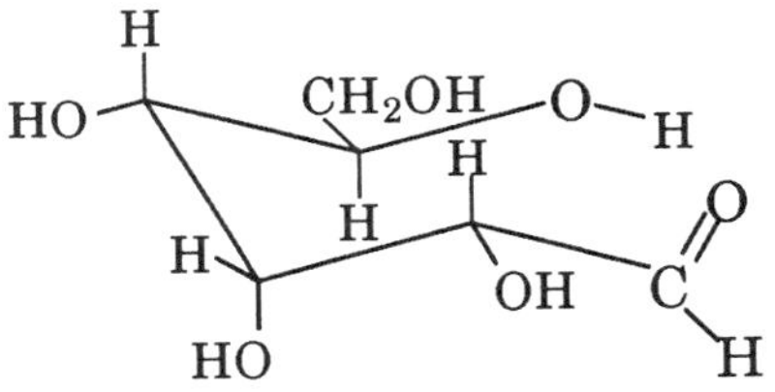

Open form of allose

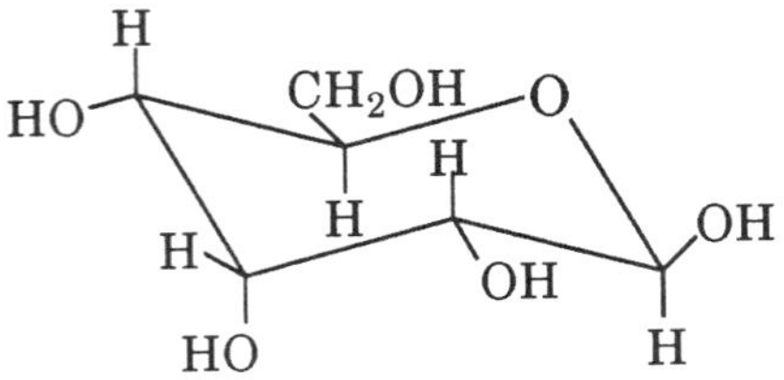

β-Allose

8.28   As molecules of the open-chain form are oxidized, the equilibrium continuously shifts to make more of the open chain form from the cyclic forms.

8.29   As the beta form is used, molecules of the other forms continuously change into it as the equilibria shift.

8.30

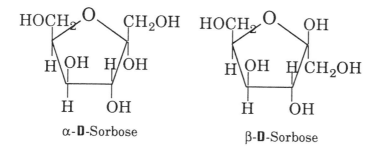

8.31   Something else: β-2-deoxyfructose

8.32

8.33   No, an OH group is required at position 4 to make possible the formation of the five-membered ring.

8.34   Sorbose is epimeric with fructose at positions 3 and 4.

α-**D**-Sorbose   β-**D**-Sorbose

8.35

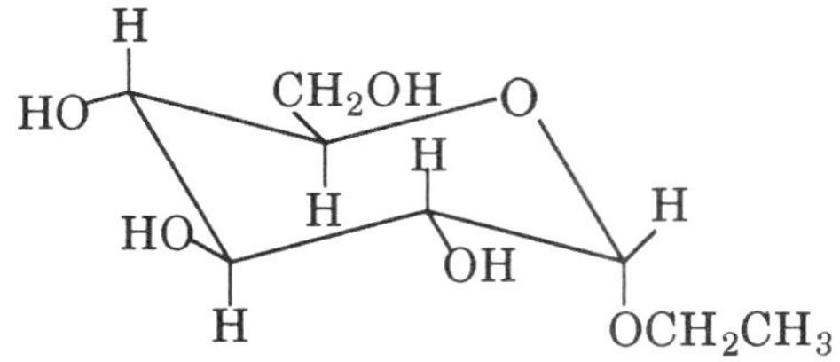

Ethyl α-glucoside

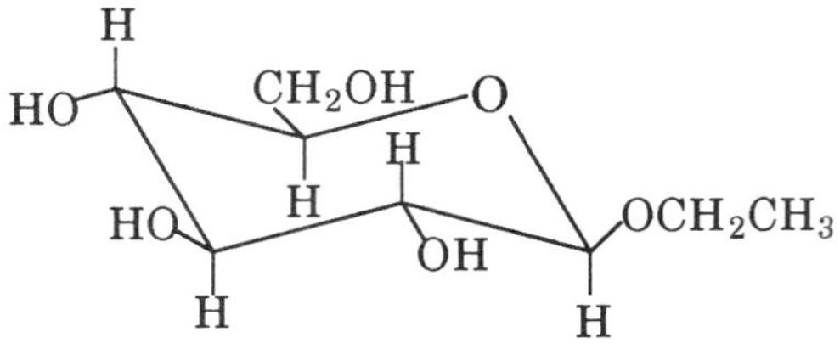

Ethyl β-glucoside

Another kind (diastereomers, Special Topic 7.1).

8.36

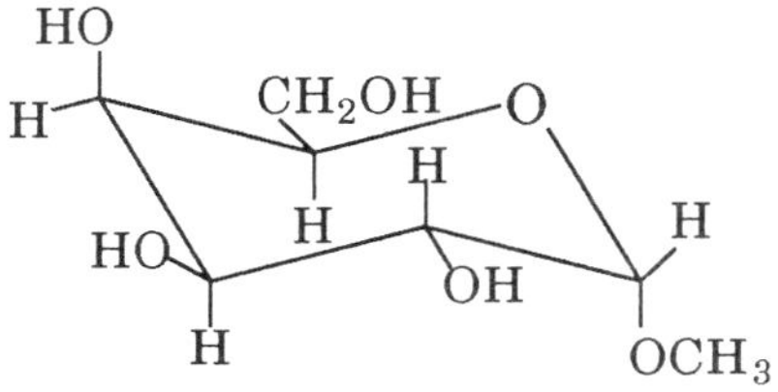

Methyl α-galactoside

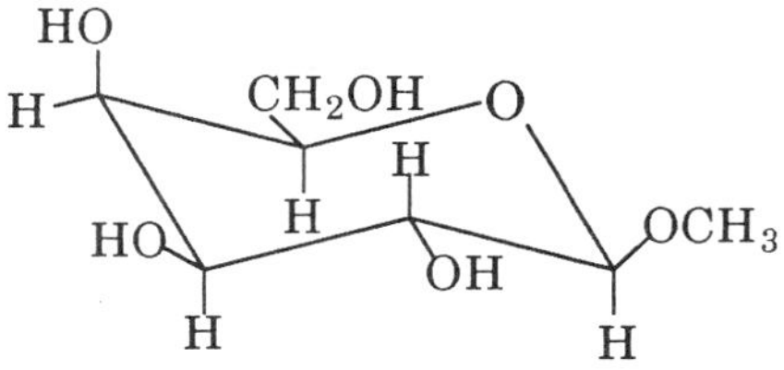

Methyl β-galactoside

These are cis-trans isomers in the sense that we have used this concept, because the OCH_3 group is on opposite sides of the rings. However, the term is just not used in connection with glycosides.

8.37   Maltose, lactose, and sucrose.

8.38   A 50 : 50 mixture of glucose and fructose.

8.39   Its molecules have no hemiacetal or hemiketal group at which ring-opening and ring-closing can occur.

8.40   (a) Yes, see arrow
        (b) Yes, see enclosure

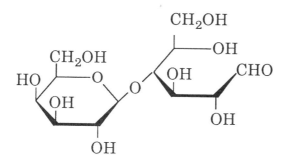

(c)   A β(1→4) bridge
(d)   Yes, it has the hemiacetal system so the open form of the corresponding ring (on the right) has an aldehyde group.
(e)   Maltose has an α(1→4) bridge between the two rings.
(f)   Two glucose molecules

8.41   (a)   No, it has no hemiacetal or hemiketal system and so cannot give a Tollens' or a Benedict's test.
(b)   No, for the same reason given in (a).
(c)   Two molecules of glucose

8.42

8.43

8.44   Amylose and amylopectin in starch; glycogen; cellulose

8.45   The oxygen bridges in amylose are $\alpha(1\rightarrow4)$ and in cellulose they are $\beta(1\rightarrow4)$.

8.46   They are polymers of α-glucose and have $\alpha(1\rightarrow4)$ bridges.

8.47   Amylose is an entirely linear polymer of α-glucose in which all of the oxygen bridges are $\alpha(1\rightarrow4)$, and amylopectin has branching in which $\alpha(1\rightarrow6)$ bridges link amylose-like strands to other amylose strands.

8.48   Humans lack the enzyme for catalyzing this reaction.

8.49   A test for starch. The reagent is a solution of iodine in aqueous potassium iodide. It is used to test for starch, and a positive test is the immediate appearance of a blue-black color.

8.50   They are very similar except that glycogen is more branched.

8.51   To store glucose units.

8.52   Solar energy is the energy that makes possible the synthesis of glucose from low-energy, simple molecules.

8.53 $$nCO_2 + nH_2O + \text{solar energy} \xrightarrow[\text{plant enzymes}]{\text{chlorophyll}} (CH_2O)_n + nO_2$$

8.54   Chlorophyll. Green

8.55   The oceans. By algae and phytoplankton

8.56   Oxygen is synthesized by photosynthesis and then consumed by the animals and plants that use the products of photosynthesis. As the living systems use these products, they release CO_2 and H_2O (and minerals), which is again re-used in photosynthesis.

8.57   Less nudging of electron-clouds occurs in the chair forms.

8.58   The electron clouds associated with groups in equatorial positions are farther apart than when they are in axial positions.

8.59

(a)

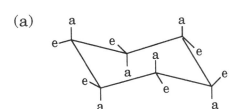

(b)

(c)

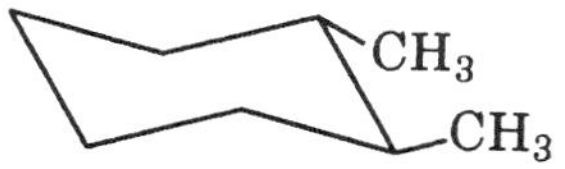

CH₃

CH₃

(d)

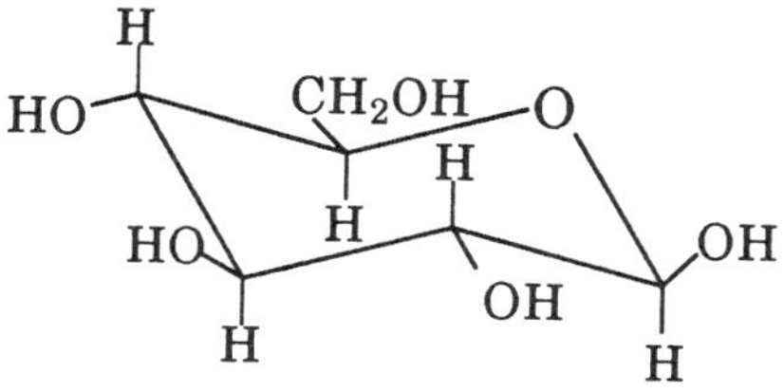

H

HO

CH₂OH   O

H

HO   H

OH

OH

H

H

β-Glucose
(all substituents are equatorial)

(e)

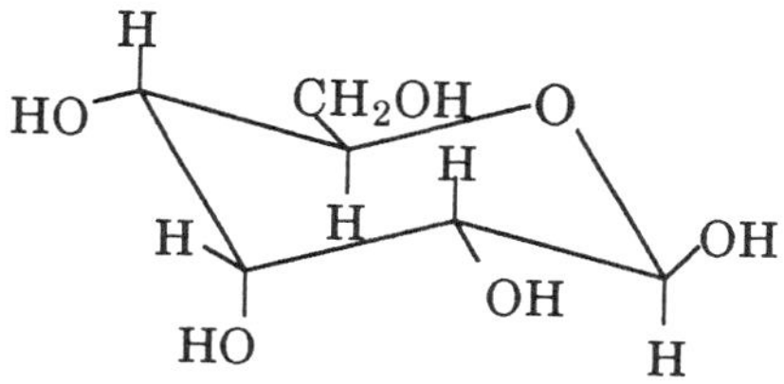

H

HO

CH₂OH   O

H

H   H

OH

OH

HO

H

β-Allose
(one substituent axial)

8.60   Over a period of a week some of the sucrose (a nonreducing sugar) hydrolyzes to give glucose and fructose, both being reducing sugars that give a positive Tollens' test.

8.61   An enzyme (amylase) in the saliva catalyzes the hydrolysis of enough of the starch so that the resulting solution fails to give the iodine test.

8.62   5

8.63   To form a cyclic hemiacetal, the ring would be limited to four atoms. Although four-membered rings are known, they are difficult to form because of the unfavorable bond angle (90° as compared to the normal angle of 109.5°).

Chapter 9

Practice Exercises, Chapter 9

1.

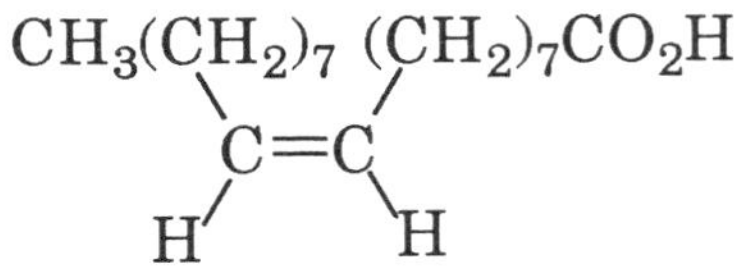

$$CH_3(CH_2)_7 \quad (CH_2)_7CO_2H$$
$$\underset{H}{\overset{}{\diagdown}} C = C \underset{H}{\overset{}{\diagup}}$$

2. $CH_3(CH_2)_{26}CO_2(CH_2)_{25}CH_3$

3.

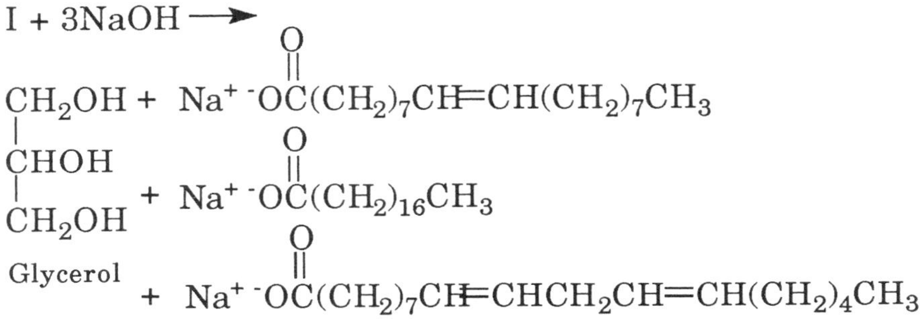

$I + 3NaOH \longrightarrow$

$CH_2OH + Na^+ \; {}^-OC(CH_2)_7CH{=}CH(CH_2)_7CH_3$
$\underset{|}{CHOH}$
$\underset{|}{CH_2OH} + Na^+ \; {}^-OC(CH_2)_{16}CH_3$

Glycerol

$+ Na^+ \; {}^-OC(CH_2)_7CH{=}CHCH_2CH{=}CH(CH_2)_4CH_3$

4.

$I + 3H_2 \xrightarrow{\text{catalyst}}$

$CH_2-O-\overset{O}{\overset{\|}{C}}(CH_2)_{16}CH_3$
$\underset{|}{CH}-O-\overset{O}{\overset{\|}{C}}(CH_2)_{16}CH_3$
$CH_2-O-\overset{O}{\overset{\|}{C}}(CH_2)_{16}CH_3$

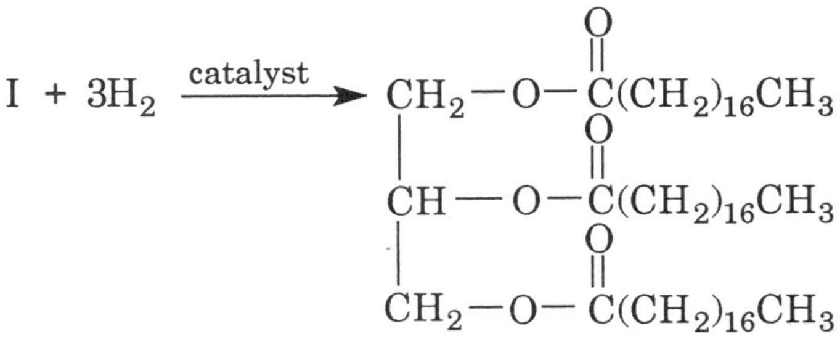

Review Exercises, Chapter 9

9.1   It is not obtainable from living plants or animals.

9.2   It is extractable from animal and plant sources by relatively nonpolar solvents.

9.3   It is soluble in water, and it isn't present in plant or animal sources.

9.4   It is present in undecomposed plant or animal materials and is extractable by relatively nonpolar solvents.

9.5   Palmitic acid, $CH_3(CH_2)_{14}CO_2H$
       Stearic acid, $CH_3(CH_2)_{16}CO_2H$

9.6

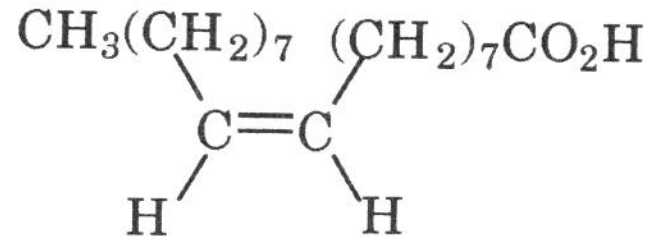

$CH_3(CH_2)_7$ $(CH_2)_7CO_2H$

C=C

H   H

Oleic acid

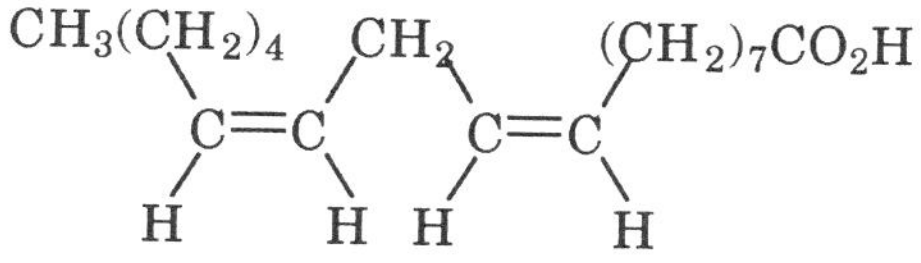

$CH_3(CH_2)_4$ CH_2 $(CH_2)_7CO_2H$

C=C   C=C

H   H H   H

Linoleic acid

CH_3CH_2 CH_2 CH_2 $(CH_2)_7CO_2H$

C=C   C=C   C=C

H   H H   H H   H

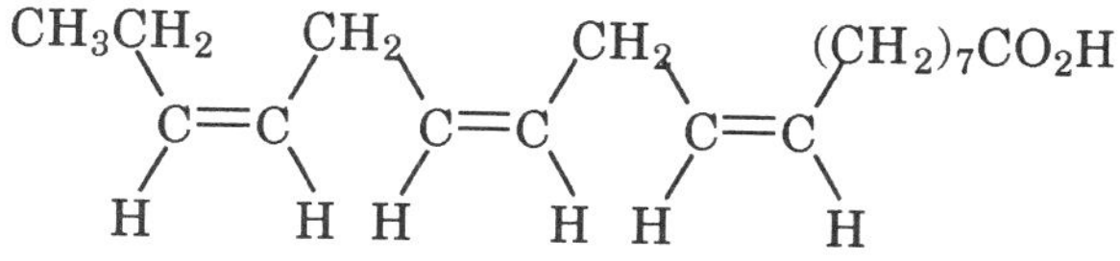

Linolenic acid

9.7   (a) $CH_3(CH_2)_{14}CO_2H + NaOH \rightarrow CH_3(CH_2)_{14}CO_2{}^-Na^+ + H_2O$

      (b) $CH_3(CH_2)_{14}CO_2H + CH_3OH \xrightarrow{\text{HCl}} CH_3(CH_2)_{14}CO_2CH_3 + H_2O$

9.8   The organic products of the reactions are the following.

    (a) 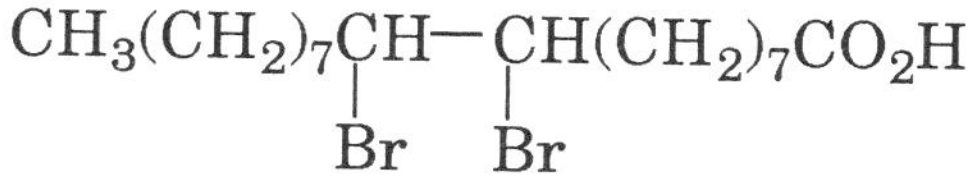 $CH_3(CH_2)_7CH-CH(CH_2)_7CO_2H$

                    Br   Br

    (b) $CH_3(CH_2)_7CH=CH(CH_2)_7CO_2{}^-K^+$
    (c) $CH_3(CH_2)_{16}CO_2H$
    (d) $CH_3(CH_2)_7CH=CH(CH_2)_7CO_2CH_2CH_3$

9.9   **A.** **B** is branched and has an uneven number of carbon atoms.

9.10   They are fatty acids with 20 carbons per molecule, including five-membered rings.

9.11

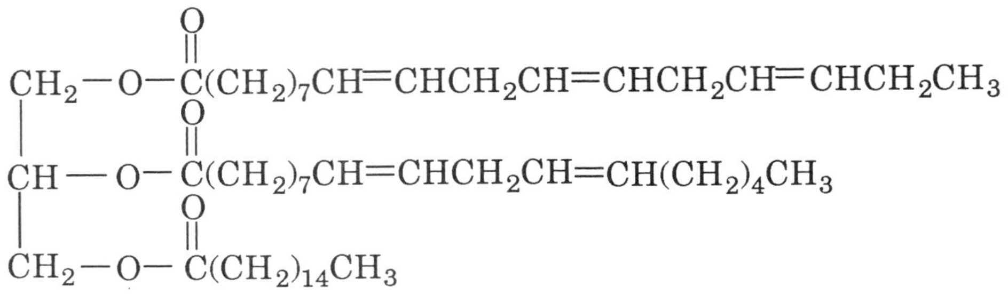

$$CH_2-O-\overset{\overset{\displaystyle O}{\|}}{C}(CH_2)_7CH=CHCH_2CH=CHCH_2CH=CHCH_2CH_3$$

$$CH-O-\overset{\overset{\displaystyle O}{\|}}{C}(CH_2)_7CH=CHCH_2CH=CH(CH_2)_4CH_3$$

$$CH_2-O-\overset{\overset{\displaystyle O}{\|}}{C}(CH_2)_{14}CH_3$$

9.12

$$CH_2-O-\overset{\overset{\displaystyle O}{\|}}{C}(CH_2)_{16}CH_3$$

$$CH-O-\overset{\overset{\displaystyle O}{\|}}{C}(CH_2)_7CH=CH(CH_2)_7CH_3$$

$$CH_2-O-\overset{\overset{\displaystyle O}{\|}}{C}(CH_2)_{14}CH_3$$

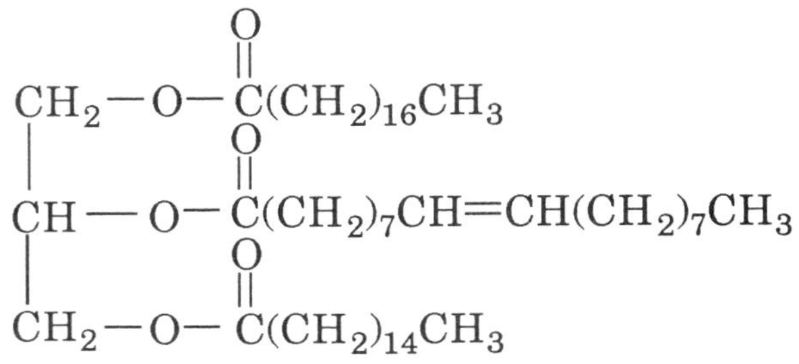

9.13

$$HOCH_2\underset{\underset{\displaystyle OH}{|}}{C}HCH_2OH + HO_2C(CH_2)_7CH=CHCH_2CH=CH(CH_2)_4CH_3$$

$$+ \ HO_2C(CH_2)_{12}CH_3 \ + \ HO_2C(CH_2)_7CH=CH(CH_2)_7CH_3$$

9.14

$$HOCH_2\underset{\underset{\displaystyle OH}{|}}{C}HCH_2OH + Na^+ \ {}^-O_2C(CH_2)_7CH=CHCH_2CH=CH(CH_2)_4CH_3$$

$$+ Na^+ \ {}^-O_2C(CH_2)_{12}CH_3 + Na^+ \ {}^-O_2C(CH_2)_7CH=CH(CH_2)_7CH_3$$

9.15   More than one structure is possible because the three different acyl groups can be joined in different orders to the glycerol unit. One possible structure is the following.

$$
\begin{array}{l}
\text{CH}_2\text{—O—}\overset{\displaystyle O}{\overset{\|}{\text{C}}}(\text{CH}_2)_{10}\text{CH}_3 \\
| \\
\text{CH—O—}\overset{\displaystyle O}{\overset{\|}{\text{C}}}(\text{CH}_2)_7\text{CH}=\text{CHCH}_2\text{CH}=\text{CH(CH}_2)_4\text{CH}_3 \\
| \\
\text{CH}_2\text{—O—}\overset{\displaystyle O}{\overset{\|}{\text{C}}}(\text{CH}_2)_7\text{CH}=\text{CH(CH}_2)_7\text{CH}_3
\end{array}
$$

9.16   Only one structure is possible if the molecule is to be chiral (have a tetrahedral stereocenter indicated by the asterisk).

$$
\begin{array}{l}
\text{CH}_2\text{—O—}\overset{\displaystyle O}{\overset{\|}{\text{C}}}(\text{CH}_2)_7\text{CH}=\text{CH(CH}_2)_7\text{CH}_3 \\
| \\
\overset{*}{\text{CH}}\text{—O—}\overset{\displaystyle O}{\overset{\|}{\text{C}}}(\text{CH}_2)_{10}\text{CH}_3 \\
| \\
\text{CH}_2\text{—O—}\overset{\displaystyle O}{\overset{\|}{\text{C}}}(\text{CH}_2)_{10}\text{CH}_3
\end{array}
$$

9.17   There are more alkene double bonds per molecule in vegetable oils than in animal fats.

9.18   The triacylglycerol molecules that are present have several alkene units per molecule, so the substances are more "polyunsaturated" than the animal fats.

9.19   Hydrogenation.

9.20   Butter melts on the tongue:, lard and tallow do not.

9.21 $CH_3(CH_2)_{16}CO_2(CH_2)_{17}CH_3$

9.22   **C. A** is ruled out because *both* the acid and alcohol portions of the wax molecule are usually long-chain. **B** is ruled out because *both* of these portions are likely to have an even number of carbons.

9.23   The molecules of both types give glycerol and phosphoric acid plus a fatty acid when fully hydrolyzed.

9.24   The phosphate ester unit; the negative charge is on oxygen. A nitrogen atom carries the positive charge.

9.25 Molecules of both types are derivatives of sphingosine (rather than glycerol). Sphingomyelin molecules have a phosphate ester unit but those of the cerebrosides have a monosaccharide unit instead.

9.26 The OH (alcohol) groups on the sugar unit.

9.27 Their molecules bear electrical charges at different locations.

9.28 Cell membranes.

9.29 Sphingomyelins and cerebrosides

9.30 Glycosidic links. The linkage involves the hemiacetal carbon of the sugar ring. The glycosidic link is more easily hydrolyzed (being an acetal, not an ordinary ether).

9.31

(a)

$$CH_2-O-\overset{\overset{O}{\|}}{C}(CH_2)_7CH=CHCH_2CH=CHCH_2CH=CHCH_2CH_3$$
$$\overset{|}{\underset{*}{C}}H-O-\overset{\overset{O}{\|}}{C}(CH_2)_7CH=CH(CH_2)_7CH_3$$
$$CH_2-O-\overset{\overset{O}{\|}}{\underset{\underset{O^-}{|}}{P}}OCH_2CH_2\overset{+}{N}(CH_3)_3$$

(b) A glycerophospholipid, because it is based on glycerol, not sphingosine.

(c) Yes, the asterisk in the structure of part (a) marks the tetrahedral stereocenter.

(d) A lecithin, because its hydrolysis would give 2-(trimethylamino)ethanol.

9.32

(a)

$$CH_2-O-\overset{\overset{O}{\|}}{C}(CH_2)_{10}CH_3$$
$$\overset{|}{\underset{*}{C}}H-O-\overset{\overset{O}{\|}}{C}(CH_2)_7CH=CH(CH_2)_7CH_3$$
$$CH_2-O-\overset{\overset{O}{\|}}{\underset{\underset{O^-}{|}}{P}}OCH_2CH_2NH_2$$

(b) A phosphoglyceride, because it is based on glycerol.

(c) Yes, the asterisk marks a tetrahedral stereocenter.

(d) A cephalin, because its hydrolysis would give ethanolamine.

9.33   The anion of cholic acid.

9.34   Vitamin D_3.

9.35   Estradiol, progesterone, testosterone, and androsterone are four.

9.36   Cholesterol.

9.37   Cholesterol. As surface-active agents (detergents) to aid in the digestion of lipids and the absorption of hydrophobic molecules from the digestive tract.

9.38   In various lipoprotein complexes.

9.39   The membrane consists mostly of two layers of phospholipid molecules whose hydrophobic parts intermingle with each other between the layers and whose hydrophilic parts face toward aqueous solutions whether they are inside the cell or outside.

9.40   The hydrophobic tails intermesh with each other between the two layers of the bilayer.

9.41   Proteins

9.42   The water-avoiding properties of the hydrophobic units and the water-attracting properties of the hydrophilic units.

9.43   The provide conduits for the movements of small ions and molecules and they furnish recognition sites for hormones.

9.44   Arachidonic acid

9.45   Aspirin inhibits the synthesis of prostaglandins. Since prostaglandins enhance a fever, aspirin reduces a fever by this action.

9.46   It has a double bond at the third carbon counting from the ω-carbon, the one most remote from the carboxyl group.

9.47   Marine oils

9.48   Some evidence suggests that they protect one against the heart disease associated with elevated cholesterol levels.

9.49   Detergent. Soap is just one example of a detergent.

9.50   A mixture of the sodium or potassium salts of long-chain fatty acids.

9.51   A synthetic detergent, because it works better in hard water.

9.52   The detergent properties are those of an anion.

9.53   The hydrophobic tails of the detergent ions become embedded in the grease layer and the hydrophilic heads stick out into the wash solution. As the grease breaks up, its tiny globules become pincushioned with detergent ions, and the effect of the charges is to help bring the globules into a colloidal dispersion.

9.54   (a) Yes

(b) Yes

(c) Ester groups

(d)

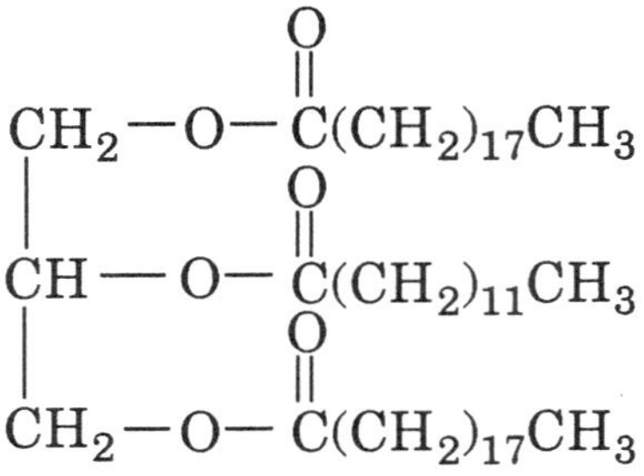

$$
\begin{array}{l}
\quad\quad\quad\quad\quad\quad O \\
\quad\quad\quad\quad\quad\quad \| \\
CH_2{-}O{-}C(CH_2)_{17}CH_3 \\
| \quad\quad\quad\quad\quad O \\
| \quad\quad\quad\quad\quad \| \\
CH{-}O{-}C(CH_2)_{11}CH_3 \\
| \quad\quad\quad\quad\quad O \\
| \quad\quad\quad\quad\quad \| \\
CH_2{-}O{-}C(CH_2)_{17}CH_3
\end{array}
$$

(e) No. Its fatty acid units have odd numbers of carbon atoms.

9.55   (a) Yes; it has both hydrophobic sections and polar groups (OH).

(b) No, steroids have *three* six-membered rings plus a five-membered ring.

Chapter 10

Practice Exercises, Chapter 10

1.

Glycine $^+NH_3CH_2CO_2{}^-$

Alanine $^+NH_3CHCO_2{}^-$
$$\quad\quad\quad |$$
$$\quad\quad CH_3$$

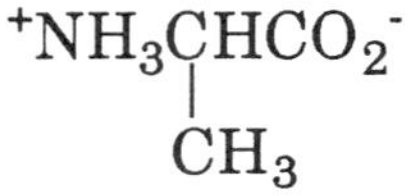

Lysine $^+NH_3CHCO_2{}^-$
$$\quad\quad\quad |$$
$$\quad\quad CH_2CH_2CH_2CH_2NH_2$$

Glutamic acid $^+NH_3CHCO_2{}^-$
$$\quad |$$
$$\quad CH_2CH_2CO_2H$$

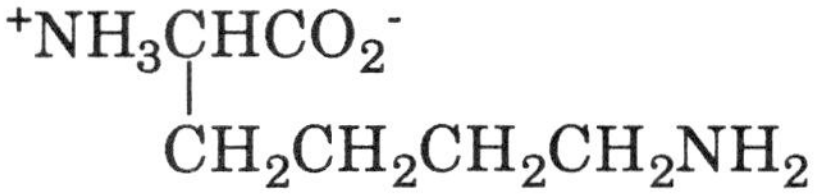

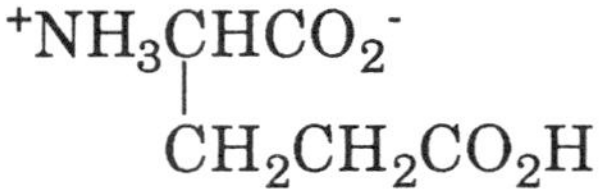

2.

(a)

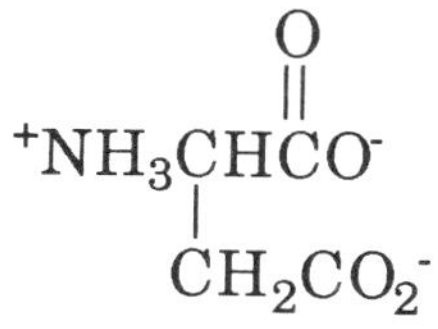

$$\overset{O}{\underset{\overset{|}{CH_2CO_2^-}}{\underset{|}{+NH_3CHCO^-}}}$$

(b)

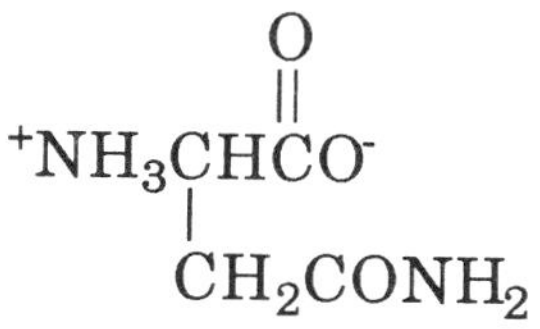

$$\overset{O}{\underset{\overset{|}{CH_2CONH_2}}{\underset{|}{+NH_3CHCO^-}}}$$

3.

$$\overset{O}{\underset{\overset{|}{CH_2CH_2CH_2NHCNH_2}}{+NH_3CHCO^-}} \quad \overset{+NH_2}{\underset{}{}}$$

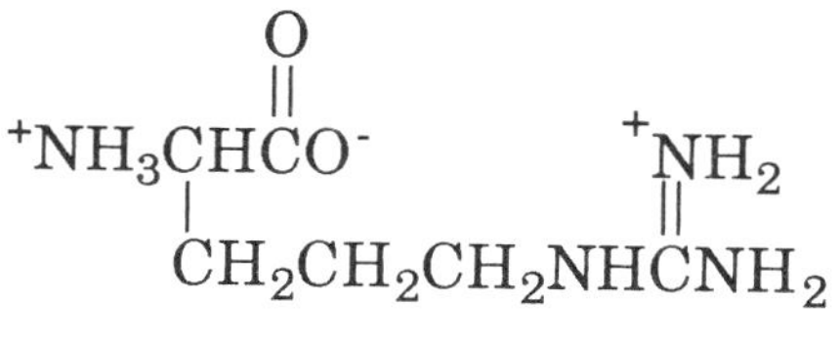

4. Hydrophilic; neutral. (The side chain has an amide group, not an amino group.)

5.

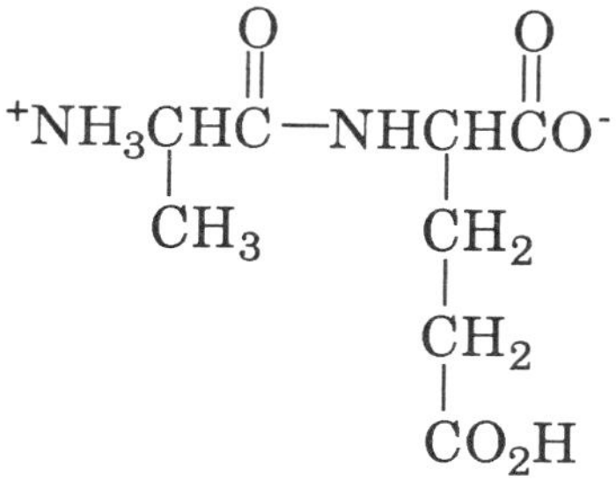

$$\overset{O \quad O}{\underset{\overset{|}{CH_3} \quad \overset{|}{CH_2}}{+NH_3CHC-NHCHCO^-}}$$
$$\underset{\overset{|}{CH_2}}{}$$
$$\underset{\overset{|}{CO_2H}}{}$$

Ala-Glu

$$\overset{O \quad O}{\underset{\overset{|}{CH_2} \quad \overset{|}{CH_3}}{+NH_3CHC-NHCHCO^-}}$$
$$\underset{\overset{|}{CH_2}}{}$$
$$\underset{\overset{|}{CO_2H}}{}$$

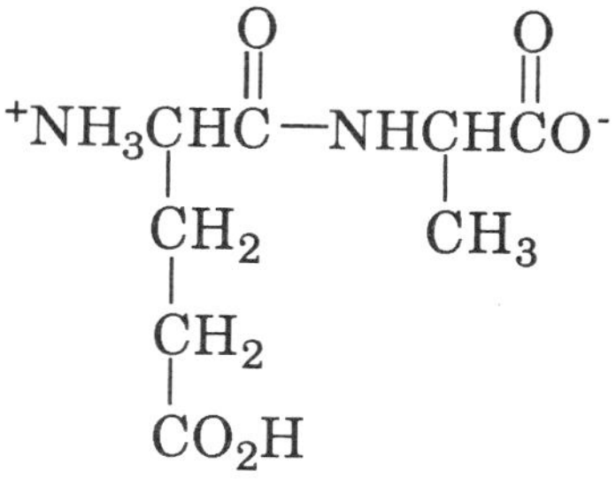

Glu-Ala

Review Exercises, Chapter 10

10.1 **B**. Its NH_3^+ group is not on the same carbon that holds the CO_2^- group.

10.2

(a)

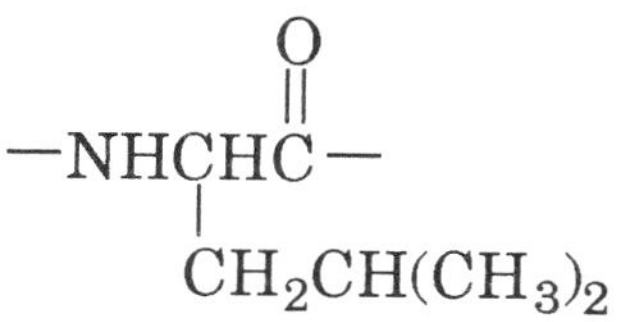

$$\overset{O}{\underset{\overset{|}{CH_2CH(CH_3)_2}}{-NHCHC-}}$$

(b) Leucine, Leu

(c) Hydrophobic

10.3 $^+NH_3CH_2CO_2H$

10.4

$$NH_2\underset{\underset{CH_3}{|}}{C}HCO_2^-$$

10.5

$$NH_2\underset{\underset{CH_3}{|}}{C}HCO_2CH_2CH_3$$

The polarity of this molecule is much, much less than the polarity of the dipolar ionic form of alanine. Thus, the ester molecules stick together with weaker forces than present between alanine molecules, and the ester has a lower melting point than alanine.

10.6   The ester has an NH_2 group as the proton acceptor, whereas alanine itself has the CO_2^- group as the acceptor. The CO_2^- group in alanine is made a weak acceptor by the electron withdrawal of the adjacent NH_3^+ group of the dipolar ion. (The withdrawal of electron density from a proton-accepting site renders this site less able to take and hold a proton.)

10.7   **A**. It has amine-like groups that can both donate hydrogen bonds to water molecules and accept them. (**B** has an alkyl group side chain, which is hydrophobic.)

10.8   **B**. Its side chain is hydrocarbon-like, whereas **A** has an OH group.

10.9   (a)   At a pH of 10. At the more basic pH, all protons that can be donated to base from the amino acid have left the molecule leaving it with a net charge of 1–.

(b)   To the anode.

10.10   Hydrophobic groups in the water-insoluble compound are unable to establish hydrogen bonds to water molecules. Thus the water molecules continue to "stick" to each other forcing the hydrophobic molecule to stay gathered together and insoluble.

10.11   Oxidizing agent

10.12   In the presence of additional acid, the following equilibrium shifts to the right in accordance with Le Châtelier's principle. This neutralizes the extra acid.

$$^+NH_3CH_2CO_2^- + H^+ \rightleftharpoons {}^+NH_3CH_2CO_2H$$

In the presence of additional base, the following equilibrium shifts to the right in accordance with Le Châtelier's principle. This neutralizes the extra base.

$$^+NH_3CH_2CO_2^- + OH^- \rightleftharpoons NH_2CH_2CO_2^- + H_2O$$

10.13

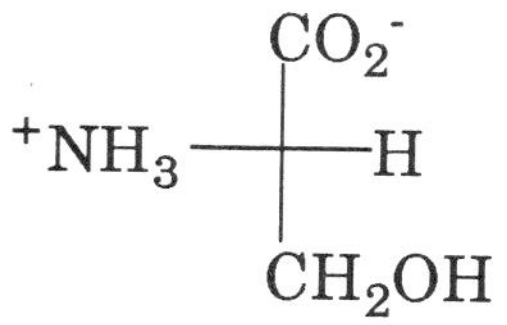

$$\begin{array}{c} CO_2^- \\ | \\ {}^+NH_3 \text{---} H \\ | \\ CH_2OH \end{array}$$

10.14   **A** has an amide bond not to the amino group of the α–position of an amino acid unit but to an amino group of a side chain (that of lysine). **B** has a proper peptide bond.

10.15

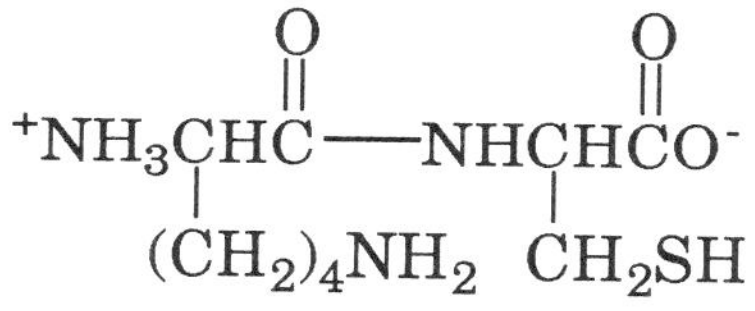

Lys-Cys 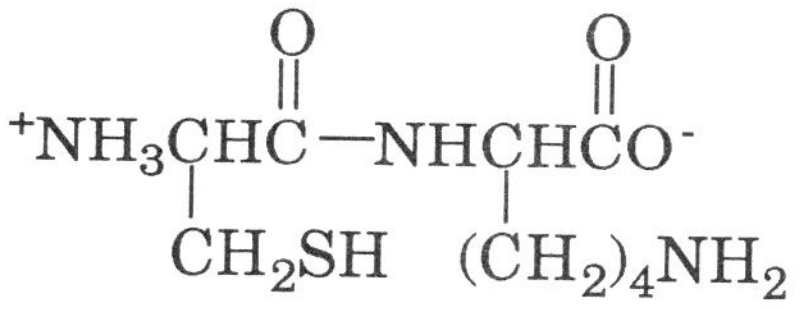   Cys-Lys

10.16

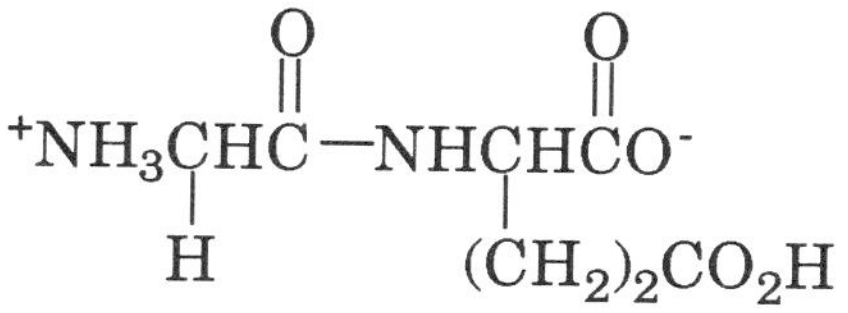

Gly-Glu 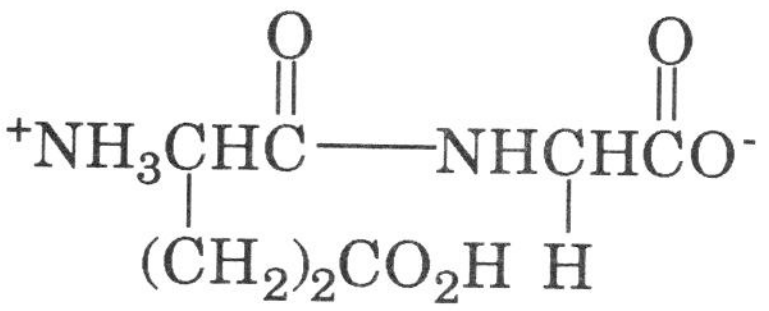   Glu-Gly

10.17   Lys-Glu-Cys   Glu-Cys-Lys Cys-Lys-Glu
        Lys-Cys-Glu   Glu-Lys-Cys Cys-Glu-Lys
10.18   Gly-Cys-Ala   Cys-Ala-Gly Ala-Gly-Cys
        Gly-Ala-Cys   Cys-Gly-Ala Ala-Cys-Gly

10.19

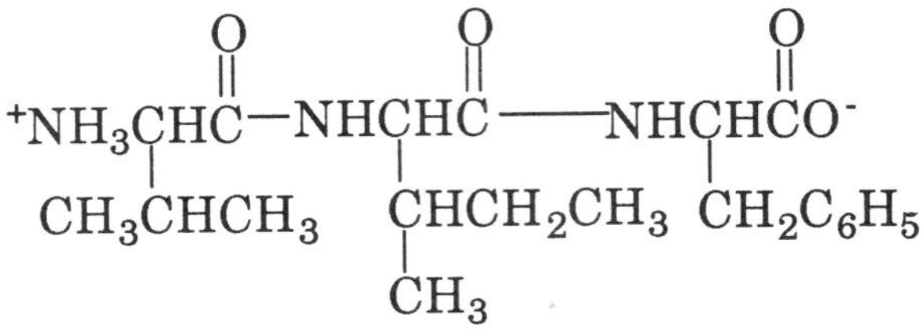

$$\overset{+}{N}H_3\underset{\underset{CH_3CHCH_3}{|}}{C}HC\overset{\overset{O}{||}}{}-NH\underset{\underset{\underset{CH_3}{|}}{CHCH_2CH_3}}{C}HC\overset{\overset{O}{||}}{}-NH\underset{\underset{CH_2C_6H_5}{|}}{C}HCO^-$$

10.20

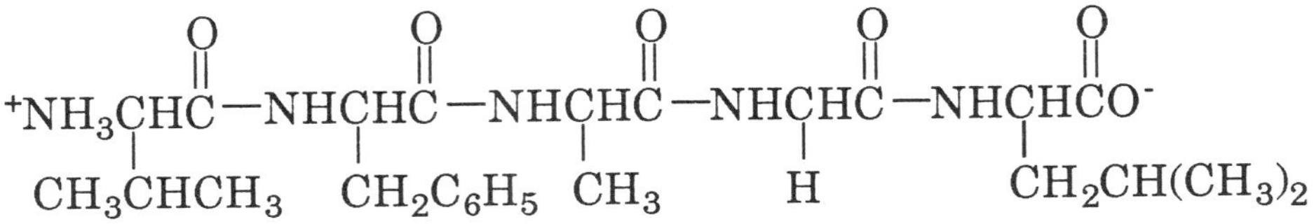

$$\overset{+}{N}H_3\underset{\underset{CH_3CHCH_3}{|}}{C}HC\overset{\overset{O}{||}}{}-NH\underset{\underset{CH_2C_6H_5}{|}}{C}HC\overset{\overset{O}{||}}{}-NH\underset{\underset{CH_3}{|}}{C}HC\overset{\overset{O}{||}}{}-NH\underset{\underset{H}{|}}{C}HC\overset{\overset{O}{||}}{}-NH\underset{\underset{CH_2CH(CH_3)_2}{|}}{C}HCO^-$$

10.21

$$\overset{+}{N}H_3\underset{\underset{CH_2CO_2H}{|}}{C}HC\overset{\overset{O}{||}}{}-NH\underset{\underset{(CH_2)_4NH_2}{|}}{C}HC\overset{\overset{O}{||}}{}-NH\underset{\underset{(CH_2)_2CO_2H}{|}}{C}HC\overset{\overset{O}{||}}{}-NH\underset{\underset{\underset{OH}{|}}{CHCH_3}}{C}HC\overset{\overset{O}{||}}{}-NH\underset{\underset{CH_2}{|}}{C}HCO^-$$

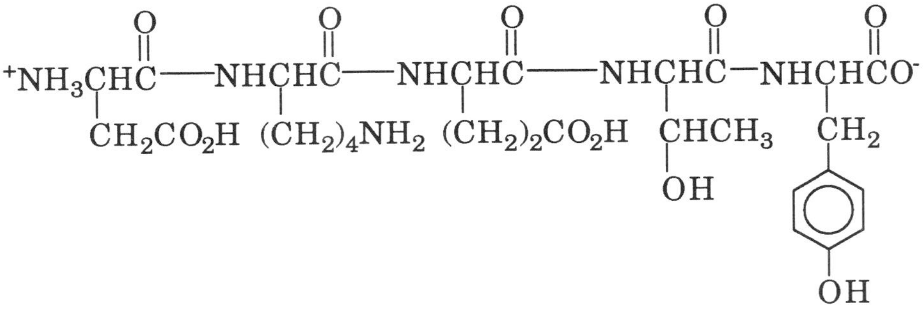

10.22 (a) **A**
       (b) **B**; it has only hydrophilic side chains. All those in **A** are hydrophobic.
10.23 (a) **D**; its side chains are hydrophobic whereas those of **C** are hydrophilic.
10.24

       Gly-Cys-Ala
              |
       Gly-Cys-Ala

10.25   It is the sequence of four atoms of the polypeptide "backbone" that starts with an α-carbon and goes through the carbonyl carbon and its attached nitrogen to the next α-position (see Figure 10.3). These four atoms lie in the same plane.

10.26   The side chain on the first α-position of the peptide group is cis to the H atom on the N atom and trans to the side chain on the second α-position.

10.27   Primary structure

10.28   A protein with the overall structure and shape as it is found in the living system.

10.29   Tertiary

10.30   It forms *after* a polypeptide with a cysteine side chain has been put together, so it forms after the primary structure has become set.

10.31   Quaternary

10.32   Reduce

10.33   It is a right-handed helix stabilized by hydrogen bonds between carbonyl oxygen atoms and H atoms on N atoms farther down the helix. The side chains project to the outside of the helix.

10.34   A left-handed helix structure.

10.35   It aids in the hydroxylation of proline and lysine residues without which collagen is not adequately made.

10.36   It consists of three left-handed collagen helices wound together as a right-handed, cable-like triple helix. Between the strands occur molecular "bridges." A fibril forms when individual triple helices overlap lengthwise.

10.37   Covalent linkages fashioned from lysine side chains.

10.38   Hydrogen bonds. The side chains project above and below the "sheets."

10.39   No, they represents portions of the secondary structure of a polypeptide and often both features are present.

10.40   Hydrophobic and hydrophilic interactions.

10.41   The force of attraction between a site bearing a full negative charge (e,g., a CO_2^- group on a glutamic acid or aspartic acid side chain) and a site with a full positive charge (e.g., a NH_3^+ group on a lysine side chain).

10.42   In the development of tertiary structure.

10.43   It consists of more than two polypeptides associated together in a specific way, each with primary, secondary, and tertiary structure.

10.44   (a)   Myoglobin is single stranded; hemoglobin has four subunits.
        (b)   Myoglobin is in muscle tissue; hemoglobin is in red cells.
        (c)   Both have the heme unit.
        (d)   Myoglobin accepts and stores O_2 molecules carried into tissue by hemoglobin molecules.

10.45

$$
\overset{\text{O}}{\underset{\underset{\text{CH}_2\text{OH}}{|}}{\overset{\|}{^+\text{NH}_3\text{CHCO}^-}}} + \overset{\text{O}}{\underset{\underset{\text{CH}_3}{|}}{\overset{\|}{^+\text{NH}_3\text{CHCO}^-}}} + \overset{\text{O}}{\underset{\underset{\text{CH}_3\text{CHCH}_3}{|}}{\overset{\|}{^+\text{NH}_3\text{CHCO}^-}}}
$$

$$
+ \;\; \overset{\text{O}}{\underset{\underset{(\text{CH}_2)_4\text{NH}_2}{|}}{\overset{\|}{^+\text{NH}_3\text{CHCO}^-}}} + {}^+\text{NH}_3\text{CH}_2\text{CO}_2^-
$$

10.46 When pH = pI, the protein molecules carry equal numbers of opposite charges and so have net charges of 0. The oppositely charged sites of separate molecules are able to attract each other and form huge clusters of protein molecules that drop out of solution.

10.47 Digestion is the hydrolysis of peptide bonds to give a mixture of amino acids. Denaturation is the disorganization of the overall shape of a protein without necessarily the breakup of peptide bonds.

10.48 The reducing agent cleaves disulfide groups to SH units, and on renaturation by oxidation the same disulfide groups form again.

10.49 A change in the value of something, like concentration, from one place to another.

10.50 Plasma

10.51 Cell fluid

10.52 Cell fluid

10.53 It moves sodium and potassium ions through membranes against their concentration gradients in order to reestablish these gradients.

10.54 Metabolic, energy-consuming reactions involving membrane components drive the transport of substances through membranes.

10.55 Gap junctions are tubules made of proteins and fastened between cells that provide avenues for the direct movements of ions and molecules from one cell to another.

10.56 Ca^{2+}

10.57 A unit of a carbohydrate molecule.

10.58 An oligosaccharide is made from more than two monosaccharide units, and it never has the thousands of such units commonly present in polysaccharide molecules.

10.59 It is the generic name of all polysaccharides.

10.60   D-glucosamine

10.61   A shock-absorbing gel-like material made of glycosaminoglycans and found in cartilage and other extracellular spaces that hold fibrous proteins.

10.62   The molecules of fibrous proteins (collagen and elastin) give tensile strength; ground substance provides resiliency and shock-absorbancy.

10.63   The resiliency of ground substance depends on the hydrogen bonds increasing the "stickiness" of the molecules of ground substance and their abilities to hold large amounts of water as water of hydration.

10.64   Because of the several OH groups per monosaccharide unit, the oxygen bridges between monosaccharide units can be formed in a large number of ways.

10.65   Fibrous proteins are insoluble in water; globular proteins are more soluble.

10.66   Collagen changes to gelatin when boiled in water.

10.67   They both have strengthening functions in tissue; both are fibrous proteins. The action of hot water on collagen turns it to gelatin, but elastin is unaffected in this way.

10.68   The globulins are less soluble in water than the albumins and they need the presence of dissolved salts to dissolve.

10.69   Fibrin is the protein that forms a blood clot. Fibrinogen is changed to fibrin by the clotting mechanism.

10.70   A lipoprotein

10.71   In a β-subunit, a valine residue with an isopropyl side chain has replaced a glutamic acid residue with a $CH_2CH_2CO_2H$ side chain. This affects the shape of the hemoglobin molecule.

10.72   Deoxygenated hemoglobin precipitates inside the red cell.

10.73   The distorted red cells are harder to pump and they can clump together to plug capillaries.

10.74   A membrane-bound protein to which a hormone (or neurotransmitter) molecule can bind and thus initiate some action in the cell.

10.75   RU 486 prevents the implantation of a fertilized ovum in the uterus by blocking the action of progesterone, a hormone needed to prepare the uterus for the implantation.

10.76   The linkages of oligosaccharide unit occur largely at protein surfaces, not in their interiors. Loss of the oligosaccharide thus leaves the protein shape intact.

10.77

$$^+NH_3CHC\overset{O}{\overset{\|}{}}-NHCHC\overset{O}{\overset{\|}{}}-NHCHC\overset{O}{\overset{\|}{}}-NHCHC\overset{O}{\overset{\|}{}}-NHCHC\overset{O}{\overset{\|}{}}O^-$$

with CH_3 side chains.

10.78   (a)   Yes. What is shown is a tripeptide written backwards from the conventional way.
         (b)   Phe
         (c)   Phe-Ala-Cys (N-terminus to C-terminus)
         (d)   No effect

Chapter 11

Practice Exercises, Chapter 11

1.   (a)   sucrose   (b) glucose   (c) protein   (d) an ester
2.   Feedback inhibition

Review Exercises, Chapter 11

11.1   (a)   A catalyst.
        (b)   It consists of a protein.
11.2   Each enzyme catalyzes a reaction for a specific substrate or a specific kind of reaction.
11.3   (a)   An apoenzyme is the wholly polypeptide part of the enzyme.
        (b)   A cofactor is a non-polypeptide molecule or ion needed to make the complete enzyme.
        (c)   A coenzyme is one kind of cofactor, an organic molecule.
11.4 $CO_2 + H_2O \rightleftharpoons HCO_3^- + H^+$
        Carbonic anhydrase provides one of the greatest rate enhancements of all enzymes.
11.5   It catalyzes the rapid reestablishment of the equilibrium after it has been disturbed.
11.6   Nicotinamide.
11.7   Riboflavin.
11.8

$$\underset{CH_3\overset{\overset{\displaystyle OH}{|}}{C}HCH_3}{} + NAD^+ \longrightarrow \underset{CH_3\overset{\overset{\displaystyle O}{\|}}{C}CH_3}{} + NAD{\cdot}H + H^+$$

11.9 $H^+ + NADH + FAD \rightarrow NAD^+ + FADH_2$
11.10   By a phosphate ester unit. NADPH
11.11   (a)   An oxidation   (b)   The transfer of a methyl group
        (c)   A reaction with water   (d)   An oxidation-reduction equilibrium

11.12   Lactose is a disaccharide and the substrate for the enzyme, lactase.

11.13   Hydrolysis is a kind of reaction catalyzed by a hydrolase enzyme.

11.14   Enzymes of identical function but with slight differences in structure.

11.15   CK(MM) in skeletal muscle; CK(BB) in brain; and CK(MB) in heart muscle.

11.16   Active site.

11.17   By the necessity of the fitting of the substrate molecule to the surface of the enzyme much as a key must fit to a particular lock.

11.18   During the formation of the enzyme-substrate complex, the substrate induces modifications in the enzyme's shape to enable a better fit .

11.19   One

11.20   (a) $V \propto [E_o]$

(b) $V \propto [S]$

11.21   The value of $[S]$ at which the reaction rate is one-half of the maximum rate.

11.22   At the another site. *Allosteric* describes an action induced at a site on an enzyme molecule at some distance from the active site.

11.23   The enzyme has more than one active site and that the (slower) activation of one site automatically causes the activation of the other(s).

11.24   As the substrate binds to one active site it induces changes in the shape of the enzyme that enable all active sites to become active, so the rate of the reaction suddenly increases rapidly.

11.25   An effector binds allosterically to the enzyme (by binding at a place other than any of the catalytically active sites) and induces changes in shape that activate these sites.

11.26   Calmodulin and troponin, the latter being in muscle cells.

11.27   In the cytosol: 10^{-7} mol/L; in the fluid just outside the cell: 10^{-3} mol/L. Ca^{2+} ions that enter the cell are pumped back out after their work is over. The calcium channels through the membrane are kept shut until an appropriate signal arrives.

11.28   At higher Ca^{2+} concentration, $Ca_3(PO_4)_2$ would precipitate.

11.29   No, most Ca^{2+} is held by calmodulin or troponin.

11.30 Ca^{2+} converts them to activated effectors.

11.31   An enzyme might be activated or a muscle might be induced to contract. After such action, Ca^{2+} is pumped back out of the cell.

11.32   When a zymogen is cleaved properly, an active enzyme emerges. Trypsinogen is the zymogen for trypsin.

11.33   It is a compound released from cells of the small intestine when food material moves into this part of the intestinal tract. Enteropeptidase converts trypsinogen to trypsin, a protease.

11.34   A proteolytic, blood-clot dissolving enzyme. It normally circulates in its inactive form, plasminogen.

11.35   The activation of certain enzymes.

11.36   The inhibitor is a non-substrate molecule resembling the true substrate enough to enable the binding of the inhibitor to the enzyme. By thus occupying the active site, the enzyme's work is inhibited.

11.37   Molecules of the product of enzyme action take up active sites on one of the enzymes used to make it and thus inhibit further enzyme action.

11.38   Because it shuts down a pathway when it is no longer needed but lets the pathway occur when it is needed.

11.39   Competitive inhibition involves blocking an active site; allosteric inhibition involves blocking the site where an effector normally works to activate the enzyme.

11.40   (a)   It binds to a metal ion cofactor and so deactivates the enzyme.
        (b)   It denatures enzymes by combining with their SH groups.
        (c)   They deactivate enzymes of the nervous system

11.41   Antimetabolites are compounds that interfere with the metabolism of disease-causing bacteria. Antibiotics are those antimetabolites that are made by microorganisms.

11.42   It inhibits an enzyme needed for the growth of bacteria.

11.43   The levels of these enzymes increase in blood as the result of a disease or injury to particular tissues, which causes tissue cells to release their enzymes.

11.44   The CK(MB) band originates in the leakage of this isoenzyme only from damaged heart muscle.

11.45   CK(MM).

11.46   Of the five LD isoenzymes, LD_1 normally is less concentrated than LD_2. The "flip" is the reversal of this relationship. LD_1 shows up as *more* concentrated than LD_2. This flip is observed in patients who have suffered a myocardial infarction.

11.47   Urease is immobilized on an electrode used to determine the concentration of urea in some fluid. The urease catalyzes the hydrolysis of urea to the ammonium ion, the species actually measured.

11.48   To catalyze the breakdown of heparin that is used to prevent blood clots during hemodialysis.

11.49   Urease immobilized on an electrode for urea in blood catalyzes the hydrolysis of the urea, and the electrode picks up the change in concentration of NH_4^+.

11.50   APSAC, streptokinase, and tissue plasminogen activator (rtPA). rtPA occurs in human blood and it can be manufactured by recombinant DNA technology.

11.51   Fibrin, which dissolves as a result of catalysis by rtPA.

11.52   (a) Adrenal glands   (b) Nerve cells

11.53   These concepts describe how receptors recognize molecules of hormones or neurotransmitters.

11.54   They are primary chemical messengers.

11.55 Receptor
11.56 Cyclic AMP and inositol phosphate
11.57 G-protein
11.58 When activated, adenylate cyclase catalyzes the formation of cyclic AMP (from ATP), which then activates an enzyme inside the target cell.
11.59 It is an enzyme activator.
11.60 The cyclic AMP is hydrolyzed to AMP.
11.61 It is started in the same general way, up to and including the step in which the G-protein does its work.
11.62 One helps keep the cellular glucose level high and the other helps to bring the Ca^{2+} level of the cytosol up.
11.63 Steroids, polypeptides, simple amino compounds, and local hormones (prostaglandins).
11.64 They are hydrocarbon-like and so slip through a hydrocarbon-like lipid bilayer.
11.65 (a) Glucose (b) amino acids (c) metal ions
11.66 The work where they are made or very close by.
11.67 It is hydrolyzed back to acetic acid and choline. The enzyme is cholinesterase. Nerve poisons inactivate this enzyme.
11.68 It inactivates the receptor protein for acetylcholine.
11.69 It blocks the receptor protein for acetylcholine.
11.70 It prevents the synthesis of acetylcholine.
11.71 They catalyze the deactivation of neurotransmitters such as norepinephrine and thus reduce the level of signal-sending activity that depends on such neurotransmitters.
11.72 Iproniazid inhibits the monoamine oxidases and thus lets norepinephrine work at a higher level of activity.
11.73 They inhibit the reabsorption of norepinephrine by the presynaptic neuron and thus reduce the rate of its deactivation by the monoamine oxidases.
11.74 Norepinephrine, acting as a hormone, serves as a backup to its acting as a neurotransmitter should some injury disrupt the latter action.
11.75 Dopamine.
11.76 They bind to dopamine receptors in the postsynaptic nerve and inhibit the action of dopamine.
11.77 They accelerate the release of dopamine from the presynaptic neuron.
11.78 Degenerated neurons can use L-DOPA to make dopamine.
11.79 BDNF protects those cells that make dopamine from further degeneration, stimulates the cells to recover, and protects susceptible cells (at least in animal studies).
11.80 GABA (gamma-aminobutyric acid), whose signal-inhibiting activity is enhanced by Valium and Librium.

11.81   Enkephalin molecules enter pain-signalling neurons and inhibit the release of substance P, a neurotransmitter that helps to send pain signals. Thus enkephalin, like an opium-drug, inhibits pain.

11.82   It moderates pain signals.

11.83   By reducing the flow of Ca^{2+} into cells of heart muscles, the heart beats with reduced vigor.

11.84   A substance made by a signal-receiving nerve cell that moves back to the signal-sending cell to strengthen the connection between the two cells.

11.85   They consist of very tiny, gaseous molecules that easily slip through cell membranes.

11.86 $H{:}^-$ and H^+. $H{:}^-$ goes to NAD^+. H^+ is handled by the buffer.

11.87 $H{:}^-$ and H^+. Both go to FMN or FAD to form $FMNH_2$ or $FADH_2$.

11.88   The positive charge on the nicotinamide ring (the "N" of NAD).

11.89 H_2O

11.90   A disease-causing microorganism or virus.

11.91   Cellular immunity uses T lymphocytes or T cells to handle viruses that have entered cells as well as parasites, fungi, and foreign tissue. Humoral immunity uses B lymphocytes or B cells to handle bacterial infections and viral agents while they are at work outside cells.

11.92   Cellular immunity, by attacking the helper T cells.

11.93   An antigen is any molecular species or any pathogen that induces the immune system to make antibodies and to give the immune system a molecular-cellular memory for the antigen. An antibody is a glycoprotein that is able to take antigen molecules out of circulation enabling white cells to destroy the antigens.

11.94   The presence of glycoproteins with highly individual oligosaccharide units. This enables a lock-and-key kind of specificity between antibody and antigen.

11.95   Humoral immunity; antibodies.

11.96   It works to generate an antibody when introduced into the presence of type B blood.

11.97   In the specific structures of the oligosaccharide units on the glycolipids of red blood cell membranes.

11.98   A red cell with an oligosaccharide that functions as an antibody in the presence of the introduced type B blood.

11.99   Type O red cells carry the H antigens and the blood carries both anti-A and anti-B antibodies, so only type O blood can be received. But type O blood can be donated to people of any type because the H antigen has no enemies in other types of blood.

11.100   (a)   Ess shape (cf. Figure 11.5)

    (b)   See Figure 11.4.

11.101   Competitive inhibition.

11.102 (a) The lock-and-key theory, perhaps as modified by induced fit.

(b) Add water to the alkene group and then oxidize the resulting 2° alcohol to a ketone. (An enzyme would have to guide the addition of the water molecule to give the specific 2° alcohol needed.)

11.103

$$
\begin{array}{cccc}
\begin{array}{c} CO_2^- \\ ^+NH_3\!-\!\!\!-\!H \\ CH_3\!-\!\!\!-\!H \\ CH_2CH_3 \end{array}
&
\begin{array}{c} CO_2^- \\ H\!-\!\!\!-\!NH_3^+ \\ H\!-\!\!\!-\!CH_3 \\ CH_2CH_3 \end{array}
&
\begin{array}{c} CO_2^- \\ ^+NH_3\!-\!\!\!-\!H \\ H\!-\!\!\!-\!CH_3 \\ CH_2CH_3 \end{array}
&
\begin{array}{c} CO_2^- \\ H\!-\!\!\!-\!NH_3^+ \\ CH_3\!-\!\!\!-\!H \\ CH_2CH_3 \end{array}
\end{array}
$$

11.104   165.1 mg of isoleucine

Chapter 12

Review Exercises, Chapter 12

12.1   Interstitial fluid and blood.

12.2   Saliva, gastric juice, pancreatic juice, and intestinal juice.

12.3   The distension of the stomach caused by entering food causes the hormone gastrin to be secreted from cells of the gastric lining of the stomach. Gastrin stimulates the release of gastric juice.

12.4   Molecules of a competitive inhibitor occupy the active sites of an enzyme. Cimetidine shuts down the K^+–H^+ pump and prevents the secretion of gastric juice, which gives the ulcer time to heal in a relatively acid-free environment.

12.5   Cholecystokinin modulates the release of food materials from the stomach into the upper intestinal tract. Secretin causes the pancreas to release bicarbonate ion for the neutralization of the gastric juice that accompanies food matter into the upper intestinal tract.

12.6   (a) α-Amylase.

(b) Pepsinogen and gastric lipase.

(c) α-Amylase, lipase, nuclease, trypsinogen, chymotrypsinogen, procarboxypeptidase, and proelastase

(d) No enzymes.

(e) Amylase, aminopeptidase, sucrase, lactase, maltase, lipase, nucleases, enteropeptidase.

12.7   (a)   Pepsin from its zymogen in gastric juice:, trypsin, chymotrypsin, and elastase from zymogens in pancreatic juice.
       (b)   Lipases provided in gastric juice, pancreatic juice, and intestinal juice.
       (c)   Amylases in saliva, pancreatic juice, and intestinal juice.
       (d)   Sucrase in intestinal juice.
       (e)   Carboxypeptidase from its zymogen in pancreatic juice; aminopeptidase from its zymogen in intestinal juice.
       (f)   Nucleases in pancreatic juice and intestinal juice.

12.8   (a)   Amino acids.
       (b)   Glucose, fructose, and galactose.
       (c)   Fatty acids and monoacylglycerols (plus some diacylglycerols).

12.9   (a)   Peptide (amide) bonds in proteins.
       (b)   Acetal systems in carbohydrates.
       (c)   Ester groups in triacylglycerols.

12.10   It catalyzes the conversion of trypsinogen to trypsin. Then trypsin catalyzes the conversion of other zymogens to chymotrypsin, carboxypeptidase, and elastin. Thus enteropeptidase turns on enzyme activity for three major protein-digesting enzymes.

12.11   They would catalyze the digestion of proteins that make up part of the pancreas to the serious harm of this organ.

12.12   They are surface active agents that help to break up lipid globules, wash lipids from the particles of food, and aid in the absorption of fat-soluble vitamins.

12.13   (a)   Lubricates the food.
       (b)   Protects the stomach lining from gastric acid and pepsin.

12.14   (a)   HCl   (b)   Enteropeptidase
       (c)   Trypsin   (d)   Trypsin
       (e)   Trypsin

12.15   It helps to coagulate the protein in milk so that this protein stays longer in the stomach where it can be digested with the aid of pepsin.

12.16   This enzyme is inactive at the high acidity of the digesting mixture in the adult stomach, but the acidity of this mixture in the infant's stomach is less.

12.17   Pancreatic juice delivers its zymogens and enzymes into the duodenum, whereas the enzymes of the intestinal juice work within cells of the intestinal wall.

12.18   Dilute sodium bicarbonate released from the pancreas. This raises the pH of the chyme to the optimum pH for the action of the enzymes that will function in the duodenum.

12.19   They recombine to molecules of triacylglycerol during their migration from the intestinal tract toward the lymph ducts.

12.20   The flow of bile normally delivers colored breakdown products from hemoglobin in the blood, and these products give the normal color to feces. When no bile flows, no colored products are available to the feces.

12.21   Cholesterol has no hydrolyzable groups.

12.22   The concentration of soluble proteins is greater in blood.

12.23   The serum-soluble proteins (albumins, mostly).

12.24   Fibrinogen is a protein in blood that is changed to fibrin, the insoluble protein of a blood clot, by the clotting mechanism.

12.25   Albumin molecules transport hydrophobic molecules such as fatty acids and cholesterol, and albumins contribute as much as 75-80% of the osmotic effect of the blood.

12.26 Na^+ is in blood plasma and other extracellular fluids; K^+ is chiefly in intracellular fluids. The two ions help to maintain osmotic pressure relationships; be part of the regulatory mechanisms for acid-base balance; and participate in the smooth working of the muscles and the nervous system.

12.27   145 meq/L

12.28   135 meq/L

12.29   Such injuries allow the contents of cells to spill out and enter circulation.

12.30   5.0 meq/L

12.31   Hyponatremia.

12.32   Hypermagnesemia and cardiac arrest.

12.33   The activation of several enzymes.

12.34   In bones and teeth

12.35   Activates enzymes and initiates muscle contraction

12.36   Hypercalcemia.

12.37   Hypomagnesemia.

12.38   Chloride ion, Cl^-.

12.39   100-106 meq/l. Loss of Cl^- causes retention of HCO_3^-, a base.

12.40   Blood pressure and osmotic pressure. Blood pressure tends to force blood fluids out of the blood vesseland osmotic pressure tends to force fluids back. The return of fluids to the blood from the interstitial compartment on the arterial side is overbalanced by the blood pressure. The net effect on the arterial side is a diffusion of fluids from the blood.

12.41   Blood pressure and osmotic pressure. The natural return of fluids to the blood on the venous side from the interstitial compartment is not balanced by the now reduced blood pressure on the venous side, so fluids return to the blood from which they left on the arterial side.

12.42   Serum proteins are lost from the blood, which upsets the osmotic pressure of the blood. Water leaves the blood for the interstitial compartment, and the blood volume drops. Loss of blood delivery to the brain leads to the symptoms of shock.

12.43   (a)   Blood proteins leak out which allows water to leave the blood and enter interstitial spaces throughout various tissues.
(b)   Blood proteins are lost to the blood by being consumed which also leads to the loss of water from the blood and its appearance in interstitial compartments.
(c)   Capillaries are blocked at the injured site reducing the return of blood in the veins, so fluids accumulate at the site.

12.44   Oxygen and carbon dioxide.

12.45   Hemoglobin.

12.46   The first oxygen molecule to bind changes the shapes of other parts of the hemoglobin molecule and makes it much easier for the remaining three oxygen molecules to bind. This ensures that all four oxygen-binding sites of each hemoglobin molecule will leave the lungs fully loaded with oxygen.

12.47 $HHb + O_2 \rightleftharpoons HbO_2^- + H^+$
(a) To the left.   (b) To the left.
(c) To the right.   d) To the left.
(e) To the left.   (f) To the right.

12.48

(a) $HHb + O_2 \longleftarrow HbO_2^- + H^+$

$CO_2 + H_2O \longrightarrow HCO_3^- + H^+$

Isohydric shift in metabolizing tissue

(b) $HHb + O_2 \longrightarrow HbO_2^- + H^+$

$CO_2 + H_2O \longleftarrow HCO_3^- + H^+$

Isohydric shift in alveolus

12.49   It generates H^+ needed to convert HCO_3^- to CO_2 and H_2O and to convert $HbCO_2^-$ to HHb and CO_2.

12.50   Waste CO_2 combines with water to give HCO_3^- and the H^+ that is needed to react with HbO_2^- to form HHb and release O_2.

12.51   It helps to shift the following equilibrium to the left:
$$HHb + O_2 \rightleftharpoons HbO_2^- + H^+$$

12.52   It is found in red cells.
   (a)   It catalyzes the conversion of HCO_3^- and H^+ to CO_2 and H_2O.
   (b)   It catalyzes the conversion of CO_2 and H_2O to HCO_3^- and H^+.
   It can do both because it accelerates *both* the forward and the reverse reactions in the equilibrium:

$$CO_2 + H_2O \rightleftharpoons HCO_3^- + H^+$$

   Other factors, such as the value of the partial pressure of carbon dioxide, determine whether the forward or the reverse reaction is favored.

12.53   The migration of BPG molecules out of hemoglobin molecules as the first oxygen molecules enter the hemoglobin helps the remaining oxygen molecules to bind more readily.

12.54   It migrates into a cavity within the hemoglobin molecule and helps to change the shapes of subunits so that oxygen molecules are more easily ejected.

12.55   The body makes more hemoglobin and red cells. This allows the body to pick up oxygen more readily. The body also makes more BPG, which helps the system release oxygen where needed.

12.56   For oxygenation:
$$HHb\text{-}BPG + O_2 + HCO_3^- \rightarrow HbO_2^- + BPG + CO_2 + H_2O$$
   For deoxygenation:
$$HbO_2^- + BPG + CO_2 + H_2O \rightarrow HHb\text{-}BPG + O_2 + HCO_3^-$$

12.57   As HCO_3^- in the serum and as $HbCO_2^-$ (carbaminohemoglobin) in red cells.

12.58   Oxygen affinity is lowered. Where the partial pressure of CO_2 is relatively high (as in actively metabolizing tissue) there is a need for oxygen, so the lowering effect of CO_2 on oxygen affinity helps to release O_2 precisely where O_2 is most needed.

12.59   The exchange of a chloride ion for a bicarbonate ion between a red blood cell and blood serum. This brings Cl^- inside the red cell when it is needed to help deoxygenate HbO_2^-.

12.60   Myoglobin can take oxygen from oxyhemoglobin and thus ensure that the oxygen needs of myoglobin-containing tissue are met.

12.61   Fetal hemoglobin can take oxygen from the oxyhemoglobin of the mother's blood and thus ensure that the fetus gets needed oxygen.

12.62

Condition	pH	pCO_2	[HCO_3^-]
Normal	7.35-7.45	35-40 mm Hg	25-30 meq/L
Metabolic acidosis	↓ 7.20	↓ 30 mm Hg	↓ 14 meq/L
Metabolic alkalosis	↑ 7.45	↑ >45 mm Hg	↑ >29 meq/L
Respiratory acidosis	↓ 7.10	↑ 68 mm Hg	↑ 40 meq/L
Respiratory alkalosis	↑ 7.54	↓ 32 mm Hg	↓ 20 meq/L

12.63 The pH of the blood decreases in both but both pCO_2 and $[HCO_3^-]$ increase in respiratory acidosis and both decrease in metabolic acidosis.

12.64 Hyperventilation is observed in metabolic acidosis, and HCO_3^- can be given intravenously to neutralize excess acid. Hyperventilation is also observed in respiratory alkalosis (because the patient can't help hyperventilating), and CO_2 is given (by rebreathing one's own air) to keep up the supply of H_2CO_3, which can neutralize excess base.

12.65 Hypoventilation is observed in metabolic alkalosis, and isotonic ammonium chloride can be given to neutralize the excess base. Involuntary hypoventilation is observed in respiratory acidosis, and isotonic sodium bicarbonate might be given to neutralize excess acid.

12.66 In metabolic acidosis, because it helps to blow out CO_2 and thereby to reduce the level of H_2CO_3 in the blood and simultaneously raise the pH.

12.67 In respiratory alkalosis. The involuntary loss of CO_2 reduces the level of H_2CO_3 in the blood and thereby reduces the level of H^+.

12.68 In metabolic alkalosis.

12.69 In respiratory acidosis.

12.70 The kidneys work to remove acids from the blood, but to remove them they must also remove water from the blood. If too much water is taken in this way, then the blood obtains more water by taking it from interstitial and intracellular compartments.

12.71 (a) Respiratory alkalosis   (b) Metabolic alkalosis
    (c) Respiratory acidosis   (d) Respiratory acidosis
    (e) Metabolic acidosis   (f) Respiratory alkalosis
    (g) Metabolic acidosis   (h) Metabolic alkalosis
    (i) Respiratory acidosis   (j) Respiratory acidosis

12.72 (a) Hyperventilation   (b) Hypoventilation
    (c) Hypoventilation   (d) Hypoventilation
    (e) Hyperventilation   (f) Hyperventilation
    (g) Hyperventilation   (h) Hypoventilation
    (i) Hypoventilation   (j) Hypoventilation

12.73 CO_2 is removed at an excessive rate, which removes carbonic acid, so the blood becomes more alkaline and the pH of the blood rises (alkalosis).

12.74 Hypoventilation in emphysema lets the blood retain carbonic acid, and the pH decreases.

12.75 The loss of acid with the loss of the stomach contents results in a loss of acid from the blood, which means an increase in the blood's pH and alkalosis.

12.76 The loss of alkaline fluids from the duodenum and lower intestinal tract leads to a loss of base from the bloodstream, too. The result is a decrease in the blood's pH and thus acidosis.

12.77 Hypocapnia

12.78   Hypercapnia

12.79   The blood has become more concentrated in solutes.

12.80   It acts to prevent the loss of water via the urine by letting the hypophysis secrete vasopressin whose target cells are in the kidneys. The retention of water helps to keep the blood's osmotic pressure from increasing further. The thirst mechanism is also activated to bring in more water to dilute the blood.

12.81   Aldosterone is secreted from the adrenal cortex, and it instructs the kidneys to retain sodium ion in the blood.

12.82   The kidneys secrete a trace of renin into the blood. This catalyzes the conversion of angiotensinogen into angiotensin I, which catalyzes the formation of angiotensin II, a neurotransmitter and powerful vasoconstrictor. With constricting of the capillaries, the blood pressure has to increase to keep the delivery of blood going.

12.83   The rate of diuresis increases.

12.84   They transfer hydrogen ions into the urine and put bicarbonate ions into the bloodstream.

12.85   A molecule of a monoacylglycerol retains a considerable hydrophobic unit (the fatty acyl "tail") and so enters the hydrophobic region of a membrane more easily than a molecule like glycerol with three hydrophilic OH groups.

12.86   Without the bile salts in the bile, normally obtained from the gall bladder, there is less emulsifying action to aid in the digestion of triacylglycerols with the usual long fatty acyl side chains. The shorter chain molecules are a little more soluble in the digestive medium.

12.87   The blood produced while at a high altitude has a higher concentration of hemoglobin and of BPG. This aids in their ability to use oxygen during a race.

12.88   These animals can store more oxygen in heart muscle, which helps them to go longer without breathing.

Chapter 13

Practice Exercises, Chapter 13

1.	(a)	Proline	(b)	Arginine
	(c)	Glutamic acid	(d)	Lysine
2.	(a)	Serine	(b)	CT (chain termination)
	(c)	Glutamic acid	(d)	Isoleucine

Review Exercises, Chapter 13

13.1   (a)   Cytosol   (b)   Protoplasm
        (c)   Cytoplasm   (d)   Ribosome
        (e)   Mitochondrion   (f)   Deoxyribonucleic acid
        (g)   Chromatin   (h)   Histone
        (i)   Gene

13.2   *Chromatin* is a substance made up of nucleosomes—complexes of proteins (called histones) around which are wrapped coiled molecules of DNA—connected by DNA strands. *Chromosomes* are rod-like bodies made microscopically observable when chromatin thickens at the onset of the process that leads to cell division.

13.3   The cell nucleus.

13.4   Duplicate copies of DNA are made.

13.5   Nucleic acids

13.6   Nucleotides

13.7   Ribose and deoxyribose.

13.8   (a) Adenine, A. Thymine, T. Guanine, G. Cytosine, C.
        (b) Adenine, A. Uracil, U. Guanine, G. Cytosine, C.

13.9   The main chains all have the same phosphate-deoxyribose-phosphate-deoxyribose repeating system.

13.10   In the sequence of bases attached to the deoxyribose units of the main chain.

13.11   The main chains all have the same phosphate-ribose-phosphate-ribose repeating system.

13.12   DNA occurs as a double helix, and it alone has the base thymine (T). RNA alone has the base uracil (U). (The remaining three bases, A, G, and C, are the same in both DNA and RNA.) DNA molecules have one less OH group per pentose unit than RNA molecules.

13.13   A and T pair to each other, so they must be in a 1:1 ratio regardless of the species. Similarly, G and C pair and so must be in a 1:1 ratio.

13.14   Bases that project from the twin spirals of DNA chains fit to each other on opposite strands by hydrogen bonds. The geometries and functional groups are such that only A and T can pair (or only A and U can pair when RNA is involved), and only G and C can pair.

13.15   Hydrophobic interactions stabilize the helices. (Hydrogen bonds hold two helices together.)

13.16

$$5' \rightarrow 3'$$

Given segment: AGGCTGA

Opposite segment: TCCGACT

$$3' \leftarrow 5'$$

13.17   The base pairings of A with T and G with C.

13.18   In cells of higher animals, a series of separated segments of a DNA molecule comprise one gene, each segment called an *exon* and each separated by DNA segments called *introns*.

13.19   The introns are b, d, and f, because they are the longer segments.

13.20   A sequence of base triplets comprising a gene corresponds to and specifies a sequence of amino acid residues of a polypeptide.

13.21   It consists of several proteins plus rRNA, and it serves as the assembly area for the synthesis of polypeptides.

13.22   It is heterogeneous nuclear RNA. Its sequence of bases is complementary to a sequence of bases on DNA, those of both the exons and introns. It is processed to make mRNA whose base sequence is complementary only to the exons of the DNA.

13.23   Heterogeneous nuclear RNA (hnRNA)

13.24   A codon is a specific triplet of bases that corresponds to a specific amino acid residue in a polypeptide, and mRNA consists of a continuous sequence of codons.

13.25   A triplet of bases found on tRNA and which is complementary to a codon found on mRNA.

13.26   ATA. A codon is RNA material, and T does not occur in RNA.

13.27   (a)   Phenylalanine   (b)   Serine
        (c)   Threonine   (d)   Aspartic acid

13.28   Writing them in the 5' to 3' direction:
        (a)   AAA   (b)   GGA
        (c)   UGU   (d)   AUC

13.29   (a) $5' \rightarrow 3'$
            UUUCUUAUAGAGUCCCCAACAGAU
        (b) $5' \rightarrow 3'$
            UUUUCCACAGAU
        (c)   Phe-Ser-Thr-Asp

13.30   B < A < D < C

13.31   Translation is the mRNA-directed synthesis of a polypeptide. Transcription is the DNA-directed synthesis of mRNA.

13.32   (a)   A large number of sequences are possible because three of the specific amino acid residues are coded by more than one codon. The possibilities are indicated by:

Met—Ala—Trp—Ser—Tyr
AUG GCU UGG UCU UAU   (5' → 3')
       GCC   UCC UAC
       GCA   UCA
       GCG   UCG

    (b)   CAU (5'→ 3') or if (3' → 5'), then UAC

13.33   The starting point is a specific triplet that forces the acceptance of all remaining sets of three, in order, to be seen as specific codons.

13.34   How polypeptide synthesis can be controlled by the use of repressors.

13.35   Structural gene. The *regulator gene* is coded to direct the synthesis of a polypeptide repressor. The *operator gene* is the site at which the repressor binds and blocks the operation of the strutural gene.

13.36   A substance whose molecules can block DNA-directed polypeptide synthesis.

13.37   It cancels the effect of a repressor.

13.38   Some block the expression of a gene by interfering with DNA-directed polypeptide synthesis.

13.39   The four-letter DNA language and the twenty-letter amino acid language.

13.40   The codons specify the same amino acids in all organisms. No.

13.41   By causing chemical changes in the substances involved with the genetic apparatus so that cell division is no longer controlled.

13.42   At high enough, well-focused doses they destroy the genetic apparatus of a cancer cell and render it incapable of dividing.

13.43   A substance that mimics the effects of radiations in cells.

13.44   All viruses have nucleic acid, and many also contain a protein.

13.45   A molecular feature on the surface of a virus particle fits by a flexible lock-and-key mechanism to a specific glycoprotein on the membrane of one specific kind of host cell.

13.46   A protein part of a virus catalyzes the digestion of part of the cell's membrane. This opens a hole for the virus particle or its nucleic acid to enter the cell.

13.47   RNA replicase is able to direct the synthesis of RNA from the "directions" encoded on RNA. A normal host cell does not contain RNA replicase, so the virus either must bring it along or it must direct its synthesis inside the host cell.

13.48   Reverse transcriptase, a DNA polymerase enzyme, directs the synthesis of viral DNA that subsequently is used to code for the synthesis of more viral RNA. Reverse transcriptase is able to use RNA information to make DNA.

13.49   (+)mRNA

13.50   It must not only make (+) RNA to direct multiplication, it must also make the (–)RNA that new virus particles require.

13.51   RNA replicase

13.52   That the cell is able to move information backward, from RNA to DNA. (Normally information always flows from DNA to RNA.)

13.53   A silent gene is a unit provided by a virus particle to a host cell but which does not take over the genetic apparatus of the cell until activated.

13.54   Cancer-causing

13.55   Human immunodeficiency virus

13.56   The T-lymphocyte. It's part of the body's immune system, so when it is destroyed, the body has no defense against infectious diseases and certain kinds of cancer.

13.57   The theory is that AZT molecules will bind to reverse transcriptase and inhibit the work of this enzyme in HIV.

13.58   A circular molecule of super-coiled DNA found in bacteria.

13.59   Restriction enzyme

13.60   One source is the plasmid and another is new material added to the bacterium or yeast.

13.61   Polypeptides of critical value to human medicine or technology.

13.62   The use of recombinant DNA to make genes and the products of genes.

13.63   A defect in a gene.

13.64   The synthesis of a transmembrane protein that lets chloride ion pass through the membranes of mucous cells in the lungs and the digestive tract. The reduced movement of Cl^- out of the cell means that less water is outside of the cell, and the mucous is thereby thickened and made more viscous. Breathing is thereby impaired.

13.65   To insert correct DNA by some mechanism into cells lacking such DNA and so to repair the defect.

13.66   Retroviruses can bring about the insertion of DNA into a chromosome of a host cell.

13.67   The surface carries the recognition molecules needed to find the appropriate host cells.

13.68   A gene needed for the metabolism of phenylalanine is defective. This leads to an increase in the level of phenylpyruvic acid in the blood, which causes brain damage. A diet very low in phenylalanine is prescribed.

13.69   All of the cells of an individual have the entire genome.

13.70   It is the synthesis of clones of DNA. DNA samples at crime scenes generally involve tiny amounts of DNA, and DNA typing depends on making enough cloned DNA for the next steps of the procedure.

13.71   The common core segments of minisatellites. These DNA segments can be picked up by radioactively labeled and complementary DNA segments. Then the patterns of dark lines or spots—the "bar code"—produced on X ray film by the radiations are obtained.

13.72   A family of proteins (small glycoproteins) that inhibit viruses.

13.73   People do not develop a second viral infection when another is in progress, suggesting that the first infection produces something that provides the protection against another infection at the same time.

13.74   Viruses.

13.75   By recombinant DNA technology

13.76   (a)   By one methyl group
         (b)   No
         (c)   Both uracil and thymine can form a base pair with adenine.

13.77

(a)

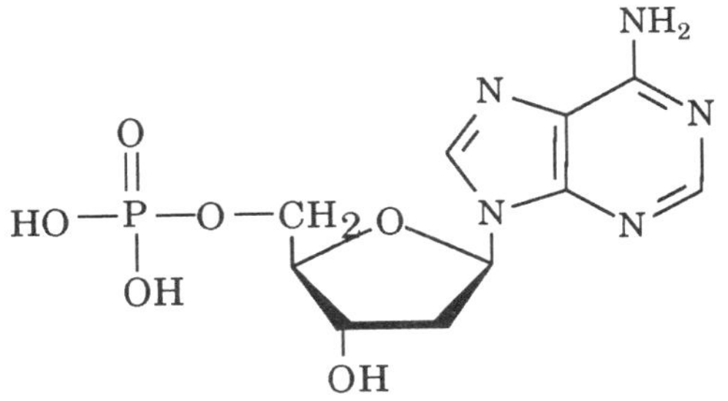

(b)

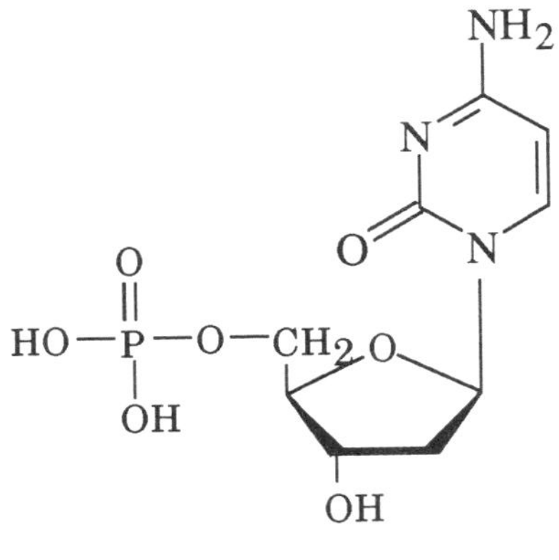

13.78   (a)   A dinucleotide; it has two side chain bases and two ribose units.
         (b)   Of RNA, because the sugar units are those of ribose.
         (c)   At the top.
         (d)   AC

Chapter 14

Practice Exercise, Chapter 14

1.   (a) Yes (b) Yes (c) No

Review Exercises, Chapter 14

14.1   All of them, but chiefly fatty acids and carbohydrates.

14.2   Carbon dioxide and water.

14.3   Combustion produces just heat. The catabolism of glucose uses about half of the energy to make ATP and the remainder appears as heat.

14.4   Carbon dioxide and water.

14.5

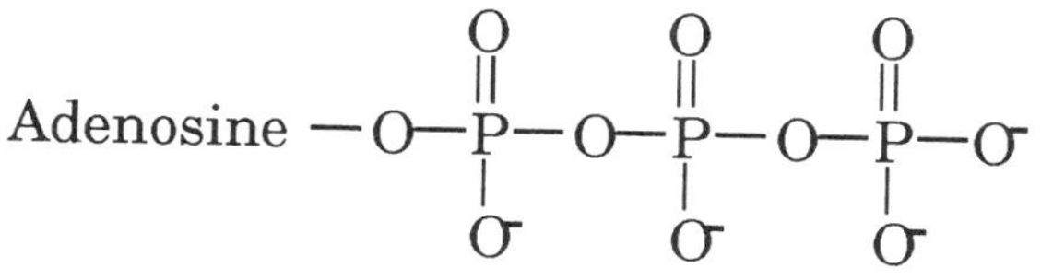

14.6

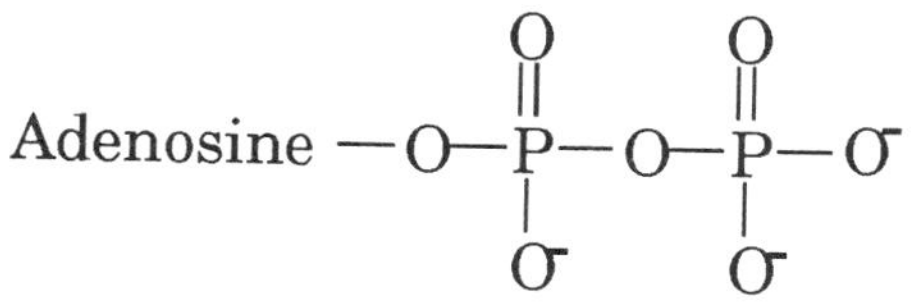

14.7   The singly and doubly ionized forms of phosphoric acid, $H_2PO_4^- + HPO_4^{2-}$.

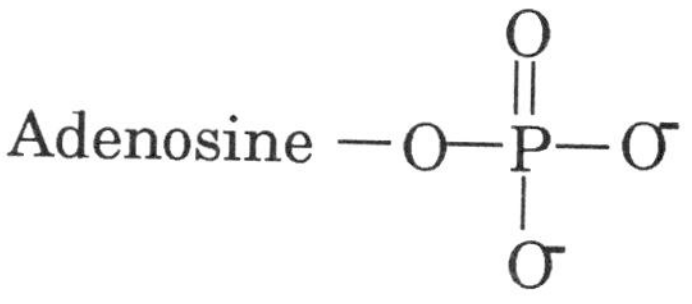

14.8   The relative potential that phosphate X has for donating a phosphate unit in the synthesis of another organophosphate compound. The higher the number is in a negative sense, the higher is this potential.

14.9   The first two have phosphate group transfer potentials equal to or higher than that of ATP, whereas this potential is lower than that of ATP for glycerol-3-phosphate.

14.10   The left.

14.11   (a) Yes   (b) Yes   (c) No   (d) No
14.12   The substrate is directly phosphorylated by the *transfer* of a phosphate group from an organic donor.
14.13   It stores phosphate group energy and transfers phosphate to ADP to remake the ATP consumed by muscular work.
14.14   As the supply of ATP decreases, the supplies of ADP and P_i increase, and this development launches events that make creatine phosphate so that more ATP can be made.
14.15   (a)   The aerobic synthesis of ATP.
        (b)   The synthesis of ATP when a tissue operates anaerobically.
        (c)   The supply of metabolites for the respiratory chain.
        (d)   The supply of metabolites for the respiratory chain and for the citric acid cycle.
14.16   The disappearance of ATP by some energy-demanding process and the simultaneous appearance of ADP + P_i.
14.17   An increase in its supply of ADP. The need to convert ADP back to ATP is met by metabolism, which requires oxygen.
14.18   The citric acid cycle.
14.19   (a) $E < B < C < A < D$
        (b) $B < C < A < D$
14.20   It starts with glycolysis and since this is aerobic as stated, it ends with the respiratory chain.
14.21   A fatty acid
14.22   It starts with glucose and ends with lactate.
14.23   When a cell is temporarily low on O_2 for running the respiratory chain (as a source of ATP), the cell can continue to make ATP anyway.
14.24   When the respiratory chain starts up, the citric acid cycle must start up to keep the chain going.
14.25

        An acetyl unit:

$$\overset{\displaystyle O}{\overset{\displaystyle \|}{CH_3C}}$$

14.26   Two
14.27   Two
14.28   ADP. The cycle helps the cell make ATP from ADP, so activation by ADP is logical.
14.29   (a) 15   (b) 12
14.30   9

14.31   An oxidation; the reactant loses hydrogen.

14.32   An enzyme like isocitrate dehydrogenase, because the reaction is a de-hydrogenation (not an addition of water to a double bond, catalyzed by an enzyme like aconitase).

14.33   A pair of electrons on the left of the arrow.

14.34   Respiratory chain.

14.35

$$\underset{\displaystyle \overset{\displaystyle OH}{\underset{|}{}}}{CH_3CHCO_2^-} + NAD^+ \longrightarrow \underset{\displaystyle \overset{\displaystyle O}{\underset{||}{}}}{CH_3CCO_2^-} + NAD{:}H + H^+$$

(a) $\underset{\displaystyle \overset{\displaystyle OH}{\underset{|}{}}}{CH_3CHCO_2^-}$

(b) NAD^+

14.36

$^-O_2CH_2CH_2CO_2^-$ ⟶ FAD

$^-O_2CCH{=}CHCO_2^-$ ⟶ FADH$_2$

14.37 $NAD^+ < FMN < FeSP < CoQ$

14.38   It catalyzes the reduction of oxygen to water.

14.39   It is a riboflavin-containing coenzyme that in its reduced form, $FADH_2$, passes electrons and H^+ into the respiratory chain.

14.40   Across the inner membrane of the mitochondrion. The value of $[H^+]$ is higher on the outer side of the innermembrane than in the mitochondrial matrix.

14.41   The flow of protons across the inner mitochondrial membrane.

14.42   If the membrane is broken, then the simple process of diffusion defeats any mitochondrial effort to set up a gradient of H^+ ions across the membrane, but the chain itself can still operate.

14.43 $MH_2 + nH^+ + f(1,2)O_2 \rightarrow$
$M + H_2O + nH^+$ (outside the inner mitochondrial membrane)

14.44   Oxidative phosphorylation is the kind made possible by the energy released from the operation of the respiratory chain. Substrate phosphorylation arises from the direct transfer of a phosphate unit from a higher to a lower energy phosphate.

14.45   A gradient of positive charge. The migration of *any* cation away from the region of higher positive charge density or the migration of *any* anion toward this region will be spontaneous.

14.46   An enzyme catalyzes the formation of ATP from ADP and P_i, but the new ATP sticks tightly to the enzyme. But this enzyme is at the end of the channel for protons in the inner mitochondrial membrane, and as protons flow through they change the enzyme so that it expels ATP.

14.47   Acetone. Unlike propane, acetone has an electronegative oxygen that causes electron density to shift out of C–H bonds tending to make the H atom more acidic.

14.48

$$CH_3CH_2\overset{\overset{\displaystyle O}{\displaystyle \|}}{C}\overset{\underset{\displaystyle CH_3}{|}}{CH}CO_2CH_2CH_3$$

14.49   (a) $CH_3CH{=}O$
      (b)   Acetic acid, CH_3CO_2H (or acetate ion, $CH_3CO_2^-$)
      (c)   α-Ketoglutarate

14.50   Pyruvate

14.51   (a)   4
      (b)   The nucleus of a hydrogen atom, because $H{:}^-$ transfers.
      (c)   Dehydrogenase

Chapter 15

Review Exercises, Chapter 15

15.1   Glucose, fructose, and galactose.

15.2   After a few steps, the metabolic pathways of galactose and fructose merge with the pathway of glucose.

15.3   The concentration of reducing monosaccharides, chiefly glucose, in the blood is called the blood sugar level. The normal fasting level is the blood sugar level after several hours of fasting.

15.4   70-110 mg/dL (3.9-6.1 mmol/L). (Note. Various references give slightly different ranges of values.)

15.5   99 mg/dL

15.6   (a)   Glucose in urine.
       (b)   A low blood sugar level.
       (c)   A high blood sugar level.
       (d)   The conversion of glycogen to glucose.
       (e)   The concentration of something (e.g., glucose) in blood above which that solute appears in the urine.
       (f)   The synthesis of glycogen from glucose.
15.7   The synthesis of glucose in the body from smaller molecules. Hypoglycemia normally activates gluconeogenesis.
15.8   The lack of glucose means the lack of the one nutrient most needed by the brain.
15.9   It increases.
15.10 Muscle tissue.
15.11 The enzyme adenylate cyclase.
15.12 $D < C < E < A < B$
15.13 10^4
15.14 Glucose might be changed back to glycogen as rapidly as it is released from glycogen, and no glucose would be made available to the cell.
15.15 The activated form of protein kinase.
15.16 Glucose-1-phosphate is the end product and phosphoglucomutase catalyzes its change to glucose-6-phosphate.
15.17 Liver, but not muscles, has the enzyme glucose-6-phosphatase that catalyzes the hydrolysis of glucose-6-phosphate. This frees glucose for release from the liver to the bloodstream.
15.18 It is a polypeptide hormone made in the alpha cells of the pancreas and released into circulation when the blood sugar level drops. At the liver it activates adenylate cyclase, which leads to glycogenolysis and the release of glucose into circulation.
15.19 Glucagon, because it works better at the liver than epinephrine in initiating glycogenolysis, and when glycogenolysis occurs at the liver there is a mechanism for releasing glucose into circulation.
15.20 It stimulates the release of glucagon, which leads to the release of glucose into circulation.
15.21 It is a polypeptide hormone released from the beta cells of the pancreas in response to an increase in the blood sugar level, and it acts most effectively at adipose tissue.
15.22 An increase in the blood sugar level.
15.23 Too much insulin leads to a sharp decrease in the blood sugar level and therefore a decrease in the supply of glucose, the chief nutrient for the brain.
15.24 Somatostatin is a polypeptide hormone released by the hypothalamus, and it acts at the pancreas to inhibit the release of glucagon and slow down the release of insulin.

15.25   The body's ability to manage dietary glucose without letting the blood sugar level swing too widely from its normal fasting level.

15.26   By its conversion to glucose-6-phosphate, which either enters a pathway that makes ATP or is converted to glycogen for storage.

15.27   Glucose absorbed from the blood stream circulates and some is removed by muscle tissue, some by the liver. As needed, glucose is catabolized by glycolysis, which generates ATP and pyruvate (in an aerobic situation) or lactate (in an anaerobic situation). When excess lactate is made, some lactate is further catabolized in the liver to provide the energy to convert the remaining lactate back to glucose. This may be stored as liver glycogen or it may reenter circulation, be removed at muscles, and thus replenish glycogen reserves in muscle tissue.

15.28   Some is catabolized in the liver to provide the energy to convert the rest, via gluconeogenesis, to glucose.

15.29   To test for the possibility of diabetes mellitus. An adult patient is given a drink that has 75 g of glucose. For children, 1.75 g of glucose per kilogram of body weight is given. Then the blood sugar level is measured at regular intervals.

15.30   (a)   The blood sugar level initially rises rapidly, but then drops sharply and slowly levels back to normal.
(b)   The blood sugar level, already high to start, rises much higher and never sharply drops. It only very slowly comes back down.

15.31   An over-release of epinephrine (as in a stressful situation) that induces an over-release of glucose.

15.32 $C_6H_{12}O_6 + 2\ ADP + 2P_i \rightarrow 2C_3H_5O_3^- + 2H^+ + 2ATP$

15.33   Glycolysis can operate and make ATP even when the oxygen supply is low, so a tissue in oxygen debt can continue to function.

15.34   Glyceraldehyde-3-phosphate is in the direct pathway of glycolysis, so changing dihydroxyacetone phosphate into it ensures that all parts of the original glucose molecule are used in glycolysis.

15.35   (a)   It undergoes oxidative decarboxylation and becomes the acetyl group in acetyl CoA.
(b)   Its keto group is reduced by NADH to a 2° alcohol group in lactate, which enables the NADH to be reoxidized to NAD^+ and then reused for more glycolysis.

15.36   It is reoxidized to pyruvate, which undergoes oxidative decarboxylation to the acetyl group in acetyl CoA. This enters the citric acid cycle.

15.37   (a)   17 ATP   (b)   18 ATP

15.38   PH forms; the body uses this reducing agent to make fatty acids.

15.39   It makes glucose out of smaller molecules obtained by the catabolism of fatty acids and amino acids.

15.40   All are catabolized, and parts of some of their molecules are used to make fatty acids and, thence, fat.

15.41   (a) Alanine   (b   Aspartic acid

15.42   α-Ketoglutarate is an intermediate in the citric acid cycle (Figure 14.2) and so is changed eventually to oxaloacetate (normally the acceptor of acetyl units at the start of the citric acid cycle). Gluconeogenesis also can use oxaloacetate to make "new" glucose.

15.43   Succinyl units are in the citric acid cycle, which ends with the formation of oxaloacetate. and the later can be used in gluconeogenesis.

15.44   (a) Glucose 6-phosphatase. Glucose cannot leave the liver, so the liver enlarges as glycogen reserves become high. The blood sugar level decreases, and the blood levels of pyruvate and lactate increase.
(b) An enzyme for breaking 1,6-glycosidic bonds. Not as much glycogen can be used. Symptoms develop like those of Von Gierke's disease, but milder.
(c) Phosphorylase. Glucose-1-phosphate cannot bes obtained from glycogen. Reduced physical activity follows.
(d) Enzymes for making the branches in glycogen. Liver failure occurs.

15.45   Type I diabetics cannot make insulin.

15.46   Type I

15.47   Type II

15.48   1.   Presence of genetic defects.
2.   Some triggering incident occurs, like a viral infection.
3.   Particular antibodies associated with diabetes appear in the blood.
4.   Gradual loss of ability to secrete insulin.
5.   Full-fledged diabetes. Hyperglycemia.
6.   Destruction of β-cells is complete.

15.49   The β-cells of the pancreas.

15.50   Changes occur in body molecules that the immune system reads as new antigens. The antibodies then made destroy body tissue.

15.51   One explanation: insulin receptors "wear out." Another: the β-cells of the pancreas "wear out." Still another: amylin (a recently discovered protein component of what is released by the β-cells) suppresses glucose uptake.

15.52   The basement membrane of capillaries.

15.53   Its aldehyde group reacts with amino groups on proteins and genes to form $C{=}N$ systems by means of which the glucose units are tied to other molecules.

15.54   It decreases. The equilibrium involving glucose, an amino group, and glucosylated hemoglobin shifts back to regenerate glucose.

15.55   They change to more permanent Amadori compounds, which constitute more permanent changes in cell molecules.

15.56   The sorbitol made by the hydrogenation of glucose stays inside the eye lens where it draws water osmotically. This produces a swelling of the lens, pressure, glaucoma, and eventually blindness.

15.57

(a)

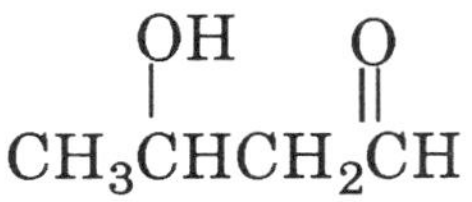

$$\underset{\underset{\displaystyle OH}{|}}{CH_3}\!CHCH_2CH$$

OH   O
 | ||
CH₃CHCH₂CH

(b)

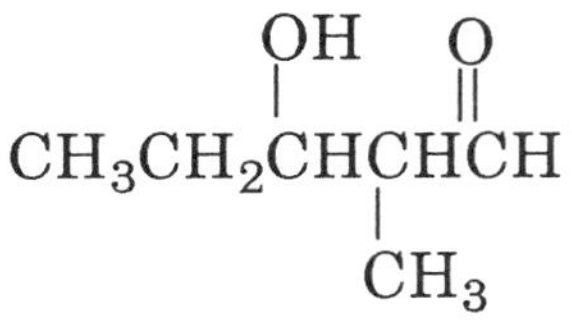

OH   O
 | ||
CH₃CH₂CHCHCH
        |
        CH₃

(c)

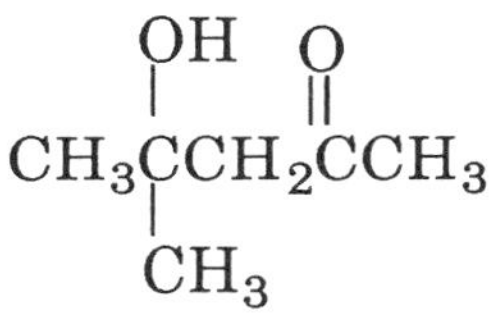

OH   O
 | ||
CH₃CCH₂CCH₃
 |
 CH₃

15.58

O
||
C₆H₅CH₂CH

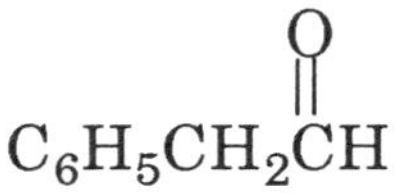

15.59   (a)   Yes, either pyruvate or lactate containing carbon-13 may reentercirculation.
(b)   Yes, either pyruvate or lactate containing carbon-13 might be absorbed by the liver from the blood stream.
(c)   Yes, glucose with carbon-13 atoms might be made via gluconeogenesis from either pyruvate or lactate containing carbon-13.

15.60   (a)   15 ATP
(b)   3 ATP
(c)   3 Glucose, (because 6 ATP are needed to make each molecule of glucose by gluconeogenesis).

Chapter 16

Review Exercises, Chapter 16

16.1   Fatty acids and monoacylglycerols (plus some diacylglycerols).

16.2   They become reconstituted into triacylglycerols.

16.3   They transport lipids received from the digestive tract to the liver.

16.4   They unload some of their triacylglycerol.

16.5   They are absorbed.

16.6   Some cholesterol has originated in the diet and some has been synthesized in the liver.

16.7   (a)   Very low density lipoprotein complex.
        (b)   Intermediate density lipoprotein complex.
        (c)   Low density lipoprotein complex.
        (d)   High density lipoprotein complex.

16.8   Triacylglycerol.

16.9   The loss of the less dense triacylglycerol leaves a higher concentration of the more dense cholesterol.

16.10   The liver.

16.11   Cholesterol.

16.12   The synthesis of steroids and the fabrication of cell membranes.

16.13   IDL and LDL.

16.14   When the receptor proteins are reduced in number, the liver cannot remove cholesterol from the blood, so the blood cholesterol level increases.

16.15   Return to the liver any cholesterol that extrahepatic tissue cannot use.

16.16   The concentration of the higher density cholesterol is greater in HDL

16.17   The grams of a particular tissue or fluid needed to store one kilocalorie of reserve energy.

16.18   3 < 1 < 2

16.19   Fasting and diabetes.

16.20   Insulin suppresses the lipase needed to hydrolyze triacylglycerols in storage prior to the release of their fatty acids into circulation.

16.21   E < D < A < C < B

16.22   The operation of the β-oxidation pathway feeds electrons and protons directly to the respiratory chain, and it makes acetyl CoA that the citric acid cycle catabolizes as it also sends electrons and protons into the respiratory chain.

16.23   The citric acid cycle processes the acetyl units manufactured by the β-oxidation pathway and so fuels the respiratory chain.

16.24   Epinephrine and glucagon. They help keep a normal blood sugar level.

16.25   An increase in the blood sugar level triggers the release of insulin which inhibits the release of fatty acids from adipose fat.

16.26   It is changed to dihydroxyacetone phosphate, which enters the pathway of glycolysis.

16.27   They are joined to coenzyme A as fatty acyl CoA.

16.28

$$CH_3CH_2CH_2\overset{O}{\overset{||}{C}}SCoA + FAD \longrightarrow CH_3CH=CH\overset{O}{\overset{||}{C}}SCoA + FADH_2$$

$$CH_3CH=CH\overset{O}{\overset{||}{C}}SCoA + H_2O \longrightarrow CH_3\overset{OH}{\overset{|}{C}}HCH_2\overset{O}{\overset{||}{C}}SCoA$$

$$CH_3\overset{OH}{\overset{|}{C}}HCH_2\overset{O}{\overset{||}{C}}SCoA + NAD^+ \longrightarrow CH_3\overset{O}{\overset{||}{C}}CH_2\overset{O}{\overset{||}{C}}SCoA + NAD{\cdot}H + H^+$$

$$CH_3\overset{O}{\overset{||}{C}}CH_2\overset{O}{\overset{||}{C}}SCoA + CoASH \longrightarrow 2CH_3\overset{O}{\overset{||}{C}}SCoA$$

No more turns of the β-oxidation pathway are possible.

16.29   FADH2 passes its hydrogen into the respiratory chain and is changed back to FAD.

16.30   NADH passes its hydrogen into the respiratory chain and is changed back to NAD$^+$.

16.31   Steps 1-3 succeed in oxidizing the beta position of the fatty acyl unit to a ketone group.

16.32   (a)   Inside mitochondria.
        (b)   Cytosol.

16.33   An acetyl unit in acetyl CoA is activated for the overall synthesis.

$$CH_3\overset{O}{\overset{||}{C}}SCoA + HCO_3^- + ATP \longrightarrow {}^-O\overset{O}{\overset{||}{C}}CH_2\overset{O}{\overset{||}{C}}SCoA + 2H^+ + ADP + P_i$$

The malonyl unit is transferred to the acyl carrier protein, ACP.

$$\text{-OCCH}_2\text{CSCoA} + \text{ACP} \longrightarrow \text{-OCCH}_2\text{CS-ACP} + \text{CoA}$$

Acetyl E is made by attaching acetyl CoA to the E unit of the enzyme.

$$\text{CH}_3\text{CSCoA} + E \longrightarrow \text{CH}_3\text{CS-}E + \text{CoA}$$

The acetyl group transfers from acetyl E to malonyl ACP.

$$\text{CH}_3\text{CS-}E + \text{-OCCH}_2\text{CS-ACP} \rightarrow \text{CH}_3\text{CCH}_2\text{CS-ACP} + \text{CO}_2 + E$$

The keto group of acetoacetyl ACP is reduced.

$$\text{CH}_3\text{CCH}_2\text{CS-ACP} + \text{NADPH} + \text{H}^+ \longrightarrow$$

$$\text{CH}_3\text{CHCH}_2\text{CS-ACP} + \text{NADP}^+$$

The 2° alcohol group is removed by dehydration.

$$\text{CH}_3\text{CHCH}_2\text{CS-ACP} \longrightarrow \text{CH}_3\text{CH=CHCS-ACP} + \text{H}_2\text{O}$$

The alkene group is reduced by NADPH.

$$CH_3CH\!\!=\!\!CH\overset{\displaystyle O}{\overset{\|}{C}}S\!\!-\!\!ACP + NADPH + H^+ \longrightarrow$$

$$CH_3CH_2CH_2\overset{\displaystyle O}{\overset{\|}{C}}S\!\!-\!\!ACP + NADP^+$$

16.34   The pentose phosphate pathway of glucose catabolism.
16.35   Mevalonate.
16.36   Cholesterol inhibits the synthesis of HMG-CoA reductase.
16.37   Oxaloacetate
16.38   Oxaloacetate is the carrier of acetyl groups in the citric acid cycle, so its loss means that the acetyl CoA level increases.
16.39

$$CH_3\overset{\displaystyle O}{\overset{\|}{C}}CH_2\overset{\displaystyle O}{\overset{\|}{C}}SCoA$$

16.40   A proton or hydrogen ion, H$^+$. If the level of hydrogen ion increases, the problem is acidosis.
16.41

$$CH_3\overset{\displaystyle O}{\overset{\|}{C}}CH_2CO_2H \qquad CH_3\overset{\displaystyle OH}{\overset{|}{C}}HCH_2CO_2H \qquad CH_3\overset{\displaystyle O}{\overset{\|}{C}}CH_3$$

Acetoacetic acid   β-Hydroxybutyrate   Acetone

16.42   An above normal concentration of the ketone bodies in the blood.
16.43   An above normal concentration of the ketone bodies in the urine.
16.44   Enough acetone vapor in exhaled air to be detected by its odor.
16.45   Ketonemia + ketonuria + acetone breath.
16.46   Metabolic acidosis brought on by an increase in the level of the ketone bodies in the blood.
16.47   Acetoacetic acid.
16.48   Diuresis is accelerated to remove ketone bodies from the blood, and their removal requires the simultaneous removal of water, so the urine volume increases.

16.49   Their *over*-production leads to acidosis.

16.50   Amino acids are catabolized at a faster than normal rate to participate in gluconeogenesis, and their nitrogen is excreted largely as urea.

16.51   Each Na^+ ion that leaves corresponds to the loss of one HCO_3^- ion, the true base, because HCO_3^- neutralizes acid generated as the ketone bodies are made. And for every negative ion that leaves with the urine a positive ion, mostly Na^+, has to leave to ensure electrical neutrality.

16.52   A disease in which arterial walls become clogged by plaques—mixtures of collagen, elastic fibers, triacylglycerols, cholesterol and cholesterol esters.

16.53   A genetically caused sharply elevated level of blood cholesterol for which both parents bore the defective genes.

16.54   LDL, because its cholesterol contributes significantly to plaque deposits.

16.55   HDL

16.56   A low level of HDL means that cholesterol is not being returned to the liver in significant amounts.

16.57   Apolipoprotein(a)

16.58   Brown cells carry out fatty acid catabolism within themselves. White cells send the fatty acids to the liver.

16.59   The inner mitochondrial membrane permits a flow of protons across it only at those sites that connect to the enzyme responsible for making and releasing ATP.

16.60   It becomes heat.

16.61   The generation of body heat by catabolism.

16.62

(a) $CH_3CH_2CH_2CH_2CH_2\overset{O}{\overset{\|}{C}}SCoA + FAD \longrightarrow$

$CH_3CH_2CH_2CH=CH\overset{O}{\overset{\|}{C}}SCoA$

(b) $CH_3CH_2CH_2CH=CH\overset{O}{\overset{\|}{C}}SCoA + H_2O \longrightarrow$

$CH_3CH_2CH_2\overset{OH}{\underset{\|}{C}}HCH_2\overset{O}{\overset{\|}{C}}SCoA$

(c) $\underset{\displaystyle CH_3CH_2CH_2\overset{\displaystyle OH}{\underset{\displaystyle |}{C}}HCH_2\overset{\displaystyle O}{\underset{\displaystyle ||}{C}}SCoA}{} + NAD^+ \longrightarrow$

$$CH_3CH_2CH_2\overset{\displaystyle O}{\underset{\displaystyle ||}{C}}CH_2\overset{\displaystyle O}{\underset{\displaystyle ||}{C}}SCoA + NADH + H^+$$

(d) $CH_3CH_2CH_2\overset{\displaystyle O}{\underset{\displaystyle ||}{C}}CH_2\overset{\displaystyle O}{\underset{\displaystyle ||}{C}}SCoA + CoASH \longrightarrow$

$$CH_3CH_2CH_2\overset{\displaystyle O}{\underset{\displaystyle ||}{C}}SCoA + CH_3\overset{\displaystyle O}{\underset{\displaystyle ||}{C}}SCoA$$

16.63   (a)   7
      (b)   6
      (c)   6
      (d)   112. A table like Table 16.2 would be the following.

Intermediate	Maximum No. of ATP from Each	Total No. of ATP Possible fromEach Intermediate as Acetyl CoA Forms
6 FADH$_2$	2	12
6 NADH	3	18
7 CH$_3$COSCoA	12	84
	Sum	114
Deduct 2 high-energy phosphate bonds for activating the myristyl group		-2
Net ATP produced for each myristyl group as it changes to acetyl CoA		112

Chapter 17

Review Exercises, Chapter 17

17.1   The entire collection of nitrogen compounds found anywhere in the body.
17.2   To synthesize protein.
         To synthesize nonprotein compounds of nitrogen.
         To synthesize nonessential amino acids.
         To contribute to the synthesis of ATP.
17.3   Infancy.
17.4   They are catabolized. Some are converted to fatty acids.
17.5   Glutamic acid (glutamate)
17.6

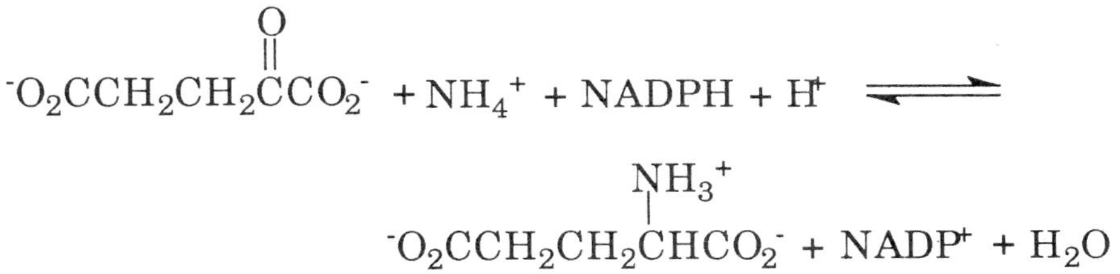

$$^-O_2CCH_2CH_2\overset{\displaystyle O}{\overset{\|}{C}}CO_2^- + NH_4^+ + NADPH + H^+ \rightleftharpoons$$

$$^-O_2CCH_2CH_2\overset{\displaystyle NH_3^+}{\overset{|}{C}H}CO_2^- + NADP^+ + H_2O$$

17.7

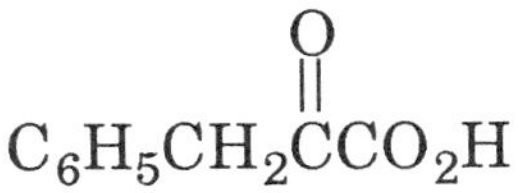

$$C_6H_5CH_2\overset{\displaystyle O}{\overset{\|}{C}}CO_2H$$

17.8

$$(CH_3)_2CH\overset{\displaystyle O}{\overset{\|}{C}}CO_2H$$

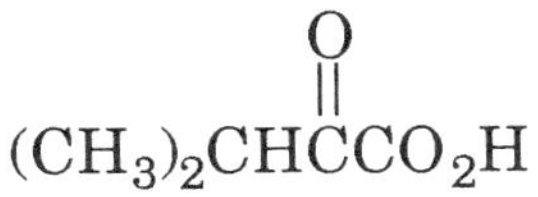

17.9   The body can use it to make glucose by gluconeogenesis.
17.10   Lysine can be used to make fatty acids but not glucose.
17.11   Ketogenic

17.12

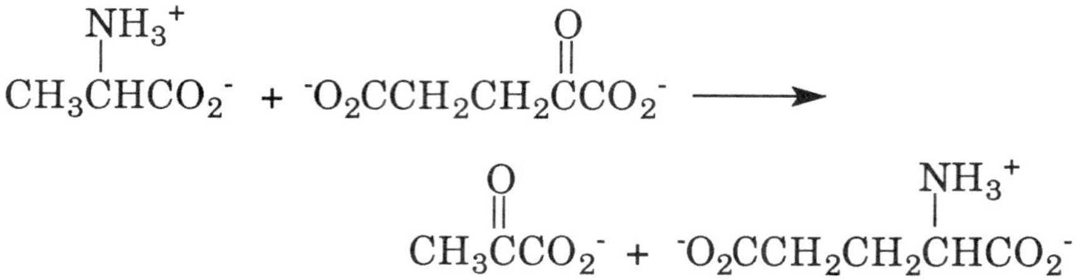

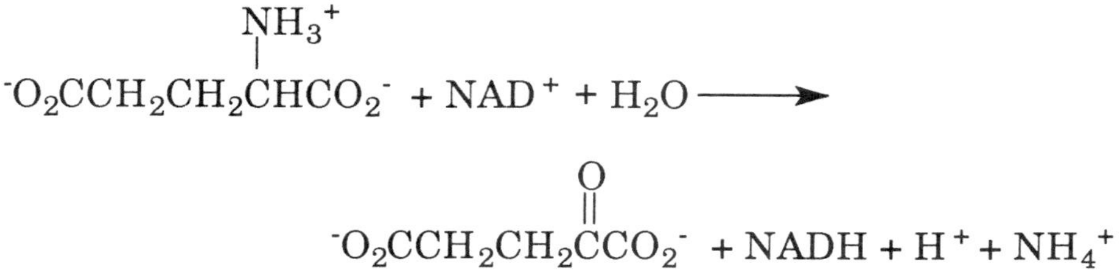

$$\underset{\overset{|}{NH_3^+}}{CH_3CHCO_2^-} + {}^-O_2CCH_2CH_2\overset{\overset{O}{\|}}{C}CO_2^- \longrightarrow$$

$$CH_3\overset{\overset{O}{\|}}{C}CO_2^- + {}^-O_2CCH_2CH_2\underset{\overset{|}{NH_3^+}}{CHCO_2^-}$$

$$\underset{\overset{|}{NH_3^+}}{{}^-O_2CCH_2CH_2CHCO_2^-} + NAD^+ + H_2O \longrightarrow$$

$${}^-O_2CCH_2CH_2\overset{\overset{O}{\|}}{C}CO_2^- + NADH + H^+ + NH_4^+$$

17.13 $4 < 1 < 5 < 2 < 3$
17.14 $5 < 2 < 1 < 3 < 4$

17.15   Yes: glucose $\dfrac{\text{aerobic}}{\text{glycolysis}} >$ pyruvate $\dfrac{\text{transamination}}{} >$ alanine

17.16

$$CH_3CH_2\overset{\overset{O}{\|}}{C}CO_2^-$$

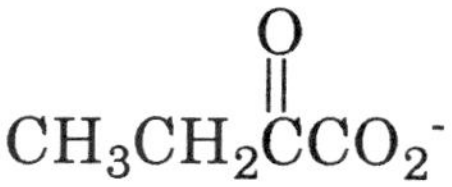

17.17

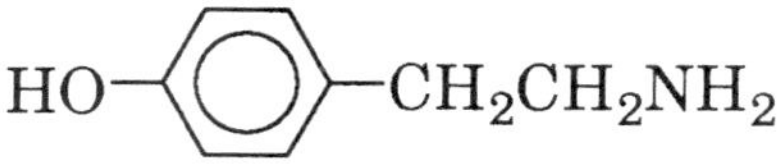

Tyramine

17.18

17.19   To synthesize glucose by means of gluconeogenesis

17.20   (a)   Originally, the amino groups of amino acids.
        (b)   Carbon dioxide.

17.21   An above normal concentration of ammonium ion in the blood. Infants improve on a low-protein diet.

17.22 $2NH_3 + H_2CO_3 \rightarrow NH_2CONH_2 + 2H_2O$

17.23   Hyperammonemia. Step 2 consumes carbamoyl phosphate, which is made using ammonia. If carbamoyl phosphate levels rise, a backup occurs to cause ammonia levels to increase.

17.24   The purine bases of nucleic acids, adenine and guanine.

17.25   Sodium urate.

17.26 $3 < 2 < 1 < 4 < 5 < 7 < 6$

17.27   It gives the skin a yellowish color.

17.28   (a)   Bile pigments form faster than the liver can clear them.
        (b)   Hepatic disease can prevent the liver from removing bilirubin from the blood.

17.29   Bile, carrying bile pigments, moves through the bile ducts. If it cannot move, the pigments back up into the blood.

17.30   Serine $\rightarrow$ pyruvate $\rightarrow$ acetyl CoA $\rightarrow$ (fatty acid synthesis)
                                                        $\rightarrow$ palmitic acid

17.31   The exclusively ketogenic amino acids, those not also glucogenic, cannot produce net extra oxaloacetate.

Chapter 18

Review Exercises, Chapter 18

18.1   It identifies the nutrients needed for health, determines the amounts required, and finds foods that are good sources of the nutrients.

18.2   It is any compound needed for health.

18.3   Foods are complex mixtures of nutrients.

18.4   Dietetics is the application of the findings of nutrition to the feeding of individuals whether ill or well.

18.5   To allow for individual differences among people and to ensure that practically all people can thrive.

18.6   1.   People with chronic diseases.
2.   People who must take special medications.
3.   Prematurely born infants.
4.   Pregnant women.
5.   Lactating women.
6.   People involved in strenuous physical activity.
7.   People exposed for prolonged periods to high temperatures.

18.7   No food contains all of the essential nutrients, and there might still be nutrients yet to be discovered but which are routinely provided by a varied diet.

18.8   (a)   The body must make its own glucose, which can lead to a buildup of harmful substances.
(b)   It lacks the essential fatty acids and it makes the absorption of the fat-soluble vitamins more difficult.

18.9   Linoleic acid. "Essential" means that it must be provided in the diet. If linoleic acid is absent in the diet, the prostaglandins are not made at a sufficient rate.

18.10   Linolenic acid and arachidonic acid.

18.11   Arachidonic acid

18.12   The body can make several amino acids itself.

18.13   Not all of the protein is digested, and of what is digested, not all is absorbed into circulation.

18.14   It breaks them down and eliminates the products.

18.15   Coefficient of digestibility $= \dfrac{(N \ \text{in} \ \text{food} \ \text{eaten} - N \ \text{in} \ \text{feces})}{N \ \text{in} \ \text{food} \ \text{eaten}}$
The numerator, (N in food eaten − N in feces), is the food nitrogen that is actually absorbed into circulation.

18.16   From an animal source.

18.17   Mill them into flour, but this also lowers their vitamin and mineral content.

18.18   The proportions of essential amino acids available from it.

18.19   Human milk protein is the best, but whole egg protein is very close.

18.20   The essential amino acid most poorly supplied by the protein.

18.21   Corn protein is low in tryptophan and lysine.

18.22   (a) $1.3 \times 10^2 \, g$
(b) $1.7 \times 10^3 \, g$
(c) $6.1 \times 10^3 \, kcal$
(d)   Very likely not, since $1.7 \times 10^3 \, g$ is nearly 3 lb.

Vitamin	Source(s)	Problem(s) If Deficient
A	carrots	night blindness, deterioration of mucous membranes, blindness
D	dairy products, fatty fish	poor bone development, rickets
E	vegetable oils	edema and anemia (in infants), accelerated hemolysis, possible heart disease
K	green leafy vegetables	increased susceptibility to hemorrhages
C	citrus fruits, potatoes, leafy vegetables, tomatoes	scurvy, possibly increased susceptibility to colds
Choline	meats, egg yolk, cereals, legumes	none known in humans; fatty liver and kidney disease in animals
Thiamine	lean meats, legumes whole grains	beri beri
Riboflavin	milk, meat	problems with tissue around the mouth, nose and tongue; skin scaling; impaired wound healing
Niacin	meat, whole grains	pellagra
Folate	green, leafy vegetables:, liver, kidneys	megaloblastic anemia
B6	meat, wheat, yeast	possibly disturbances in the central nervous system; hyperchromic microcytic anemia
B12	meat and dairy products	pernicious anemia
Pantothenic acid milk	not observed clinically	liver, kidney, egg yolk, skim in humans
Biotin	egg yolk, liver, tomatoes, yeast	seldom observed; anorexia, nausea, pallor, dermatitis, depression

18.24   They are needed in much more than trace amounts, and they come from proteins.

18.25   Vitamin B_{12}

18.26   No single vegetable source has a balanced supply of essential amino acids.

18.27   Vitamins A, D, E, and K.

18.28   Vitamin D.

18.29   Vitamin D.

18.30   Vitamin A.

18.31   Vitamin A.

18.32   Vitamin K

18.33   A species that has an unpaired electron in the valence shell of one of its atoms. Free radicals attack essential components of cells and they accelerate the aging process.

18.34   Vitamins C and E

18.35   Partially oxidized cholesterol, produced by the action of free radicals, attracts white blood cells, and this can cause the buildup of plaque.

18.36   Vitamin C.

18.37   Vitamin C, choline, thiamin, riboflavin, niacin, folate, vitamin B_6, vitamin B_{12}, pantothenic acid, and biotin

18.38   Vitamin C

18.39   Vitamin C, thiamin, riboflavin, niacin, and folate.

18.40   Thiamin

18.41   Niacin

18.42   Folate

18.43   The quantity needed per day. Minerals are needed in the amount of more than 100 mg/day and trace elements in the amount of less than 20 mg/day.

18.44   Calcium, Ca^{2+}; phosphorus, P_i (Chiefly, the mix of HPO_4^{2-} and $H_2PO_4^-$ plus some PO_4^{3-} that exists at body pH and in bone.); magnesium, Mg^{2+}; sodium, Na^+; potassium, K^+; and chloride, Cl^-.

18.45

Trace Element	Functions
Fluorine (F^-)	sound teeth
Chromium (Cr^{3+})	glucose metabolism; work of insulin
Manganese (Mn^{2+})	nerve function, sound bones, reproduction
Iron (Fe^{2+})	heme, many enzyme
Cobalt (Co^{2+})	vitamin B_{12}
Copper (Cu^{2+})	in proteins and enzymes
Zinc (Zn^{2+})	enzymes, nucleic acids, bones; prevention of dwarfism; possible protection against heart disease
Selenium	possible protection against heart disease
Molybdenum	nucleic acid metabolism
Iodine (I^-)	synthesis of thyroid hormones

18.46   Goiter

18.47 2.13×10^3 L air

18.48   (a)   79 g
(b) 3.0×10^2 g
(c) 8.5×10^2 kcal
(d)   Because 3.0×10^2 g of peanuts is about 2/3 lb, a child could probably get this much down.

NOTES

NOTES

NOTES

NOTES

NOTES

NOTES

NOTES

NOTES

NOTES

NOTES

NOTES

NOTES

NOTES

NOTES

NOTES

NOTES

NOTES